Lecture Notes in Physics

Volume 848

For further volumes:
http://www.springer.com/series/5304

The Lecture Notes in Physics

The series Lecture Notes in Physics (LNP), founded in 1969, reports new developments in physics research and teaching—quickly and informally, but with a high quality and the explicit aim to summarize and communicate current knowledge in an accessible way. Books published in this series are conceived as bridging material between advanced graduate textbooks and the forefront of research and to serve three purposes:

- to be a compact and modern up-to-date source of reference on a well-defined topic
- to serve as an accessible introduction to the field to postgraduate students and nonspecialist researchers from related areas
- to be a source of advanced teaching material for specialized seminars, courses and schools

Both monographs and multi-author volumes will be considered for publication. Edited volumes should, however, consist of a very limited number of contributions only. Proceedings will not be considered for LNP.

Volumes published in LNP are disseminated both in print and in electronic formats, the electronic archive being available at springerlink.com. The series content is indexed, abstracted and referenced by many abstracting and information services, bibliographic networks, subscription agencies, library networks, and consortia.

Proposals should be sent to a member of the Editorial Board, or directly to the managing editor at Springer:

Christian Caron
Springer Heidelberg
Physics Editorial Department I
Tiergartenstrasse 17
69121 Heidelberg/Germany
christian.caron@springer.com

Christian Beck
Editor

Clusters in Nuclei, Vol.2

 Springer

Christian Beck
Dept. de Recherches Subatomiques
Institut Pluridiciplinaire Hubert Curien
Université de Strasbourg
Bat. 20
23, rue du Loess BP28
67037 Strasbourg, Cedex 2
France

ISSN 0075-8450
ISBN 978-3-642-24706-4
DOI 10.1007/978-3-642-24707-1
Springer Heidelberg New York Dordrecht London

e-ISSN 1616-6361
e-ISBN 978-3-642-24707-1

Library of Congress Control Number: 2010932330

Printed on acid-free paper

Springer is part of Springer Science+Business Media (www.springer.com)

The second volume of the new Series Lecture Notes in Physics **Clusters in Nuclei** *is dedicated to the memory of both Professor David Alan Bromley (1926–2005) and Professor Nikola Cindro (1931–2001).*

Christian Beck

Preface

This is the second volume in a series of *Lecture Notes in Physics* entitled "Clusters in Nuclei" based on the well known Cluster Conferences that have been running since decades, on two recent **State Of The Art in Nuclear Cluster Physics** Workshops, as well as on successfull Theoretical Winter Schools, traditionally held on the Campus of the Université de Strasbourg.

A great deal of research work has been done in the field of alpha clustering and in cluster studies of light neutron-rich nuclei. The scope of this new Series of lecture notes is to deepen our knowledge of the field of nuclear cluster physics which is one of the domains of heavy-ion nuclear physics facing the greatest challenges and opportunities.

The purpose of this second volume of *Lecture Notes in Physics* **Clusters in Nuclei**, is to promote the exchange of ideas and discuss new developments in "Clustering Phenomena in Nuclear Physics and Nuclear Astrophysics" from both the theoretical and experimental points of views. It is aimed to retain the pedagogical nature of our earlier Theoretical Winter Schools and should provide a helpful reference for young researchers entering the field and wishing to get a feel of contemporary research in a number of areas.

The various aspects of the main topics in this second volume of **Clusters in Nuclei** are divided into six chapters, each highlighting new ideas that have emerged in recent years:

- Microscopic Cluster Models
- Neutron Halo and Breakup Reactions
- Breakup reaction models for two- and three-cluster projectiles
- Clustering Effects within the Dinuclear Model
- Nuclear Alpha-Particle Condensates
- Clusters in Nuclei: Experimental Perspectives

The first chapter entitled **Microscopic Cluster Models** by Descouvemont and Dufour shows how clustering aspects can be fairly well described by microscopic cluster models such as the Resonating Group Method or the Generator Coordinate Method (GCM). For the sake of pedagogy, the formalism is presented in simple conditions by assuming spinless clusters and single-channel calculations. Extensions of the GCM to multicluster and multichannel calculations are compared to different three-alpha descriptions of ^{12}C proposed as typical illustrative examples.

The second chapter of Nakamura on **Neutron Halo and Breakup Reactions** connects the phenomenological aspects of neutron halos to experimental results collected for breakup reactions at intermediate/high energies. Nakamura investigates the breakup reactions as playing significant role in elucidating exotic structures along the neutron drip line. This study is be very important for further investigations of drip-line nuclei towards heavier mass regions available with the new-generation Radioactive Ion Beam (RIB) facilities: SPIRAL2, FAIR etc. in Europe, RIBF and KoRIA in Asia and FRIB in the US.

Breakup reaction models for two- and three-cluster projectiles are deeply discussed in Chap. 3 by Baye and Capel to provide a precise reaction picture coupled to a fair description of the projectile. The projectile is assumed to possess a cluster structure revealed by the dissociation process. This cluster structure is described by a few-body Hamiltonian involving effective forces between the clusters. Within this assumption, various reaction models are reviewed.

The Chap. 4 entitled **Clustering Effects within the Dinuclear Model** by Adamian, Antonenko and Scheid describes clustering of two nuclei as a dinuclear system (DNS) following Volkov ideas. The problems of fusion dynamics in the production of superheavy nuclei, of the quasifission process and of multi-nucleon transfer between nuclei are revisited within the DNS concept. Similarly, ternary fission processes are discussed within the scission-point picture.

Yamada, Schuck and their collaborators are trying in Chap. 5 (**Nuclear Alpha-Particle Condensates**) to definitively demonstrate that a typical alpha-particle condensate is the Hoyle state of ^{12}C, which plays a crucial role for the synthesis of light-mass elements in the universe. It is conjectured that alpha-particle condensate states also exist in heavier n alpha nuclei in qualitative agreement with the experimental observations presented in Vol.1 by von Oertzen in his lecture notes.

Finally, the last chapter **Clusters in Nuclei: Experimental Perspectives** proposed by Papka treats most of experimental aspects of nuclear cluster states studies from traditional techniques to the most recent developments and emerging methods. Many aspects of acceleration, including high-intensity, low-energy and radioactiveion beams are detailed in the context of nuclear clustering. The interest in combining radioactive beams and active targets are also addressed in terms of new perspectives and possible fields of future investigations.

Each forthcoming volume will also contain lectures covering a wide range of topics from nuclear cluster theory to experimental applications that have gained a renewed interest with available RIB facilities and modern detection techniques. We stress that the contributions in this volume and the following ones are not review articles and so are not meant to contain all the latest results or to provide an exhaustive coverage of the field but are written instead in the pedagogical style of graduate lectures and thus have a reasonable long 'shelf life'.

The edition of this book could not have been possible without stimulous discussions with Profs. Greiner, Horiuchi, and Schuck. Our appreciation goes to all our co-lectures for their valuable contributions. We acknowledge also all the referees for their comments on the chapters that are included in this volume. I would like here to thank, more particularly, Prof. Poenaru for his constant helpfull suggestions from the beginning to the end. Special thanks go Dr. Christian Caron and all the members of his Springer-Verlag team (in particular, Mrs Angela Schulze-thomin and Gabriele Hakuba) for their help, fruitful collaboration and continuedsupport for this ongoing project.

Strasbourg, May 2011   Christian Beck

Contents

Chapter 1
Microscopic Cluster Models

P. Descouvemont and M. Dufour

Abstract We present an overview of microscopic cluster models, by focusing on the Resonating Group Method (RGM) and on the Generator Coordinate Method (GCM). The wave functions of a nuclear system are defined from cluster wave functions, with an exact account of antisymmetrization between all nucleons. For the sake of pedagogy, the formalism is mostly presented in simple conditions, i.e. we essentially assume spinless clusters, and single-channel calculations. Generalizations going beyond these limitations are outlined. We present the GCM in more detail, and show how to compute matrix elements between Slater determinants. Specific examples dealing with α+nucleus systems are presented. We also discuss some approximations of the RGM, and in particular, the renormalized RGM which has been recently developed. We show that the GCM can be complemented by the microscopic variant of the R-matrix method, which provides a microscopic description of unbound states. Finally, extensions of the GCM to multicluster and multichannel calculations are discussed, and illustrated by typical examples. In particular we compare different three-α descriptions of ^{12}C.

1.1 Introduction

Clustering is a well-known effect in light nuclei [1]. Historically, the observation of clustering started with the α particle, which presents a large binding energy and therefore tends to keep its own identity in light nuclei. A description of nuclear states

P. Descouvemont (✉)
Physique Nucléaire Théorique et Physique Mathématique, C.P. 229,
Université Libre de Bruxelles (ULB), 1050 Brussels, Belgium
e-mail: pdesc@ulb.ac.be

M. Dufour
IPHC Bat27, IN2P3-CNRS/Université de Strasbourg,
BP28, 67037 Strasbourg Cedex 2, France
e-mail: Marianne.Dufour@IReS.in2p3fr

C. Beck (ed.), *Clusters in Nuclei, Vol.2*, Lecture Notes in Physics 848,
DOI: 10.1007/978-3-642-24707-1_1, © Springer-Verlag Berlin Heidelberg 2012

based on a cluster structure was first suggested by Wheeler [2] and by Margenau [3], and then extended by Brink [4]. This formulation is known as the α-cluster model, and has been widely used in the literature (see for example Ref. [5]). A typical example of α cluster states is the second 0^+ level in ^{12}C, known as the Hoyle state [6], which presents a strong $\alpha + ^8$Be cluster structure, and plays a crucial role in stellar evolution.

The cluster structure in α nuclei (i.e. with nucleon numbers $A = 4n$) was clarified by Ikeda [7] who proposed a diagram which identifies situations where a cluster structure can be observed. The α model and its extensions were utilized by many authors to investigate the properties of α-particle nuclei such as ^{8}Be, ^{12}C, ^{16}O, etc. In particular, the interest for α-cluster models was recently revived by the hypothesis of a new form of nuclear matter, in analogy with the Bose–Einstein condensates [8].

If the α particle, owing to its large binding energy, plays a central role in clustering phenomena, it soon became clear that other cluster structures can be observed. Reviews of recent developments in cluster physics can be found in Refs. [9, 10]. In many nuclei, some states present an α +nucleus structure. A well known example is ^{7}Li, well described by an $\alpha + t$ model [11]. In recent years, clustering phenomena have been observed in several nuclei such as ^{16}O, ^{18}O, ^{19}F, etc. More exotic states were suggested by Freer et al. [12] who found evidence for an ^{6}He+^{6}He rotational band in ^{12}Be. This unusual structure was subsequently supported by various calculations (see, for example [13, 14]).

An important property of clustering is that it may change from level to level in the same nucleus [15]. There are many examples: in ^{5}He, the ground state present an $\alpha + n$ structure, whereas the $3/2^+$ excited state is better described by a t+d configuration [16]. More generally, many nuclei exhibit α-cluster bands in their high-energy region. Recently "extreme" α-clustering has been reported in the ^{18}O nucleus [17].

The observation of clustering effects is the basis of cluster models, which are essentially divided into two categories: (i) non-microscopic models, where the internal structure of the clusters is neglected [18, 19], and (ii) microscopic theories where the clusters are described by shell-model wave functions [20, 21]. The Schrödinger equation is written as

$$H\Psi = E_T\Psi, \qquad (1.1.1)$$

where H is the Hamiltonian, $P\Psi$ the wave function, and E_T the total energy.

In non-microscopic approaches, the Hamiltonian of a system involving A nucleons distributed over N clusters is given by

$$H = \sum_{i=1}^{N} \frac{\mathbf{P}_i^2}{2M_i} + \sum_{i>j=1}^{N} V_{ij}(\mathbf{R}_i - \mathbf{R}_j), \qquad (1.1.2)$$

where the N clusters with masses M_i have a space coordinate $\mathbf{R}_i$ and a momentum $\mathbf{P}_i$. In this definition, V_{ij} is a nucleus–nucleus interaction which can be local or non local. It may also depend on other cluster coordinates such as the spin or the velocity. Of course, the simplest variant is a two-cluster model ($N=2$) where, after removal of the c.m. motion, the Hamiltonian only depends on the relative coordinate $\mathbf{r} = \mathbf{R}_1 - \mathbf{R}_2$. An important issue in non-microscopic models is the choice of the potentials V_{ij}. In general, these potentials are fitted on some properties of the system, such as binding energies or nucleus–nucleus phase shifts. In most cases they depend on the angular momentum between the clusters. It is well known that, to simulate the Pauli principle, this potential must satisfy some requirements. Deep potentials [19] or their supersymmetric partners [22] can partially account for antisymmetrization effects, although the associated wave functions neglect the structure of the clusters.

Non-microscopic theories can be extended to more than two clusters. For three-body models, the hyperspherical method [23] or the Faddeev approach [24] are efficient techniques. Because of their relative simplicity, at least for two-cluster variants, non-microscopic models can be directly extended to scattering states, i.e. to solutions of (1.1.2) at positive energies. This raises difficulties to properly include the asymptotic behaviour of the wave function, but is now well mastered for two-body and three-body scattering states.

The present work is devoted to microscopic cluster theories [20]. In a microscopic model, the Hamiltonian of the A-nucleon system is written as

$$H = \sum_{i=1}^{A} \frac{\mathbf{p}_i^2}{2m_N} + \sum_{i>j=1}^{A} v_{ij}(\mathbf{r}_i - \mathbf{r}_j), \tag{1.1.3}$$

where m_N is the nucleon mass (assumed to be equal for neutrons and protons), $\mathbf{r}_i$ and $\mathbf{p}_i$ are the space coordinate and momentum of nucleon i, and v_{ij} a nucleon–nucleon interaction. We explicitly mention the dependence on space coordinates, but v_{ij} may also depend on other nucleon coordinates. Until now, most microscopic cluster calculations neglect three-body forces (see however Ref. [25]).

Hamiltonian (1.1.3) is common to all microscopic theories, which explicitly treat all nucleons of the system. Examples are the shell model [26] and its "No-Core" extensions [27], the Antisymmetric Molecular Dynamics (AMD, see Ref. [28]), or the Fermionic Molecular Dynamics (FMD, see Ref. [29]). For small nucleon numbers (i.e. $A \leq 4$), efficient techniques are available to solve the Schrödinger equation with realistic nucleon–nucleon interactions (see Ref. [30] and references therein). These methods can be applied to bound as well as to continuum states. When the nucleon number is larger, some approximation must be used. The specificity of cluster models is that the wave function of the A-nucleon system, solution of the Schrödinger equation associated with (1.1.3), is described within the cluster approximation. In other words, the A nucleons are assumed to be divided in clusters, described by shell-model wave functions, and the total wave function is fully antisymmetric. For a two-cluster system with internal wave functions ϕ_1 and ϕ_2, the total wave function is written as

$$\Psi = \mathscr{A}\phi_1\phi_2 g(\boldsymbol{\rho}), \tag{1.1.4}$$

where $\mathscr{A}$ is the A-nucleon antisymmetrizor, and the radial function $g(\rho)$ depends on the relative coordinate $\boldsymbol{\rho}$. The cluster approximation is at the origin of the Resonating

Group Method (RGM) proposed by Wheeler [2] and widely used and developed by many groups (see for example [16, 20, 31]).

A significant breakthrough in microscopic cluster theories was achieved by the introduction of the Generator Coordinate Method (GCM), equivalent to the RGM, but allowing simpler and more systematic calculations [32]. The principle of the GCM is to expand the radial wave function $g(\rho)$ in a Gaussian basis. Under some restrictions, the total wave function (1.4) can then be rewritten as a combination of Slater determinants, well adapted to numerical calculations. Over the last decades, the GCM was developed in various directions: multi-cluster extensions [33–35], improved shell-model descriptions of the cluster wave functions [36], monopole distortion of the clusters [37], etc.

In nuclear spectroscopy, microscopic cluster models present a wide range of applications. They are remarkably well suited to molecular states, which are known to be strongly deformed, and present a marked cluster structure (see, for example, Refs. [13, 38–40]). The physics of exotic nuclei, and in particular of halo nuclei, is rather recent [41], and is also well described by cluster models. These nuclei are regarded as a core surrounded by external nucleons moving at large distances [42], and can be considered as cluster systems (see, for example, Refs. [43–45]). Unbound nuclei are extreme applications of cluster models, well adapted to resonances [46]. Several other applications, such as β decay [47] or charge symmetry in the Asymptotic Normalization Constant [48], have also been analyzed within microscopic cluster theories.

Microscopic cluster models have been also applied to various types of reactions: elastic, inelastic, transfer, etc. At low energies, the wavelength associated with the relative motion is large with respect to the typical dimensions of the system, and antisymmetrization effects are expected to be important. Microscopic theories have been widely applied in nuclear astrophysics (see e.g. [49, 50]), where measurements in laboratories are in general impossible at stellar energies [51–53]. This includes low-energy capture and transfer processes. Other nuclear reactions, such as nucleus–nucleus bremsstrahlung [54], have been studied in microscopic approaches. Being restricted to a limited number of cluster configurations (in general one), a microscopic cluster model is well adapted to the spectroscopy of low-lying states, and to low-energy reactions, where the level density and the number of open channels are limited.

It is of course impossible to provide an exhaustive bibliography of microscopic cluster theories. Excellent reviews can be found, for example, in Refs. [16, 20, 32, 55–57]. The paper is organized as follows. In Sect. 1.2, we discuss effective nucleon–nucleon interactions used in microscopic theories. In Sect. 1.3, we present the RGM in simple conditions: we consider systems made of two spinless clusters. We present an illustrative example with the $\alpha + n$ system. Section 1.4 is devoted to the GCM and to its link with the RGM. In Sect. 1.5 we give more specific information on the calculation of GCM matrix elements. In Sect. 1.6, we discuss some approximations and reformulations of the RGM equations. Section 1.7 is devoted to extensions of the model to multicluster and multichannel approaches. The treatment of scattering states in the GCM framework is outlined in Sect. 1.8. We discuss some applications of the RGM in Sect. 1.9. Concluding remarks are presented in Sect. 1.10.

1.2 Choice of the Nucleon–Nucleon Interaction

In the A-body Hamiltonian (1.1.3), the nucleon–nucleon interaction v_{ij} must account for the cluster approximation of the wave function. This leads to effective interactions, adapted to harmonic-oscillator orbitals. For example, using $0s$ orbitals for the α particle makes all matrix elements of non-central forces equal to zero. The effect of non-central components is simulated by an appropriate choice of the central effective interaction.

The nucleon–nucleon interaction contains Coulomb and nuclear terms and is written as

$$v_{ij}(r) = v_{ij}^C(r) + v_{ij}^N(r), \tag{1.2.1}$$

where the Coulomb term

$$v_{ij}^C(r) = \frac{e^2}{r} \left(\frac{1}{2} - t_{iz} \right) \left(\frac{1}{2} - t_{jz} \right), \tag{1.2.2}$$

is defined in the isospin formalism. For the nuclear term, most calculations performed with the RGM use central $v_{ij}^{N,c}(r)$ and spin–orbit $v_{ij}^{N,so}(r)$ interactions with

$$v_{ij}^N(r) = v_{ij}^{N,c}(r) + v_{ij}^{N,so}(r). \tag{1.2.3}$$

In general, the central part is written as a combination of N_g Gaussian form factors

$$v_{ij}^{N,c}(r) = \sum_{k=1}^{N_g} V_{0k} \exp(-(r/a_k)^2)(w_k - m_k P_{ij}^\sigma P_{ij}^\tau + b_k P_{ij}^\sigma - h_k P_{ij}^\tau). \tag{1.2.4}$$

Other potentials, such as the M3Y force [58] are defined from Yukawa form factors. However, the use of Gaussian form factors is well adapted to harmonic-oscillator orbitals. Parameters V_{0k} and a_k are given in Table 1.1 for the Volkov V2 [59] and Minnesota [60] interactions. Both forces contain one adjustable parameter (M and u, respectively). The standard values are $M=0.6$ and $u=1$, but these parameters can be slightly modified in order to reproduce an important property of the system. A typical example is the energy of a resonance or of a bound state.

The Volkov interaction involves two Gaussian functions and does not depend on spin and isospin ($b_k = h_k = 0$). With this force the deuteron binding energy is underestimated and the dineutron system is bound with the same energy. The Minnesota interaction [60] is defined by three different Gaussian functions. This force reproduces the deuteron binding energy and some properties of nucleon–nucleon scattering. It simulates the missing tensor force in the binding energy, as well as possible three-body effects, through the central term. Of course, the quadrupole moment of the deuteron, which is determined by the tensor force, is exactly zero with the Minnesota interaction.

Table 1.1 Amplitudes V_{0k} (in MeV) and ranges a_k (in fm) of the Volkov V2 and Minnesota interactions

Interaction	k	V_{0k}	a_k	w_k	m_k	b_k	h_k
Volkov V2	1	-60.65	1.80	$1-M$	M	0	0
	2	61.14	1.01	$1-M$	M	0	0
Minnesota	1	200	$1/\sqrt{1.487}$	$u/2$	$1-u/2$	0	0
	2	-178	$1/\sqrt{0.639}$	$u/4$	$1/2-u/4$	$u/4$	$1/2-u/4$
	3	-91.85	$1/\sqrt{0.465}$	$u/4$	$1/2-u/4$	$-u/4$	$u/4-1/2$

In Ref. [40], we have extended the Volkov $V2$ interaction by introducing Bartlett and Heisenberg components. This development was motivated by the need for more flexible interactions, able to reproduce thresholds in transfer reactions. This force is referred to as the EVI (Extended Volkov Interaction) interaction.

In most calculations a spin–orbit term is included. We take it as defined in Ref. [61] (see also [60]),

$$v_{ij}^{N,so}(\mathbf{r}) = -\frac{S_0}{\hbar^2 r_0^5}\big((\mathbf{r}_i - \mathbf{r}_j) \times (\mathbf{p}_i - \mathbf{p}_j)\big) \cdot (\mathbf{s}_i + \mathbf{s}_j)\exp\big(-(r/r_0)^2\big), \qquad (1.2.5)$$

where S_0 is the amplitude (expressed in MeV.fm^5), and $\mathbf{s}_i$ is the spin of nucleon i. Standard values of S_0 are $S_0 \approx 30\,\text{MeV.fm}^5$, which provides a fair approximation of the $1/2^- - 3/2^-$ energy splitting in ^{15}N. We use a range $r_0 = 0.1$ fm, which is equivalent to a zero-range force.

As mentioned earlier, cluster models make use of effective nucleon–nucleon forces. In contrast, *ab initio* models [29, 62] aim at determining exact solutions of the Schrödinger equation (1.1.1), without the cluster approximation. For instance, the No-Core Shell Model (NCSM) is based on very large one-center harmonic-oscillator (HO) bases and effective interactions [63], derived from realistic forces such as Argonne [64] or CD-Bonn [65]. These interactions are adapted for finite model spaces through a particular unitary transformation. Wave functions are then expected to be accurate, but states presenting a strong clustering remain difficult to describe with this model. Indeed, in spite of considerable advances in computer facilities, the calculations remain limited by the size of the model space. Realistic interactions are adjusted to reproduce properties of the nucleon–nucleon system with a high precision. The necessity to introduce a $3N$ force or more ($4N$, ...) is now established in order to get highly accurate spectra [66]. However, genuine expressions of these potentials remain under study [66].

1.3 The Resonating Group Method

1.3.1 The RGM Equation

Let us consider A nucleons with coordinates $\mathbf{r}_i$, assumed to be divided in two clusters with A_1 and A_2 nucleons. The center of mass (c.m.) of each cluster is given by

$$\mathbf{R}_{cm,1} = \frac{1}{A_1} \sum_{i=1}^{A_1} \mathbf{r}_i,$$

$$\mathbf{R}_{cm,2} = \frac{1}{A_2} \sum_{i=A_1+1}^{A} \mathbf{r}_i, \qquad (1.3.1)$$

which define the c.m. of the system $\mathbf{R}_{cm}$, and the relative coordinate $\boldsymbol{\rho}$ as

$$\mathbf{R}_{cm} = \frac{1}{A} (A_1 \mathbf{R}_{cm,1} + A_2 \mathbf{R}_{cm,2}),$$

$$\boldsymbol{\rho} = \mathbf{R}_{cm,2} - \mathbf{R}_{cm,1}. \qquad (1.3.2)$$

For each cluster, we define a set of translation-invariant coordinates

$$\boldsymbol{\xi}_{1i} = \mathbf{r}_i - \mathbf{R}_{cm,1} \quad \text{for } i = 1, \ldots, A_1,$$

$$\boldsymbol{\xi}_{2i} = \mathbf{r}_i - \mathbf{R}_{cm,2} \quad \text{for } i = A_1 + 1, \ldots, A. \qquad (1.3.3)$$

The A_1 and A_2 sets of coordinates $\boldsymbol{\xi}_{1i}$ and $\boldsymbol{\xi}_{2i}$ are not independent since we have, according to the definitions of $\mathbf{R}_{cm,1}$ and $\mathbf{R}_{cm,2}$,

$$\sum_{i=1}^{A_1} \boldsymbol{\xi}_{1i} = \sum_{i=A_1+1}^{A} \boldsymbol{\xi}_{2i} = 0. \qquad (1.3.4)$$

In the RGM, the total wave function is based on internal cluster wave functions $\phi_1(\boldsymbol{\xi}_{1i})$ and $\phi_2(\boldsymbol{\xi}_{2i})$. These internal wave functions are defined in the harmonic-oscillator model with oscillator parameter b. Here, we always assume that the oscillator parameter is common to all clusters. Going beyond this approximation introduces serious technical problems due to spurious c.m. components (see for example [16, 37, 67]). The RGM wave function is written, for two-cluster systems, as

$$\Psi(\boldsymbol{\xi}_{1i}, \boldsymbol{\xi}_{2i}, \boldsymbol{\rho}) = \mathscr{A} \phi_1(\boldsymbol{\xi}_{1i}) \phi_2(\boldsymbol{\xi}_{2i}) g(\boldsymbol{\rho}), \qquad (1.3.5)$$

where $g(\boldsymbol{\rho})$ is the relative wave function, to be determined from the Schrödinger equation (1.1.1), and $\mathscr{A}$ the antisymmetrization operator

$$\mathscr{A} = \sum_{p=1}^{A!} \varepsilon_p P_p, \qquad (1.3.6)$$

where P_p is a permutation over the A nucleons and $\varepsilon_p = \pm 1$ is the sign of this permutation. This operator not only acts inside the clusters, but also contains exchange terms between them. With this definition, the antisymmetrization operator is not exactly a projector since we have

$$\mathscr{A}^2 = A!\mathscr{A}. \tag{1.3.7}$$

In Eq. (1.3.5), we do not include the spins of the clusters, neither the relative angular momentum between the clusters. Definition (1.3.5) only contains one cluster configuration or, in other words, a single arrangement of the nucleons. More generally, several cluster wave functions (1.3.5) can be combined to improve the total wave function of the system. Here we limit ourselves to this simple case, for the sake of clarity. Various extensions will be developed in Sect. 1.7.

At first glance, the RGM wave function may appear as suitable for cluster states only, where the cluster approximation is obvious. However, owing to the antisymmetrization operator $\mathscr{A}$, the RGM (and the equivalent GCM described in Sect. 1.4) can be also applied to non-cluster states, such as shell-model or single-particle states [56].

Another remarkable advantage of the RGM wave function (1.3.5) is its direct applicability to scattering states. The main issue for scattering is to treat the asymptotic behaviour of the wave functions. At large relative distances between the colliding nuclei, antisymmetrization effects are negligible and the factorization (1.3.5) is exact without the antisymmetrization operator. This property is one of the main advantages of the RGM with respect to other microscopic approaches, such as the shell model or the FMD, where the treatment of scattering states is a serious problem, in particular to go beyond nucleon+nucleus scattering [68].

To derive the relative wave function $g(\rho)$, let us rewrite Eq. (1.3.5) as

$$\Psi = \mathscr{A}\phi_1\phi_2 g(\boldsymbol{\rho}) = \int \mathscr{A}\phi_1\phi_2\delta(\boldsymbol{\rho} - \mathbf{r})g(\mathbf{r})d\mathbf{r}, \tag{1.3.8}$$

where $\mathbf{r}$ is a parameter on which operator $\mathscr{A}$ does not act, and where the internal coordinates are implied. Then, using (1.3.8) in the Schrödinger equation (1.1.1) provides the RGM equation

$$\int [\mathscr{H}(\boldsymbol{\rho}, \boldsymbol{\rho}') - E_T \mathscr{N}(\boldsymbol{\rho}, \boldsymbol{\rho}')]g(\boldsymbol{\rho}')d\boldsymbol{\rho}' = 0. \tag{1.3.9}$$

In this equation, $\mathscr{N}$ and $\mathscr{H}$ are the (non-local) overlap and Hamiltonian kernels defined as

$$\left\{ \begin{array}{c} \mathscr{N}(\boldsymbol{\rho}, \boldsymbol{\rho}') \\ \mathscr{H}(\boldsymbol{\rho}, \boldsymbol{\rho}') \end{array} \right\} = \langle \phi_1\phi_2\delta(\boldsymbol{\rho} - \mathbf{r})| \left\{ \begin{array}{c} 1 \\ H \end{array} \right\} |\mathscr{A}\phi_1\phi_2\delta(\boldsymbol{\rho}' - \mathbf{r})\rangle, \tag{1.3.10}$$

where the integrals are performed over the internal coordinates and over the relative coordinate $\mathbf{r}$. In the Hamiltonian operator H, the kinetic energy of the center of mass (c.m.) has been subtracted. Accordingly E_T is defined with respect to the c.m. energy.

The RGM equation (1.3.9) can be simplified further by rewriting $\mathscr{A}$ and H as

$$\mathscr{A} = 1 + \mathscr{A}'$$
$$H = H_1 + H_2 + H', \tag{1.3.11}$$

where $\mathscr{A}'$ only contains exchange terms, H_1 and H_2 are the internal Hamiltonians of the clusters, and H' is given by

$$H' = -\frac{\hbar^2}{2\mu}\Delta_\rho + \sum_{i=1}^{A_1}\sum_{j=A_1+1}^{A} v_{ij}, \tag{1.3.12}$$

μ being the reduced mass $\mu = \mu_0 m_N$ with $\mu_0 = A_1 A_2/(A_1 + A_2)$. The internal energies E_1 and E_2 are given by

$$E_i = \langle \phi_i | H_i | \phi_i \rangle, \tag{1.3.13}$$

and the relative energy E is

$$E = E_T - E_1 - E_2. \tag{1.3.14}$$

In these conditions, kernels (1.3.10) can be expressed as

$$\mathscr{N}(\boldsymbol{\rho}, \boldsymbol{\rho}') = \delta(\boldsymbol{\rho} - \boldsymbol{\rho}') + \mathscr{N}_E(\boldsymbol{\rho}, \boldsymbol{\rho}')$$
$$\mathscr{H}(\boldsymbol{\rho}, \boldsymbol{\rho}') = \left(-\frac{\hbar^2}{2\mu}\Delta_\rho + V_D(\boldsymbol{\rho}) + E_1 + E_2\right)\delta(\boldsymbol{\rho} - \boldsymbol{\rho}') + \mathscr{H}_E(\boldsymbol{\rho}, \boldsymbol{\rho}') \tag{1.3.15}$$

where $\mathscr{N}_E$ and $\mathscr{H}_E$ are the exchange kernels, and where the direct potential V_D is given by

$$V_D(\boldsymbol{\rho}) = \langle \phi_1 \phi_2 | \sum_{i=1}^{A_1}\sum_{j=1}^{A_2} v_{ij} | \phi_1 \phi_2 \rangle. \tag{1.3.16}$$

The RGM equation (1.3.9) is finally written as

$$\left(-\frac{\hbar^2}{2\mu}\Delta_\rho + V_D(\boldsymbol{\rho})\right)g(\boldsymbol{\rho}) + \int K(\boldsymbol{\rho}, \boldsymbol{\rho}')g(\boldsymbol{\rho}')d\boldsymbol{\rho}' = Eg(\boldsymbol{\rho}), \tag{1.3.17}$$

with

$$K(\boldsymbol{\rho}, \boldsymbol{\rho}') = \mathscr{H}_E(\boldsymbol{\rho}, \boldsymbol{\rho}') - E_T \mathscr{N}_E(\boldsymbol{\rho}, \boldsymbol{\rho}'). \tag{1.3.18}$$

Equation (1.3.17) is the standard form of the RGM equation. It can be solved by different techniques (see for example [69]). The non-local term (1.3.18) is energy dependent and arises from exchange effects in the antisymmetrization operator (1.3.11). If $\mathscr{A}' = 0$, i.e. if antisymmetrization is neglected, the kernels $\mathscr{N}_E$ and $\mathscr{H}_E$ are equal to zero. In this simple approximation, the RGM equation is local and only involves the direct potential V_D.

1.3.2 Example: Overlap Kernel of the $\alpha+n$ System

This simple example illustrates the calculation of the overlap kernel. The extension
to the Hamiltonian kernels is given in [70]. Let us consider the α and neutron internal
wave functions

$$\phi_1 = \Phi_\alpha(\xi_1, \xi_2, \xi_3)|n_1 \downarrow n_1 \uparrow p_1 \downarrow p_1 \uparrow>$$
$$\phi_2 = |n_2 \downarrow>, \qquad\qquad\qquad\qquad\qquad (1.3.19)$$

where we have factorized the space and spin/isospin components. The spatial component Φ_α of the α-particle wave function is built from $0s$ oscillator orbitals with
parameter $\nu = 1/2b^2$. In the coordinate system (1.3.3) it is given as

$$\Phi_\alpha(\xi_1, \xi_2, \xi_3) = \frac{1}{N}\exp(-\nu\sum_{i=1}^{4}\xi_i^2), \qquad (1.3.20)$$

with the normalization factor defined by

$$\langle \Phi_\alpha \mid \Phi_\alpha \rangle = 1$$
$$= \frac{1}{N^2}\int\int\int \exp\left(-2\nu(\xi_1^2+\xi_2^2+\xi_3^2 + (\xi_1 + \xi_2 + \xi_3)^2)\right) d\xi_1 d\xi_2 d\xi_3$$
$$= \frac{1}{N^2}\left(\frac{\pi^3}{32\nu^3}\right)^{3/2}.$$
$$(1.3.21)$$

Notice that the internal wave function (1.3.20) only depends on three independent
coordinates [see Eqs. (1.3.3, 1.3.4)]. Coordinate ξ_4 is defined from (1.3.4).

Since we assume that the external neutron has a spin down, only the exchange
operator P_{15} between nucleons 1 and 5 contributes in the antisymmetrization operator
(1.3.6). Applying P_{15} on the internal and relative coordinates provides

$$P_{15}\xi_1 = \frac{3}{4}\rho + \frac{1}{4}\xi_1,$$

$$P_{15}\xi_2 = -\frac{1}{4}\rho + \frac{1}{4}\xi_1 + \xi_2,$$

$$P_{15}\xi_3 = -\frac{1}{4}\rho + \frac{1}{4}\xi_1 + \xi_3,$$

$$P_{15}\rho = -\frac{1}{4}\rho + \frac{5}{4}\xi_1. \qquad\qquad (1.3.22)$$

A simple calculation leads to

$$P_{15}\Phi_\alpha\delta(\rho - \mathbf{r}) = \Phi_\alpha\exp\left[-\frac{4\nu}{5}(r^2 - (P_{15}\rho)^2)\right]\delta(P_{15}\rho - \mathbf{r}), \qquad (1.3.23)$$

and the exchange overlap kernel is deduced from (1.3.15) as

$$\mathcal{N}_E(\boldsymbol{\rho}, \boldsymbol{\rho}') = - \int \int \int \int d\mathbf{r} d\boldsymbol{\xi}_1 d\boldsymbol{\xi}_2 d\boldsymbol{\xi}_3 \Phi_\alpha(\boldsymbol{\xi}_1, \boldsymbol{\xi}_2, \boldsymbol{\xi}_3)$$
$$\times \delta(\boldsymbol{\rho} - \mathbf{r}) P_{15} \Phi_\alpha(\boldsymbol{\xi}_1, \boldsymbol{\xi}_2, \boldsymbol{\xi}_3) \delta(\boldsymbol{\rho}' - \mathbf{r}). \tag{1.3.24}$$

The integral is first performed over $\mathbf{r}$. Then, integration over $\boldsymbol{\xi}_2$ and $\boldsymbol{\xi}_3$ provides

$$\int |\Phi_\alpha(\boldsymbol{\xi}_1, \boldsymbol{\xi}_2, \boldsymbol{\xi}_3)|^2 d\boldsymbol{\xi}_2 d\boldsymbol{\xi}_3 = \left(\frac{\sqrt{3}\pi}{6\nu}\right)^3 \exp\left(-\frac{8\nu}{3}\boldsymbol{\xi}_1^2\right). \tag{1.3.25}$$

This gives, by integrating over $\boldsymbol{\xi}_1$ and using the delta function in (1.3.23),

$$\mathcal{N}_E(\boldsymbol{\rho}, \boldsymbol{\rho}') = - \left(\frac{4}{5}\right)^3 \left(\frac{8\nu}{3\pi}\right)^{3/2} \exp\left[-\frac{4\nu}{75}(17\rho^2 + 17\rho'^2 + 16\boldsymbol{\rho} \cdot \boldsymbol{\rho}')\right]. \tag{1.3.26}$$

This result can be also found in Refs. [70, 71] for example (see also Ref. [72]). Of course it does not depend on the spin and isospin of the external nucleon.

1.4 The Generator Coordinate Method

1.4.1 Introduction

The main problem associated with the RGM is not to solve the integro-differential equation (1.3.17). This can be done, for example, by using finite-difference methods [73], or the Lagrange-mesh technique [69, 74]. In contrast, the determination of the overlap and Hamiltonian kernels (1.3.10) requires heavy analytical calculations, in particular for systems involving p-shell clusters. The non-systematic character of the RGM makes it quite difficult to apply in multicluster systems or in multichannel problems.

This limitation received an efficient solution with the introduction of the Generator Coordinate Method [16, 32, 75]. The idea underlying the GCM is to expand the radial function $g(\boldsymbol{\rho})$ (1.3.5) over Gaussian functions, centered at different locations, called the generator coordinates. This expansion allows to express the total wave function (1.3.5) as a superposition of Slater determinants. The RGM and the GCM methods are therefore equivalent, but the use of Slater determinants makes the GCM better adapted to numerical calculations. The GCM has been applied, in the last decades, to many nuclei or reactions. In particular the spectroscopy of exotic nuclei [45, 76], and reactions of astrophysical interest [77] have been investigated.

1.4.2 Slater Determinants and GCM Basis Functions

Let us consider a one-center Slater determinant $\Phi_1(\mathbf{S})$ built from A_1 orbitals. All orbitals are centered at a common location $\mathbf{S}$ as

$$\Phi_1(\mathbf{S}) = \frac{1}{\sqrt{A_1!}} \det\left\{\hat{\varphi}_1(\mathbf{S}) \ldots \hat{\varphi}_{A_1}(\mathbf{S})\right\} = \frac{1}{\sqrt{A_1!}} \mathscr{A} \hat{\varphi}_1(\mathbf{S}) \ldots \hat{\varphi}_{A_1}(\mathbf{S}), \qquad (1.4.1)$$

where the individual orbitals $\hat{\varphi}_i(\mathbf{S})$ are factorized in space, spin and isospin components. Each function $\hat{\varphi}_i(\mathbf{S})$ is therefore defined as

$$\hat{\varphi}_i(\mathbf{S}) = \varphi_i(\mathbf{r}, \mathbf{S})|m_{s_i}\rangle|m_{t_i}\rangle, \qquad (1.4.2)$$

where $|m_{s_i}\rangle$ is a spinor and $|m_{t_i}\rangle$ the isospin function. In this definition, the space, spin and isospin coordinates are implied. The radial part $\varphi_i(\mathbf{r}, \mathbf{S})$ is an harmonic-oscillator function, normalized to unity [78]. For s waves, it reads

$$\varphi_i(\mathbf{r}, \mathbf{S}) = \varphi_{0s}(\mathbf{r}, \mathbf{S}) = (\pi b^2)^{-3/4} \exp\left(-\frac{(\mathbf{r} - \mathbf{S})^2}{2b^2}\right). \qquad (1.4.3)$$

The parameter ν and oscillator energy $\hbar\omega$ are related to the oscillator parameter b as

$$\nu = 1/2b^2$$
$$\hbar\omega = \frac{\hbar^2}{m_N b^2}. \qquad (1.4.4)$$

For p waves, the radial functions are

$$\varphi_i(\mathbf{r}, \mathbf{S}) = \varphi_{1p\mu}(\mathbf{r}, \mathbf{S}) = \frac{\sqrt{2}}{b}(r_\mu - S_\mu)\varphi_{0s}(\mathbf{r}, \mathbf{S}), \qquad (1.4.5)$$

where index μ corresponds to the Cartesian coordinates (x, y, z). In the following we do not explicitly write the labels $0s$ or $1p\mu$ to the nucleon orbitals. We assume that all orbitals have a common oscillator parameter and are all centered at the same location. This is different from the AMD or FMD, where the oscillator parameters are optimized individually for each nucleon.

A drawback of the internal wave function (1.4.1) is that it is not invariant under translation. However, the Slater determinant (1.4.1) can be rewritten as

$$\Phi_1(\mathbf{S}) = \exp\left(-\frac{A_1}{2b^2}(\mathbf{R}_{cm,1} - \mathbf{S})^2\right)\phi_1, \qquad (1.4.6)$$

where ϕ_1 is the translation-invariant function defined in Sect. 1.3.1, and where $\mathbf{R}_{cm,1}$ is the c.m. coordinate (1.3.1) of the system. The factorization (1.4.6) is known as the Bethe and Rose theorem [79], and assumes that all shells below some maximum are included in the Slater determinant (1.4.1).

Let us now consider a two-center wave function defined from two cluster functions (1.4.1) located at $\mathbf{S}_1$ and $\mathbf{S}_2$. The generator coordinate is defined as $\mathbf{R} = \mathbf{S}_2 - \mathbf{S}_1$. We choose the origin of the system along the axis between $\mathbf{S}_1$ and $\mathbf{S}_2$. The location of the origin is therefore defined by a parameter λ (with $0 \leq \lambda \leq 1$). Typical values are $\lambda = 0$ which corresponds to the center of cluster 1, and $\lambda = A_2/A$, where the origin is located at the center of mass. We define the two-cluster Slater determinant as

$$
\begin{aligned}
\Phi(\mathbf{R}) &= \frac{1}{\sqrt{A!}} \det \left\{ \hat{\varphi}(-\lambda \mathbf{R}) \ldots \hat{\varphi}_{A_1}(-\lambda \mathbf{R}) \hat{\varphi}_{A_1+1}((1-\lambda)\mathbf{R}) \ldots \hat{\varphi}_A((1-\lambda)\mathbf{R}) \right\}, \\
&= \frac{1}{\sqrt{N_0}} \mathscr{A} \, \Phi_1(-\lambda \mathbf{R}) \Phi_2((1-\lambda)\mathbf{R}),
\end{aligned} \tag{1.4.7}
$$

where the nucleon coordinates are implied. The normalization factor $N_0 = \frac{A!}{A_1! A_2!}$ stems from property (1.3.7). This Slater determinant is built with A_1 orbitals at $-\lambda \mathbf{R}$ and A_2 orbitals at $(1-\lambda)\mathbf{R}$. Definition (1.4.7) can be directly extended to more than two clusters [21, 32, 80]. Obviously this basis function is not invariant under translation. However, by using the factorization (1.4.6) for both clusters, and assuming a common oscillator parameter b, Eq. (1.4.7) can be rewritten as

$$
\Phi(\mathbf{R}) = \frac{1}{\sqrt{N_0}} \Phi_{cm} \mathscr{A} \phi_1 \phi_2 \Gamma(\boldsymbol{\rho}, \mathbf{R}), \tag{1.4.8}
$$

which involves the translation-invariant functions ϕ_1 and ϕ_2. The c.m. and radial wave functions read

$$
\begin{aligned}
\Phi_{cm}(\mathbf{R_{cm}}) &= \left(\frac{A}{\pi b^2}\right)^{3/4} \exp\left(-\frac{A}{2b^2}\left[\mathbf{R}_{cm} + \mathbf{R}(\lambda - A_2/A)\right]^2\right), \\
\Gamma(\boldsymbol{\rho}, \mathbf{R}) &= \left(\frac{\mu_0}{\pi b^2}\right)^{3/4} \exp\left(-\frac{\mu_0}{2b^2}(\boldsymbol{\rho} - \mathbf{R})^2\right).
\end{aligned} \tag{1.4.9}
$$

The c.m. and radial coordinates are therefore uncoupled. The associated functions are simple Gaussian functions with oscillator parameters $b/\sqrt{A}$ and $b/\sqrt{\mu_0}$, respectively. This factorization of the c.m. motion greatly simplifies the calculation of GCM matrix elements. Let us express the Slater determinant (1.4.7) as

$$
\Phi(\mathbf{R}) = \Phi_{cm} \bar{\Phi}(\mathbf{R}), \tag{1.4.10}
$$

where $\bar{\Phi}(\mathbf{R})$ is a physical basis function, independent of the c.m. coordinate. Functions $\Phi(\mathbf{R})$, on the contrary, contain spurious c.m. components, but are well adapted to a numerical calculation since they are Slater determinants. Using (1.4.10), we have

$$
\langle \Phi(\mathbf{R}) | \Phi(\mathbf{R}') \rangle = \langle \Phi_{cm} | \Phi_{cm} \rangle \langle \bar{\Phi}(\mathbf{R}) | \bar{\Phi}(\mathbf{R}') \rangle, \tag{1.4.11}
$$

with

$$\langle \Phi_{cm} | \Phi_{cm} \rangle = \exp \left(-\frac{A(\lambda - A_2/A)^2}{4b^2} (\mathbf{R} - \mathbf{R}')^2 \right). \qquad (1.4.12)$$

The overlap between basis functions $\bar{\Phi}(\mathbf{R})$ is therefore obtained from a matrix elements between Slater determinants, corrected by a simple c.m. factor. Notice that both c.m. functions in the matrix elements (1.4.12) may involve different generator coordinates.

Matrix elements between GCM basis states $\bar{\Phi}(\mathbf{R})$ should not depend on λ. This provides a strong test of the calculations. The choice $\lambda = A_2/A$, i.e. taking the origin at the center of mass, is commonly used since c.m. correction factors are trivial. Another choice adopted in the literature is $\lambda = 0$, where all orbitals are centred at the origin of cluster 1. If the orbitals of the external clusters are orthogonalized to the core orbitals [32, 81], the calculation of matrix elements is strongly simplified (see Sect. 1.5). This technique is quite efficient when the core is a closed-shell nucleus (α, ^{16}O, ^{40}Ca, etc.) and is surrounded by $0s$ orbitals (see for example Ref. [82] for the ^{16}O+^{3}He+p three-cluster system).

Matrix elements of other operators should account for the spurious c.m. contribution. Garthenaus and Schwartz [83] have shown that the removal of the c.m. component of the wave function can be achieved by using transformed operators, obtained by a modification of the space and momentum coordinates as

$$\mathbf{r}_i \longrightarrow \mathbf{r}_i - \mathbf{R}_{cm}$$
$$\mathbf{p}_i \longrightarrow \mathbf{p}_i - \frac{1}{A} \mathbf{P}_{cm}, \qquad (1.4.13)$$

where $\mathbf{P}_{cm}$ is the c.m. momentum. The transformation-invariant forms of the kinetic energy and of the r.m.s. radius are therefore

$$T \longrightarrow T - T_{cm} = \sum_i \frac{\mathbf{p}_i^2}{2m_N} - T_{cm},$$
$$< r^2 > \longrightarrow < r^2 > - \mathbf{R}_{cm}^2 = \frac{1}{A} \sum_i \mathbf{r}_i^2 - \mathbf{R}_{cm}^2, \qquad (1.4.14)$$

and the c.m. matrix elements take the simple forms

$$\langle \Phi_{cm} | T_{cm} | \Phi_{cm} \rangle = \frac{1}{4} \hbar \omega \left(3 - \frac{A(\lambda - A_2/A)^2}{2b^2} (\mathbf{R} - \mathbf{R}')^2 \right) \langle \Phi_{cm} | \Phi_{cm} \rangle$$
$$\langle \Phi_{cm} | \mathbf{R}_{cm}^2 | \Phi_{cm} \rangle = \left(\frac{3}{2} b^2 + \frac{(\lambda - A_2/A)^2}{4} (\mathbf{R} + \mathbf{R}')^2 \right) \langle \Phi_{cm} | \Phi_{cm} \rangle. \qquad (1.4.15)$$

A similar calculation can be performed for the electric operators of rank L [84]. At the long-wavelength approximation, the translation-invariant form is defined as

$$\mathcal{M}_{LM}^E = e \sum_i \left(\frac{1}{2} - t_{iz}\right) |\mathbf{r}_i - \mathbf{R}_{cm}|^L Y_L^M (\Omega_{\mathbf{r}_i - \mathbf{R}_{cm}}). \tag{1.4.16}$$

This operator can be expanded as [78]

$$\mathcal{M}_{LM}^E = \sum_k \left(\frac{4\pi(2L+1)!}{(2k+1)!(2L-2k+1)!}\right)^{1/2} \left[\mathcal{M}_{L-k}^E(\mathbf{r}_i) \otimes \mathcal{M}_k^E(\mathbf{R}_{cm})\right]^{LM},$$
$$\tag{1.4.17}$$

where $\mathcal{M}_{L-k}^E(\mathbf{r}_i)$ is defined from (1.4.16) and where the c.m. contributions read

$$\mathcal{M}_{km}^E(\mathbf{R}_{cm}) = R_{cm}^k Y_k^m (\Omega_{\mathbf{R}_{cm}}). \tag{1.4.18}$$

Matrix elements can be obtained as in (1.4.15). However, the calculation of additional multipoles ($k < L$) can be avoided by choosing $\lambda = A_2/A$ or, in other words, by taking the c.m. as origin of the coordinate system. In that case, only $k=0$ contributes in the matrix elements of (1.4.17) and the c.m. correction is trivial. Similar developments can be performed for magnetic multipoles [85].

The factorization of the internal wave functions and of the radial part makes GCM basis functions (1.4.8) well adapted to collisions (see Sect. 1.8). If the oscillator parameters of the clusters are different, the removal of the spurious c.m. components is however a delicate problem [16, 86]. This can be tackled by using the Complex GCM [16]. In this variant the generator coordinate $\mathbf{R}$ is complex. The calculation of matrix elements is very similar, but the imaginary part provides an efficient tool to deal with different oscillator parameters of the clusters.

1.4.3 Equivalence Between RGM and GCM

In the two-cluster approximation, the total wave function of a system is defined as a superposition of GCM basis functions

$$\Psi = \Phi_{cm}^{-1} \int f(\mathbf{R}) \Phi(\mathbf{R}) d\mathbf{R}, \tag{1.4.19}$$

where $f(\mathbf{R})$ is the generator function, to be determined from the microscopic Hamiltonian (1.1.3) According to Eq. (1.4.8), wave function Ψ is invariant under translation. Using (1.4.8), wave function (1.4.19) reads

$$\Psi = \mathcal{A} \phi_1 \phi_2 g(\rho), \tag{1.4.20}$$

with

$$g(\rho) = \int f(\mathbf{R}) \Gamma(\rho, \mathbf{R}) d\mathbf{R}, \tag{1.4.21}$$

which shows the equivalent between the RGM and the GCM. The generator function is obtained from the Hill–Wheeler equation [87]

$$\int \left[H(\mathbf{R}, \mathbf{R}') - E_T N(\mathbf{R}, \mathbf{R}') \right] f(\mathbf{R}') d\mathbf{R}' = 0, \qquad (1.4.22)$$

where the GCM kernels are given by

$$\begin{Bmatrix} N(\mathbf{R}, \mathbf{R}') \\ H(\mathbf{R}, \mathbf{R}') \end{Bmatrix} = \langle \bar{\Phi}(\mathbf{R})| \begin{Bmatrix} 1 \\ H \end{Bmatrix} |\bar{\Phi}(\mathbf{R}')\rangle, \qquad (1.4.23)$$

and where c.m. components have been removed. In practice the integral in (1.4.21) is discretized over a finite set of R_n values as

$$g(\rho) \approx \sum_n f(\mathbf{R}_n) \Gamma(\rho, \mathbf{R}_n). \qquad (1.4.24)$$

For bound states, the integral equation (1.4.22) is therefore replaced by the diagonalization of a matrix (typically 10 R_n values are used). The treatment of scattering states requires an additional tool to correct for the Gaussian asymptotic behaviour of the relative function (1.4.24). This will be developed in Sect. 1.8.

The calculation of matrix elements between Slater determinants is well known [4] and will be discussed in Sect. 1.5 (see also Refs. [32, 67] for further detail). The RGM and GCM kernels [(1.3.10) and (1.4.23), respectively] can be linked to each other by integral transforms [32, 67, 88].

1.4.4 Two-Cluster Angular-Momentum Projection

Let us consider the partial-wave expansion of the GCM basis states (1.4.7)

$$\Phi(\mathbf{R}) = 4\pi \sum_{\ell m} \Phi^{\ell m}(R) Y_\ell^{m*}(\Omega_R). \qquad (1.4.25)$$

Notice that the overall normalization does not play a role, and can be chosen freely as long as it is consistently used in the calculation of the matrix elements. We use the expansion of (1.4.9)

$$\Gamma(\rho, \mathbf{R}) = 4\pi \sum_{\ell m} \Gamma_\ell(\rho, R) Y_\ell^m(\Omega_\rho) Y_\ell^{m*}(\Omega_R),$$

$$\Gamma_\ell(\rho, R) = \left(\frac{\mu_0}{\pi b^2} \right)^{3/4} \exp\left[-\frac{\mu_0}{2b^2}(\rho^2 + R^2) \right] i_\ell\left(\frac{\mu_0 \rho R}{b^2} \right), \qquad (1.4.26)$$

where $i_\ell(x)$ is a spherical Bessel function [89]. Then Eqs. (1.4.8) and (1.4.25) provide

$$\Phi^{\ell m}(R) = \frac{1}{\sqrt{N_0}} \mathscr{A} \phi_1 \phi_2 \Gamma_\ell(\rho, R) Y_\ell^m(\Omega_\rho) \qquad (1.4.27)$$

and the total wave function reads

$$\Psi^{\ell m} = \int f_\ell(R)\Phi^{\ell m}(R)dR. \tag{1.4.28}$$

The generator function $f_\ell(R)$ is obtained from the Hill–Wheeler equation involving projected GCM kernels

$$\int \left[H_\ell(R, R') - E_T N_\ell(R, R')\right]f_\ell(R')dR' = 0. \tag{1.4.29}$$

The projected overlap kernel $N_\ell(R, R')$ is derived from the expansion

$$
\begin{aligned}
N(\mathbf{R}, \mathbf{R}') &= \langle\Phi(\mathbf{R})|\Phi(\mathbf{R}')\rangle \\
&= (4\pi)^2 \sum_{\ell m}\langle\Phi_\ell(R)|\Phi_\ell(R')\rangle Y_\ell^m(\Omega_R)Y_\ell^{m*}(\Omega_{R'}) \\
&= 4\pi \sum_\ell (2\ell + 1)\langle\Phi_\ell(R)|\Phi_\ell(R')\rangle P_\ell(\cos\theta),
\end{aligned}
\tag{1.4.30}
$$

where θ is the angle between $\mathbf{R}$ and $\mathbf{R}'$. Then, by inverting (1.4.30), we find

$$N_\ell(R, R') = \langle\Phi_\ell(R)|\Phi_\ell(R')\rangle = \frac{1}{8\pi}\int_0^\pi N(\mathbf{R}, \mathbf{R}')P_\ell(\cos\theta)d\cos\theta, \tag{1.4.31}$$

and a similar equation holds for the Hamiltonian kernel $H_\ell(R, R')$. Since $N(\mathbf{R}, \mathbf{R}')$ and $H(\mathbf{R}, \mathbf{R}')$ only depend on the relative angle θ, the orientation of one generator coordinate can be chosen arbitrarily. A common choice is to take $\mathbf{R}$ along the z axis, and $\mathbf{R}'$ in the xz plane. The integration over θ can be performed numerically with a Gauss–Legendre quadrature, or analytically in some simple cases.

For two-clusters systems, the projection over parity is automatic since the angular-momentum projection provides

$$\pi = (-1)^\ell. \tag{1.4.32}$$

1.5 Matrix Elements Between Slater Determinants

1.5.1 General Presentation

Let us consider a system of A orbitals $\hat{\varphi}_i(\mathbf{S}_n)$ distributed among N clusters. The set of cluster locations is denoted as $\mathbf{S}_{\{N\}}$. As mentioned in (1.4.2) the individual orbitals involve space, spin and isospin components as

$$\hat{\varphi}_i(\mathbf{S}_n) = |\varphi_i(\mathbf{r}, \mathbf{S}_n)\rangle|m_{s_i}\rangle|m_{t_i}\rangle, \tag{1.5.1}$$

Fig. 1.1 Typical two-cluster
($\alpha + n$) and three-cluster
($^{16}O+^{3}He+p$)
configurations. The crosses
indicate the c.m. location
within the generator
coordinates

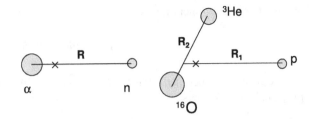

where $\varphi_i(\mathbf{r}, \mathbf{S}_n)$ is an harmonic-oscillator function. Most calculations are performed with $0s$ orbitals, but the presentation is more general; explicit definitions of s and p orbitals are given by (1.4.3) and (1.4.5). The A-nucleon Slater determinant reads

$$\Phi(\mathbf{S}_{\{N\}}) = \frac{1}{\sqrt{A!}} \det\{\hat{\varphi}_1(\mathbf{S}_1) \ldots \hat{\varphi}_A(\mathbf{S}_N)\}. \tag{1.5.2}$$

For example, the $\alpha + n$ system involves four s orbitals at $\mathbf{S}_1 = -\mathbf{R}/5$, and one $0s$ orbital at $\mathbf{S}_2 = 4\mathbf{R}/5$. The $^{19}Ne+p$ system, with ^{19}Ne described as $^{16}O+^{3}He$ [82], involves four $0s$ and twelve $1p$ orbitals at $\mathbf{S}_1 = -\mathbf{R}_1/20 - 3\mathbf{R}_2/19$, three $0s$ orbitals at $\mathbf{S}_2 = -\mathbf{R}_1/20 + 16\mathbf{R}_2/19$, and one $0s$ orbital at $\mathbf{S}_3 = 19\mathbf{R}_1/20$ (see Fig. 1.1).

The calculation of matrix elements between Slater determinants (1.5.2) is rather simple and systematic. We present here a short overview of the method, but more detail can be found in Refs. [4, 32]. The overlap is given by

$$\begin{aligned}
\langle \Phi(\mathbf{S}_{\{N\}}) | \Phi(\mathbf{S}'_{\{N\}}) \rangle &= \frac{1}{A!} \langle \mathscr{A}\hat{\varphi}_1(\mathbf{S}_1) \ldots \hat{\varphi}_A(\mathbf{S}_N) \mid \mathscr{A}\hat{\varphi}_1(\mathbf{S}'_1) \ldots \hat{\varphi}_A(\mathbf{S}'_N) \rangle \\
&= \langle \hat{\varphi}_1(\mathbf{S}_1) \ldots \hat{\varphi}_A(\mathbf{S}_N) \mid \mathscr{A}\hat{\varphi}_1(\mathbf{S}'_1) \ldots \hat{\varphi}_A(\mathbf{S}'_N) \rangle \\
&= \langle \hat{\varphi}_1(\mathbf{S}_1) \ldots \hat{\varphi}_A(\mathbf{S}_N) \mid \det \hat{\varphi}_1(\mathbf{S}'_1) \ldots \hat{\varphi}_A(\mathbf{S}'_N) \rangle \\
&= \det \mathbf{B},
\end{aligned} \tag{1.5.3}$$

where we have used (1.3.7), and where matrix $\mathbf{B}$ is given by the individual overlaps as

$$B_{ij} = \langle \hat{\varphi}_i(\mathbf{S}_i) | \hat{\varphi}_j(\mathbf{S}'_j) \rangle. \tag{1.5.4}$$

For one-body operators O_1 written as

$$O_1 = \sum_{i=1}^{A} o_1(\mathbf{r}_i), \tag{1.5.5}$$

a matrix element between Slater determinants reads

$$\langle O_1 \rangle = \langle \Phi(\mathbf{S}_{\{N\}}) | O_1 | \Phi(\mathbf{S}'_{\{N\}}) \rangle$$

$$= \sum_{ij=1}^{A} M_{ij} \langle \hat{\varphi}_i(\mathbf{S}_i) | o_1 | \hat{\varphi}_j(\mathbf{S}'_j) \rangle$$

$$= \det \mathbf{B} \sum_{ij=1}^{A} \left(\mathbf{B}^{-1} \right)_{ji} \langle \hat{\varphi}_i(\mathbf{S}_i) | o_1 | \hat{\varphi}_j(\mathbf{S}'_j) \rangle \quad \text{if } \det \mathbf{B} \neq 0, \qquad (1.5.6)$$

where M_{ij} is a cofactor of matrix $\mathbf{B}$. It is obtained from the determinant of $\mathbf{B}$ after removal of column i and line j, and multiplication by a phase factor $(-1)^{i+j}$. If $\det \mathbf{B} \neq 0$, we have

$$M_{ij} = \det \mathbf{B} \left(\mathbf{B}^{-1} \right)_{ji}. \qquad (1.5.7)$$

One-body matrix elements therefore involve a double sum over the individual orbitals. Here and in the following, we assume $\det \mathbf{B} \neq 0$, but the generalization is straightforward. Typical examples of one-body operators are the kinetic energy, the r.m.s. radius, and the electromagnetic operators.

For a two-body operators O_2 such as the nucleon–nucleon interaction

$$O_2 = \sum_{i>j=1}^{A} o_2(\mathbf{r}_i, \mathbf{r}_j) = \frac{1}{2} \sum_{i \neq j=1}^{A} o_2(\mathbf{r}_i, \mathbf{r}_j), \qquad (1.5.8)$$

a matrix element reads

$$\langle O_2 \rangle = \langle \Phi(\mathbf{S}_{\{N\}}) | O_2 | \Phi(\mathbf{S}'_{\{N\}}) \rangle$$

$$= \frac{1}{2} \sum_{ijkl=1}^{A} M_{ij,kl} \langle \hat{\varphi}_i(\mathbf{S}_i) \hat{\varphi}_j(\mathbf{S}_j) | o_2 | \hat{\varphi}_k(\mathbf{S}'_k) \hat{\varphi}_l(\mathbf{S}'_l) \rangle, \qquad (1.5.9)$$

where $M_{ij,kl}$ is a second-order cofactor of matrix $\mathbf{B}$. For $\det \mathbf{B} \neq 0$, we have

$$M_{ij,kl} = \det \mathbf{B} \left[\left(\mathbf{B}^{-1} \right)_{ki} \left(\mathbf{B}^{-1} \right)_{lj} - \left(\mathbf{B}^{-1} \right)_{kj} \left(\mathbf{B}^{-1} \right)_{li} \right]. \qquad (1.5.10)$$

In addition since the individual matrix elements satisfy the symmetry property

$$\langle \varphi_i(\mathbf{S}_i) \varphi_j(\mathbf{S}_j) | o_2 | \varphi_k(\mathbf{S}'_k) \varphi_l(\mathbf{S}'_l) \rangle = \langle \varphi_j(\mathbf{S}_j) \varphi_i(\mathbf{S}_i) | o_2 | \varphi_l(\mathbf{S}'_l) \varphi_k(\mathbf{S}'_k) \rangle, \qquad (1.5.11)$$

the following definitions are equivalent:

$$\langle O_2 \rangle = \frac{1}{2} \det \mathbf{B} \sum_{ijkl=1}^{A} \left[\left(\mathbf{B}^{-1} \right)_{ki} \left(\mathbf{B}^{-1} \right)_{lj} - \left(\mathbf{B}^{-1} \right)_{kj} \left(\mathbf{B}^{-1} \right)_{li} \right]$$

$$\times \langle \hat{\varphi}_i(\mathbf{S}_i) \hat{\varphi}_j(\mathbf{S}_j) | o_2 | \hat{\varphi}_k(\mathbf{S}'_k) \hat{\varphi}_l(\mathbf{S}'_l) \rangle$$

$$= \frac{1}{2} \det \mathbf{B} \sum_{ijkl=1}^{A} \left(\mathbf{B}^{-1} \right)_{ki} \left(\mathbf{B}^{-1} \right)_{lj} \left[\langle \hat{\varphi}_i(\mathbf{S}_i) \hat{\varphi}_j(\mathbf{S}_j) | o_2 | \hat{\varphi}_k(\mathbf{S}'_k) \hat{\varphi}_l(\mathbf{S}'_l) \rangle \right.$$

$$\left. - \langle \hat{\varphi}_i(\mathbf{S}_i) \hat{\varphi}_j(\mathbf{S}_j) | o_2 | \hat{\varphi}_l(\mathbf{S}'_l) \hat{\varphi}_k(\mathbf{S}'_k) \rangle \right]. \tag{1.5.12}$$

They involve a quadruple sum over the individual orbitals. In practice the two-body matrix elements represent the main part of the computer time. Further extensions to three-body forces can be done, but the corresponding matrix elements involve sextuple sums overs the individual orbitals.

1.5.2 Spin and Isospin Factorization

In Eq. (1.5.1) it is assumed that the individual orbitals are characterized by spin and isposin projections ($m_s = \pm 1/2$, $m_t = \pm 1/2$). In that case, the overlap matrix $\mathbf{B}$ takes the simpler form

$$\mathbf{B} = \begin{pmatrix} \mathbf{B}^{n\downarrow} & & & \\ & \mathbf{B}^{n\uparrow} & & \\ & & \mathbf{B}^{p\downarrow} & \\ & & & \mathbf{B}^{p\uparrow} \end{pmatrix}, \tag{1.5.13}$$

involving (smaller) submatrices corresponding to the nucleon types. The individual orbitals have been reordered in four groups corresponding to the spin and isospin values (notice that a phase factor (-1) may appear in the wave function when reordering the orbitals). The overlap (1.5.3) is then factorized as

$$\det \mathbf{B} = \prod_{k=1}^{4} \det \mathbf{B}^k,$$

$$B_{ij}^k = \langle \varphi_i(\mathbf{S}_i) | \varphi_j(\mathbf{S}'_j) \rangle, \tag{1.5.14}$$

where only the spatial parts of the wave functions are involved [see Eq. (1.5.1)]. In this definition, index k corresponds to the four spin/isospin projections. This means that the calculation is much faster than by using the full matrix. In many cases some of the matrices $\mathbf{B}^k$ are identical (for example in $n\alpha$ systems such as ^{8}Be or ^{12}C, the four matrices $\mathbf{B}^k$ are identical), which still simplifies the calculations.

If the one-body operator O_1 does not depend on spin and isospin (as for the kinetic energy for example), its matrix element (1.5.6) is simplified to

Table 1.2 Direct and exchange coefficients ($k_1 \geq k_2$) for the nuclear (central), Coulomb and spin–orbit interactions

$k_1 k_2$	$A_D^{k_1 k_2}$				$C_D^{k_1 k_2}$	$S_D^{k_1 k_2}$	$A_E^{k_1 k_2}$				$C_E^{k_1 k_2}$	$S_E^{k_1 k_2}$
	1	$P^\sigma P^\tau$	P^σ	P^τ			1	$P^\sigma P^\tau$	P^σ	P^τ		
$n\downarrow n\downarrow$	1	1	1	1	0	-1	1	1	1	1	0	-1
$n\downarrow n\uparrow$	1	0	0	1	0	0	0	1	1	0	0	0
$n\downarrow p\downarrow$	1	0	1	0	0	-1	0	1	0	1	0	0
$n\downarrow p\uparrow$	1	0	0	0	0	0	0	1	0	0	0	0
$n\uparrow n\uparrow$	1	1	1	1	0	1	1	1	1	1	0	1
$n\uparrow p\downarrow$	1	0	0	0	0	0	0	1	0	0	0	0
$n\uparrow p\uparrow$	1	0	1	0	0	1	0	1	0	1	0	0
$p\downarrow p\downarrow$	1	1	1	1	1	-1	1	1	1	1	1	-1
$p\downarrow p\uparrow$	1	0	0	1	1	0	0	1	1	0	0	0
$p\uparrow p\uparrow$	1	1	1	1	1	1	1	1	1	1	1	1

$$\langle O_1 \rangle = \det \mathbf{B} \sum_{k=1}^{4} \sum_{ij} (\mathbf{B}^k)_{ji}^{-1} \langle \varphi_i(\mathbf{S}_i) | o_1 | \varphi_j(\mathbf{S}_j') \rangle, \qquad (1.5.15)$$

where the spin and isospin components of the individual orbitals have been taken out. The number of terms in the summations over ij of course depends on the spin/isposin index k.

Two-body operators in general depend on spin and isospin, but Eq. (1.5.9) can also be simplified. Let us consider a central nucleon–nucleon interaction defined by Eq. (1.2.4). The matrix elements (1.5.9) and (1.5.12) can be written as

$$\begin{aligned}
\langle V^{N,c} \rangle = \frac{1}{2} \det \mathbf{B} \sum_{k_1, k_2 = 1}^{4} \sum_{ijkl} (\mathbf{B}^{k_1})_{ki}^{-1} (\mathbf{B}^{k_2})_{lj}^{-1} \\
\times \Big[A_D^{k_1 k_2} \langle \varphi_i(\mathbf{S}_i) \varphi_j(\mathbf{S}_j) | v | \varphi_k(\mathbf{S}_k') \varphi_l(\mathbf{S}_l') \rangle \\
- A_E^{k_1 k_2} \langle \varphi_i(\mathbf{S}_i) \varphi_j(\mathbf{S}_j) | v | \varphi_l(\mathbf{S}_l') \varphi_k(\mathbf{S}_k') \rangle \Big],
\end{aligned} \qquad (1.5.16)$$

where v is a Gaussian form factor, and where the direct and exchange coefficients $A_D^{k_1 k_2}$ and $A_E^{k_1 k_2}$ are defined for each operator in (1.2.4). They are given in Table 1.2, as well as the corresponding coefficients $C_D^{k_1, k_2}$ and $C_E^{k_1, k_2}$ arising from the Coulomb interaction (1.2.2). Notice that these coefficients satisfy the symmetry properties

$$A_D^{k_2 k_1} = A_D^{k_1 k_2}, \quad A_E^{k_2 k_1} = A_E^{k_1 k_2}, \qquad (1.5.17)$$

and equivalent relations hold for the Coulomb potential. Summations (1.5.16) can therefore be simplified.

1.5.3 The Spin–Orbit Interaction

Potentials considered in the previous section are scalar operators with respect to the spin. For two nucleons coupled to spin $S=0$ or 1 and projection M_S, the application of the Wigner–Eckart theorem gives

$$\langle SM_S|V^{N,c}|S'M_S'\rangle = \langle S||V^{N,c}||S'\rangle \delta_{SS'}\delta_{M_SM_S'}, \tag{1.5.18}$$

where the reduced matrix element $\langle S||V^{N,c}||S'\rangle$ does not depend on the spin projections. The spin–orbit potential is a rank-1 operator, and the previous property does not hold anymore. The calculation of the matrix elements is therefore more complicated since the overlap between wave functions with different spins is zero.

Let us consider the scalar product involved in the spin–orbit potential (1.2.5)

$$\mathbf{L}\cdot\mathbf{S} = L_zS_z + (S_+L_- + S_-L_+)/2. \tag{1.5.19}$$

Since S_z does not change the spin of the a nucleon pair, the contribution of L_zS_z can be determined as in (1.5.16)

$$\langle L_zS_z\exp(-(\frac{r}{r_0})^2)\rangle = \frac{1}{2}\det\mathbf{B}\sum_{k_1,k_2}\sum_{ijkl}(\mathbf{B}^{k_1})^{-1}_{ki}(\mathbf{B}^{k_2})^{-1}_{lj}$$

$$\times\left[S_D^{k_1k_2}\langle\varphi_i(\mathbf{R}_i)\varphi_j(\mathbf{R}_j)|L_z\exp(-(\frac{r}{r_0})^2)|\varphi_k(\mathbf{R}_k)\varphi_l(\mathbf{R}_l)\rangle\right.$$

$$\left.-S_E^{k_1k_2}\langle\varphi_i(\mathbf{R}_i)\varphi_j(\mathbf{R}_j)|L_z\exp(-(\frac{r}{r_0})^2)|\varphi_l(\mathbf{R}_l)\varphi_k(\mathbf{R}_k)\rangle\right], \tag{1.5.20}$$

where coefficients $S_D^{k_1k_2}$ and $S_E^{k_1k_2}$ are given in Table 1.1. The matrix elements of S_+L_- and S_-L_+ must be computed with the more general formula (1.5.9).

1.5.4 Matrix Elements Between Individual Orbitals

We give here matrix elements for $0s$ orbitals [32], and then discuss how to derive matrix elements involving higher shells. As mentioned previously we assume that all orbitals have the same oscillator parameter b. Notation φ_i corresponds to a $0s$ orbital centred at $\mathbf{R}_i$.

The overlap, kinetic energy and r.m.s radius are given by

$$\langle\varphi_i|\varphi_j\rangle = B_{ij} = \exp\left(-\frac{(\mathbf{R}_i-\mathbf{R}_j)^2}{4b^2}\right),$$

$$\langle\varphi_i|-\frac{\hbar^2}{2m_N}\Delta|\varphi_j\rangle = \hbar\omega\left[\frac{3}{4} - \frac{(\mathbf{R}_i-\mathbf{R}_j)^2}{8b^2}\right]B_{ij},$$

$$\langle\varphi_i|r^2|\varphi_j\rangle = \left[\frac{3}{2}b^2 + \frac{(\mathbf{R}_i+\mathbf{R}_j)^2}{4}\right]B_{ij}. \tag{1.5.21}$$

For a Gaussian form factor and for the Coulomb interaction we have

$$\langle \varphi_i \varphi_j | \exp(-\frac{(\mathbf{r}_1 - \mathbf{r}_2)^2}{a^2}) | \varphi_k \varphi_l \rangle = \left(\frac{a^2}{a^2 + 2b^2} \right)^{3/2} \exp\left(-\frac{2b^2}{a^2 + 2b^2} \mathbf{P}^2 \right) B_{ik} B_{jl},$$

$$(1.5.22)$$

and

$$\langle \varphi_i \varphi_j | \frac{1}{|\mathbf{r}_1 - \mathbf{r}_2|} | \varphi_k \varphi_l \rangle = \frac{1}{\sqrt{2}b} \frac{\mathrm{erf}(P)}{P} B_{ik} B_{jl},$$

$$(1.5.23)$$

where vector $\mathbf{P}$ is defined as

$$\mathbf{P} = \frac{1}{2\sqrt{2}b} (\mathbf{R}_i - \mathbf{R}_j + \mathbf{R}_k - \mathbf{R}_l).$$

$$(1.5.24)$$

For the spin–orbit potential, we need the matrix elements

$$\langle \varphi_i \varphi_j | \exp(-\frac{(\mathbf{r}_1 - \mathbf{r}_2)^2}{r_0^2}) \mathbf{L}_\mu | \varphi_k \varphi_l \rangle = -\frac{i}{4b^2} \frac{r_0^2}{r_0^2 + 2b^2} \left[(\mathbf{R}_i - \mathbf{R}_j) \times (\mathbf{R}_k - \mathbf{R}_l) \right]_\mu$$

$$\times \langle \varphi_i \varphi_j | \exp(-\frac{(\mathbf{r}_1 - \mathbf{r}_2)^2}{r_0^2}) | \varphi_k \varphi_l \rangle,$$

$$(1.5.25)$$

where $\mu = -1, 0, 1$. The combination of (1.2.5), (1.5.22) and (1.5.25) shows that the factors involving r_0 cancel out in the final matrix element of the spin–orbit interaction. Matrix elements involving p orbitals can be obtained by rewriting (1.4.5) as

$$\varphi_{p\mu}(\mathbf{r}, \mathbf{R}) = \sqrt{2}b \frac{d}{dR_\mu} \varphi_s(\mathbf{r}, \mathbf{R}).$$

$$(1.5.26)$$

The corresponding matrix elements are therefore obtained by differentiation of $0s$ matrix elements with respect to the generator coordinate $\mathbf{R}$. Another approach is to expand harmonic-oscillator orbitals in a Cartesian basis. Matrix elements in this basis can be computed by recurrence relations [90]. In practice the latter technique is the most efficient to include orbitals beyond the p shell.

1.5.5 Example: $\alpha + n$ Overlap Kernel

In this section we present an illustrative example with the $\alpha + n$ system, treated in a way which is adopted in numerical calculations. Similar developments are presented in Ref. [31] for the $\alpha + \alpha$ system. More general results, obtained for systems involving an α particle and an s-shell cluster will be given in the next subsection. The α particle is built with four $0s$ orbitals, whereas the external neutron can have a spin up or down. As long as the interaction does not depend on the spin, both projections are not coupled. As in Sect. 1.3.2, we define a Slater determinant with $m_s = -1/2$ for the external neutron. The GCM basis function (1.4.7) is explicitly written as

$$\Phi(\mathbf{R}) = \frac{1}{\sqrt{5!}} \times \det\{\varphi_{0s}(-\frac{\mathbf{R}}{5})n \downarrow \varphi_{0s}(-\frac{\mathbf{R}}{5})n \uparrow$$
$$\varphi_{0s}(-\frac{\mathbf{R}}{5})p \downarrow \varphi_{0s}(-\frac{\mathbf{R}}{5})p \uparrow \varphi_{0s}(\frac{4\mathbf{R}}{5})n \downarrow\}, \qquad (1.5.27)$$

where we use $\lambda = 1/5$, which makes the c.m. factor (1.4.12) equal to unity. The overlap between two Slater determinants (1.5.27) is therefore given by

$$N(\mathbf{R}, \mathbf{R}') = \langle \Phi(\mathbf{R}) | \Phi(\mathbf{R}') \rangle = \begin{vmatrix} B_{11} & B_{12} & 0 & 0 & 0 \\ B_{21} & B_{22} & 0 & 0 & 0 \\ 0 & 0 & B_{11} & 0 & 0 \\ 0 & 0 & 0 & B_{11} & 0 \\ 0 & 0 & 0 & 0 & B_{11} \end{vmatrix}$$
$$= B_{11}^3 (B_{11}B_{22} - B_{12}B_{21}), \qquad (1.5.28)$$

with

$$B_{11} = \langle \varphi_{0s}(-\frac{1}{5}\mathbf{R}) | \varphi_{0s}(-\frac{1}{5}\mathbf{R}') \rangle$$
$$B_{12} = \langle \varphi_{0s}(-\frac{1}{5}\mathbf{R}) | \varphi_{0s}(\frac{4}{5}\mathbf{R}') \rangle$$
$$B_{21} = \langle \varphi_{0s}(\frac{4}{5}\mathbf{R}) | \varphi_{0s}(-\frac{1}{5}\mathbf{R}') \rangle$$
$$B_{22} = \langle \varphi_{0s}(\frac{4}{5}\mathbf{R}) | \varphi_{0s}(\frac{4}{5}\mathbf{R}') \rangle. \qquad (1.5.29)$$

Using the single-particle overlap (1.5.21), we find

$$N(\mathbf{R}, \mathbf{R}') = \exp\left(-\frac{(\mathbf{R} - \mathbf{R}')^2}{5b^2}\right) \left[1 - \exp\left(-\frac{\mathbf{R} \cdot \mathbf{R}'}{2b^2}\right)\right]. \qquad (1.5.30)$$

The first term is the direct contribution, which stems from the diagonal of the overlap matrix. The second term is responsible for exchange effects, and is negligible at large distances. When $\mathbf{R}$ (or $\mathbf{R}'$) tends to zero, the Slater determinant (1.5.27) vanishes since two rows or columns are identical. This property is a consequence of the Pauli principle, and the total overlap (1.5.30) also vanishes.

In this simple example, projection over angular momentum is directly obtained from definition (1.4.31) as

$$N_\ell(R, R') = \frac{1}{4\pi} \exp\left(-\frac{R^2 + R'^2}{5b^2}\right) \left[i_\ell\left(\frac{2RR'}{5b^2}\right) - (-1)^\ell i_\ell\left(\frac{RR'}{10b^2}\right)\right]. \qquad (1.5.31)$$

1.5.6 GCM Kernels of $\alpha + N$ Systems

In this section we present analytical expressions for $\alpha + N$ GCM kernels, where N is a $0s$ shell nucleus (with A_2 nucleons). An extension to systems involving an ^{16}O core is given in Ref. [81]. As shown by Horiuchi [32] (see also Ref. [88]), the overlap kernel takes the general form

$$N(\mathbf{R}, \mathbf{R}') = \sum_{n=0}^{A_2} N_n f_n(\mathbf{R}, \mathbf{R}'), \qquad (1.5.32)$$

with

$$f_n(\mathbf{R}, \mathbf{R}') = \exp\left(-\frac{\mu_0 \nu}{2}(\mathbf{R} - \mathbf{R}')^2 - n\nu \mathbf{R} \cdot \mathbf{R}'\right), \qquad (1.5.33)$$

and $\nu = 1/2b^2$. Index n in (1.5.32) can be interpreted as the number of exchanged terms. The overlap kernel is therefore entirely determined from a set of integer numbers N_n. They can be obtained from algebraic calculations.

For the symmetric $\alpha + \alpha$ system ($\mu_0 = 2$), functions f_n must be symmetrized as

$$f_n(\mathbf{R}, \mathbf{R}') \rightarrow \frac{1}{2}\left(f_n(\mathbf{R}, \mathbf{R}') + f_n(\mathbf{R}, -\mathbf{R}')\right), \qquad (1.5.34)$$

which shows that terms corresponding to n and $4 - n$ are equivalent in expansion (1.5.32). After expansion on angular momentum, the symmetrized definition (1.5.34) involves even partial waves only.

For systems involving $0s$-shell orbitals, the kinetic energy matrix element between individual orbitals φ_i can be written as [see Eq. (1.5.21)]

$$\langle \varphi_i | -\frac{\hbar^2}{2m_N}\Delta|\varphi_j\rangle = \hbar\omega\left[\frac{3}{4}\langle\varphi_i|\varphi_j\rangle + \frac{\nu}{2}\frac{d}{d\nu}\langle\varphi_i|\varphi_j\rangle\right]. \qquad (1.5.35)$$

Consequently, the kinetic-energy kernel reads, after subtraction of the c.m. contribution,

$$T(\mathbf{R}, \mathbf{R}') = \hbar\omega\left[\frac{3}{4}(A - 1) + \frac{\nu}{2}\frac{d}{d\nu}\right]N(\mathbf{R}, \mathbf{R}'), \qquad (1.5.36)$$

and can be directly obtained from coefficients N_n. These coefficients are given in Table 1.3 for various systems.

For the potential kernels, we assume that the nuclear interaction is given by combinations of Gaussian functions and exchange operators O as

$$v^N(r) = V_0 \exp\left(-\frac{r^2}{a^2}\right)O, \qquad (1.5.37)$$

Table 1.3 Coefficients N_n of the overlap kernel

System	$n=0$	$n=1$	$n=2$	$n=3$	$n=4$
$\alpha + n,\ \alpha + p$	1	-1			
$\alpha + {}^3\text{H},\ \alpha + {}^3\text{He}$	1	-3	3	-1	
$\alpha + \alpha$	1	-4	6	-4	1

where O is one of the operators $1,\ P^\sigma P^\tau,\ P^\sigma,\ P^\tau$. The GCM kernel corresponding to the two-body potential (1.5.37) is given by (see Ref. [32])

$$V^N(\mathbf{R}, \mathbf{R}') = V_0 \left(\frac{a^2}{a^2 + 2b^2}\right)^{3/2} \sum_n f_n(\mathbf{R}, \mathbf{R}') \sum_{i=1}^{5} V_{ni}^N \exp[-\alpha^2 F_i^2(\mathbf{R}, \mathbf{R}')],$$

(1.5.38)

where functions $F_i(\mathbf{R}, \mathbf{R}')$ are defined as

$$F_1(\mathbf{R}, \mathbf{R}') = 0,$$
$$F_2(\mathbf{R}, \mathbf{R}') = \mathbf{R}/2\sqrt{2}b,$$
$$F_3(\mathbf{R}, \mathbf{R}') = \mathbf{R}'/2\sqrt{2}b,$$
$$F_4(\mathbf{R}, \mathbf{R}') = (\mathbf{R} + \mathbf{R}')/2\sqrt{2}b,$$
$$F_5(\mathbf{R}, \mathbf{R}') = (\mathbf{R} - \mathbf{R}')/2\sqrt{2}b,$$

(1.5.39)

and α is defined by $\alpha^2 = 2b^2/(a^2 + 2b^2)$.

The Coulomb kernel takes the general form

$$V^C(\mathbf{R}, \mathbf{R}') = \frac{e^2}{\sqrt{2}b} \sum_n f_n(\mathbf{R}, \mathbf{R}') \sum_{i=1}^{5} V_{ni}^C \frac{\text{erf}[|F_i(\mathbf{R}, \mathbf{R}')|]}{|F_i(\mathbf{R}, \mathbf{R}')|}.$$

(1.5.40)

Coefficients V_{ni}^N and V_{ni}^C are given in Tables 1.4 and 1.5, respectively. The projected kernels are directly obtained by integration over the relative angle between $\mathbf{R}$ and $\mathbf{R}'$ [see Eq. (1.4.31)].

A very simple application is the α particle with four $0s$ orbitals centred at the origin. In that case all space components in (1.5.16) take the same form. For a Gaussian potential of range a and amplitude V_0, the α nuclear and Coulomb energies are

$$V^N(\alpha) = 6(w + m) \left(\frac{a^2}{a^2 + 2b^2}\right)^{3/2} V_0$$

$$V^C(\alpha) = \sqrt{\frac{2}{\pi}} \frac{e^2}{b}.$$

(1.5.41)

Table 1.4 Coefficients V_{ni}^N of the nuclear potential for the different exchange operators. The bracketed values correspond to $i = (1,2,3,4,5)$

System	$O = 1$	$O = P^\sigma P^\tau$	$O = P^\sigma$	$O = P^\tau$
$\alpha + n, \alpha + p$				
$n = 0$	(6,0,0,4,0)	$(-6, 0, 0, 1, 0)$	(0,0,0, 2,0)	(0, 0, 0, 2, 0)
$n = 1$	$(-3, -3, -3, 0, -1)$	$(3, 3, 3, 0, -4)$	$(0, 0, 0, 0, -2)$	$(0, -1, -1, 0, -1)$
$\alpha + {}^3$H, $\alpha + {}^3$He				
$n = 0$	(9,0,0,12,0)	$(-9, 0, 0, 3, 0)$	(0,0,0,6,0)	(0,0,0,6,0)
$n = 1$	$(-12, -15, -15, -18, -3)$	$(12, 15, 15, -12, -12)$	$(0, 0, 0, -12, -6)$	$(0, 0, 0, -12, -6)$
$n = 2$	(9 ,18, 18 ,6 ,12)	$(-9, -18, -18, 9, 18)$	(0 ,0, 0, 6 ,12)	(0,0, 0, 6,12)
$n = 3$	$(-6, -3, -3, 0, -9)$	$(6, 3, 3, 0, -6)$	$(0, 0, 0, 0, -6)$	$(0, 0, 0, 0, -6)$
$\alpha + \alpha$				
$n = 0$	(12, 0, 0 ,16 ,0)	$(-12, 0, 0, 4, 0)$	(0 ,0, 0, 8, 0)	(0, 0, 0, 8, 0)
$n = 1$	$(-24, -24, -24, -36, -4)$	$(24, 24, 24, -24, -16)$	$(0, 0, 0, -24, -8)$	$(0, 0, 0, -24, -8)$
$n = 2$	(24, 48, 48, 24 ,24)	$(-24, -48, -48, 36, 36)$	$(0, 0, 0, 24, 24)$	$(0, 0, 0, 24 ,24)$
$n = 3$	$(-24, -24, -24, -4, -36)$	$(24, 24, 24, -16, -24)$	$(0, 0, 0, -8, -24)$	$(0, 0, 0, -8, -24)$
$n = 4$	(12 ,0, 0 ,0 ,16)	$(-12, 0, 0, 0, 4)$	(0 ,0 ,0 ,0,8)	(0, 0, 0, 0,8)

Table 1.5 Coefficients V_{ni}^C of the Coulomb potential (see caption of Table 1.4). V_{C1} and V_{C2} correspond to the mirror systems

System	V_{C1}	V_{C2}
$\alpha + n, \alpha + p$		
$n=0$	$(1, 0, 0, 0, 0)$	$(1,0,0,2,0)$
$n=1$	$(-1, 0, 0, 0, 0)$	$(0, -1, -1, 0, -1)$
$\alpha +^3 H, \alpha +^3 He$		
$n=0$	$(2, 0, 0, 4, 0)$	$(1, 0, 0, 2, 0)$
$n=1$	$(-2, -4, -4, -6, -2)$	$(-2, -1, -1, -4, -1)$
$n=2$	$(2, 4, 4, 2, 6)$	$(1, 2, 2, 2, 2)$
$n=3$	$(-2, 0, 0, 0, -4)$	$(0, -1, -1, 0, -1)$
$\alpha + \alpha$		
$n=0$	$(2, 0, 0, 4, 0)$	
$n=1$	$(-4, -4, -4, -10, -2)$	
$n=2$	$(4, 8, 8, 8, 8)$	
$n=3$	$(-4, -4, -4, -2, -10)$	
$n=4$	$(2, 0, 0, 0, 4)$	

1.6 Approximations of the RGM

1.6.1 Eigenvalues of the Overlap Kernel

In this Section, we consider other variants of the RGM equation. All are based on the eigenvalues of the overlap kernel [16, 31, 32, 57, 91]. Let us consider the eigenvalue problem

$$\int \mathscr{N}_\ell(\rho, \rho') \chi_{\ell n}(\rho') d\rho' = \mu_{\ell n} \chi_{\ell n}(\rho), \tag{1.6.1}$$

where $\mathscr{N}_\ell(\rho, \rho')$ is the ℓ-projected overlap kernel. In a more compact notation, Eq. (1.6.1) is rewritten as

$$\mathscr{N}_\ell \chi_{\ell n} = \mu_{\ell n} \chi_{\ell n}. \tag{1.6.2}$$

These orthogonal eigenstates $\chi_{\ell n}$ play an important role in approximations [92] and extensions [93] of the RGM. In particular, eigenstates $\chi_{\ell i}(\rho)$ corresponding to $\mu_{\ell i} = 0$ are called forbidden states. These functions are different from zero and present the property

$$\mathscr{A} \phi_1 \phi_2 \chi_{\ell i}(\rho) = 0, \tag{1.6.3}$$

i.e. they vanish from the action of the antisymmetrizor. They are typical of calculations with identical oscillator parameters. When the oscillator parameters of both clusters are different, forbidden states ($\mu_{\ell i} = 0$) are replaced by "almost" forbidden states ($\mu_{\ell i}$ small) which induce spurious states in the Schrödinger equation [16, 31, 37]. On the other hand, eigenstates with $\mu_{\ell i} \approx 1$ are weakly affected by exchange effects of the overlap kernel. The eigenvalue distribution therefore provides some insight on the importance of the antisymmetrization.

For some systems, involving a closed-shell nucleus ($\alpha, ^{16}O, ^{40}Ca, \ldots$) and an s-shell cluster, the calculation of the eigenvalues is analytical, and only depends of the quantum number [32]

$$N = 2\ell + n. \tag{1.6.4}$$

For example, the $\alpha + n$, $\alpha + t$ and $\alpha + \alpha$ eigenvalues are

$$\mu_N = \begin{cases} 1 - (-1/4)^N & \text{for } \alpha + n \\ 1 - 3(\frac{5}{12})^N + 3(-\frac{1}{6})^N - (-\frac{3}{4})^N & \text{for } \alpha + t \\ 1 - 2^{2-N} + 3\delta_{N,0} & \text{for } \alpha + \alpha \ (N \text{ even}) \end{cases} . \tag{1.6.5}$$

These eigenvalues do not depend of the oscillator parameter. The eigenfunctions are harmonic oscillator orbitals with oscillator parameter $b/\sqrt{\mu_0}$. From the example (1.6.5) we immediately see that the $\alpha + n$ system presents one forbidden state for $\ell = 0$. The reason is that the s orbital is already occupied in the α particle, and is not accessible to the external neutron. The $\alpha + \alpha$ system presents two forbidden states for $\ell = 0$, one for $\ell = 2$, and zero for $\ell \geq 4$.

In general, the eigenvalue problem (1.6.1) cannot be solved analytically. A numerical approach has been proposed by Varga and Lovas [91] who write (1.6.1) in an equivalent form

$$\mathscr{A}\phi_1\phi_2\chi_{\ell n}(\rho) = \mu_{\ell n}\phi_1\phi_2\chi_{\ell n}(\rho), \tag{1.6.6}$$

which shows that $\mu_{\ell n}$ are the eigenvalues of the antisymetrization operator. Expanding $\chi_{\ell n}(\rho)$ over a finite Gaussian basis as

$$\chi_{\ell n}(\rho) = \sum_i c_n^{\ell}(R_i)\Gamma_{\ell}(\rho, R_i) \tag{1.6.7}$$

provides the equivalent eigenvalue problem

$$\sum_i c_n^{\ell}(R_i)\left(\langle\Phi^{\ell}(R_j)|\Phi^{\ell}(R_i)\rangle - \mu_{\ell n}\langle\Gamma_{\ell}(R_j)|\Gamma_{\ell}(R_i)\rangle\right) = 0. \tag{1.6.8}$$

The first term is the overlap between two projected Slater determinants (1.4.27), whereas the second term corresponds to the direct contribution, and can be calculated analytically as [94]

$$\langle\Gamma_{\ell}(R_j)|\Gamma_{\ell}(R_i)\rangle = \exp\left[-\frac{\mu_0}{4b^2}(R_i^2 + R_j^2)\right]i_{\ell}\left(\frac{\mu_0 R_i R_j}{2b^2}\right). \tag{1.6.9}$$

Of course, the method is approximative only, but is quite simple to apply, in particular for multicluster problems [95], or in two-cluster systems with different oscillator parameters [37].

1.6.2 Reformulation of the RGM Equation

Our aim here is to reformulate the RGM equation (1.3.17) in a more transparent way. One of the reasons is that the RGM two-cluster kernels can be used in multicluster calculations [93, 96], and therefore provide a microscopic framework to multicluster calculations. After projection on angular momentum ℓ, the RGM equation (1.3.17) can be recast as

$$(T_\ell + V_\ell - E\mathcal{N}_{E\ell})g_\ell = Eg_\ell, \tag{1.6.10}$$

where T_ℓ is the kinetic energy in partial wave ℓ, and V_ℓ a potential which includes exchange contributions. Equation (1.6.10) resembles a usual two-body Schrödinger equation, but the equivalent potential depends on energy. This is not a problem in two-body calculations, but raises some ambiguities in multicluster models. In that case, cluster-cluster energies are not precisely defined. An iterative procedure has been proposed by Fujiwara et al. [97] but the method was shown to raise conceptual problems in three-body calculations [48].

An elegant and efficient method has been proposed by Suzuki et al. [96] and is briefly explained here. Let us define a modified radial function

$$\hat{g}_\ell = \mathcal{N}_\ell^{1/2} g_\ell = \sum_n \mu_{\ell n}^{1/2} \langle \chi_{\ell n}|g_\ell\rangle \chi_{\ell n}$$

$$= g_\ell - \sum_n (1 - \mu_{\ell n}^{1/2}) \langle \chi_{\ell n}|g_\ell\rangle \chi_{\ell n}. \tag{1.6.11}$$

Since $\mu_{\ell n}$ tend to unity for large n, $\hat{g}_\ell(\rho)$ and $g_\ell(\rho)$ have the same asymptotic behaviour. In addition, $\hat{g}_\ell(\rho)$ is orthogonal to the forbidden states since

$$\langle \chi_{\ell i}|\hat{g}_\ell\rangle = 0 \tag{1.6.12}$$

for $\mu_{\ell i} = 0$. Replacing g_ℓ by

$$g_\ell = \mathcal{N}_\ell^{-1/2} \hat{g}_\ell \tag{1.6.13}$$

in (1.6.10) provides

$$(T_\ell + V_\ell^{RGM})\hat{g}_\ell = E\hat{g}_\ell, \tag{1.6.14}$$

where V_ℓ^{RGM} is not local, but does not depend on energy. It is defined as

$$V_\ell^{RGM} = \mathcal{N}_\ell^{-1/2}(T_\ell + V_\ell)\mathcal{N}_\ell^{-1/2} - T_\ell$$

$$= V_\ell + W_\ell. \tag{1.6.15}$$

Of course, implicit summations in (1.6.13) and (1.6.15) only include allowed states ($\mu_{\ell n} \neq 0$). The renormalized RGM therefore contains the bare RGM potential V_ℓ, and an additional contribution W_ℓ. As shown in [96], this term can be computed by an expansion over harmonic-oscillator functions [32].

In Refs. [93, 96], a detailed comparison is performed between the use of renormalized RGM potentials in three-body calculations and fully microscopic calculations with the same nucleon–nucleon interaction. This is done for three typical three-body systems: ^{6}He $= \alpha + n + n$, ^{9}Be $= \alpha + \alpha + n$, and ^{12}C $= \alpha + \alpha + \alpha$. The $\alpha + n$ and $\alpha + \alpha$ RGM potentials W_ℓ have been obtained numerically [96]. The binding energies of the ground states are in reasonable agreement with the microscopic energies, but are slightly underestimated. This leads to the suggestion that three-body effects, missing in the renormalized RGM, should be attractive.

This technique opens various possibilities in the microscopic treatment of the nucleus–nucleus interaction. For example, three-body continuum states [98] could be treated with these potentials. In parallel, Continuum Discretized Coupled Channel (CDCC) calculations require a precise description of two and three-body projectiles [99] and could be performed with non-local RGM potentials. On the other hand, the renormalized RGM has been successfully used to compute the triton and hypertriton binding energies from nucleon–nucleon interactions based an a quark cluster model [100].

1.6.3 The Orthogonality Condition Model

The Orthogonality Condition Model (OCM) has been proposed by Saito [92]. The main idea was to simplify the RGM approach, while keeping its microscopic grounds. Let us introduce the projector

$$\Lambda_\ell = 1 - \sum_{i \in PFS} |\chi_{\ell i}\rangle \langle \chi_{\ell i}|, \qquad (1.6.16)$$

where the sum runs over the Pauli forbidden states (PFS, $\mu_{\ell i} = 0$). This provides an equivalent RGM equation [16, 92]

$$\Lambda_\ell (T_\ell + V_\ell^{OCM} - E)\hat{g}_\ell = 0, \qquad (1.6.17)$$

where the OCM potential is implicitly defined by

$$\mathcal{H}_\ell = \mathcal{N}_\ell^{1/2}(T_\ell + V_\ell^{OCM})\mathcal{N}_\ell^{1/2}. \qquad (1.6.18)$$

Equation (1.6.18) is strictly equivalent to (1.3.17). However, the potential V_ℓ^{OCM} has a non-local form. The purpose of the OCM is to include antisymmetrization effects through the operator Λ_ℓ, but to use approximations for the potential [57, 92]. Various methods have been proposed to solve the non-local equation (1.6.17) (see Refs. [92, 94] and references therein).

The OCM equation can still be simplified by replacing (1.6.17) by the local equation

$$(T_\ell + \tilde{V}_\ell - E)\tilde{g}_\ell = 0 \tag{1.6.19}$$

where $\tilde{V}_\ell$ is a local potential [18, 19]. The role of the forbidden states is simulated by additional non-physical states $\tilde{g}_i$ in the potential. Their number depends on the system and on angular momentum [see Eq. (1.6.5) for some systems]. In this way, the orthogonality condition (1.6.12) is replaced by

$$\langle \tilde{g}_i | \tilde{g}_\ell \rangle = 0. \tag{1.6.20}$$

Deep nucleus–nucleus potentials are available in the literature, in particular for $\alpha + n$ [101] and $\alpha + \alpha$ [102]. The main advantage of the local approximation is its simplicity. The calculation of bound states and phase shifts is straightforward with, for example, the Numerov method [73, 103]. All matrix elements are obtained from one-dimensional integrals. In nuclear astrophysics, many capture reactions have been investigated in this framework (see for example [94, 104]).

1.7 Recent Developments of the GCM

1.7.1 Introduction

For the sake of clarity, the formalism presented in previous sections was simplified as much as possible. In particular, we neglected the spins of the clusters, assumed single-channel problems, and limited the discussion to two-cluster systems. However, the GCM has been significantly extended in two directions, with the goal of improving the description of the system: the extension to multichannel approaches, and to multicluster calculations. We briefly review these two developments. In addition, attention has been paid on the improvement of the cluster wave functions: large shell-model bases [36], and mixing of several shell-model wave functions corresponding to different oscillator parameters [37].

1.7.2 Internal Wave Functions

Microscopic cluster calculations are performed with a shell-model description for the cluster wave functions. The standard shell-model formalism can therefore be used. However, in contrast with shell-model calculations, the definition of the internal wave functions is just a first step of cluster theories. As long as s clusters only are involved, the construction of the shell-model states is trivial. In particular, the corresponding wave functions involve a single Slater determinant (except for the deuteron, but this nucleus is poorly described by the shell model).

Going beyond s-shell clusters strongly increases the complexity of the calculations. Analytical developments are, in most cases, very difficult and need to be replaced by entirely numerical approaches. In Ref. [36], we proposed a generalized cluster model, where the cluster wave functions are defined by all p-shell configurations consistent by the Pauli principle. Formally, this can be directly extended to the sd shell or higher (see for example Refs. [105, 106] for applications involving ^{30}Ne and ^{39}Ca clusters), but raises computational problems owing to the large number of Slater determinants and of orbitals involved in the matrix elements.

In practice, cluster calculations are optimized when working in the LS coupling scheme. As shown before, this allows to factorize a Slater determinant in four nucleon types, according to the spin and isospin. In that case, each nucleon orbital (1.4.2) is defined either in a Cartesian basis as

$$\hat{\varphi}_{n_x n_y n_z, m_s m_t}(\mathbf{r}) = \varphi_{n_x}(x)\varphi_{n_y}(y)\varphi_{n_z}(z)|m_s\rangle|m_t\rangle, \qquad (1.7.21)$$

or in a spherical basis as

$$\hat{\varphi}_{n l m_\ell, m_s m_t}(\mathbf{r}) = \varphi_{n\ell}(r)Y_\ell^{m_\ell}(\Omega_r)|m_s\rangle|m_t\rangle, \qquad (1.7.22)$$

where the space, spin and isospin components have been factorized. In (1.7.22), $\varphi_{n\ell}(r)$ is an harmonic-oscillator radial function [78]. Both bases are equivalent and related to each other by a unitary transform.

An alternative is to use the jj coupling scheme, where the individual orbitals are defined as

$$\hat{\varphi}_{n\ell j m, m_t}(\mathbf{r}) = \varphi_{nl}(r)\left[Y_\ell(\Omega_r) \otimes \chi_s\right]^{jm}|m_t\rangle. \qquad (1.7.23)$$

In this option, the overlap (1.5.3) is expressed as the product of two determinants involving more orbitals (see Sect. 1.5). This property strongly increases the computation times for two-body matrix elements since they involve quadruple sums over the individual orbitals [see (1.5.12)].

Let us now define a Slater determinant built from A_1 individual orbitals. All configurations, compatible with the Pauli principle, must be included up to some excitation level N_{max}. In most cluster calculations, $N_{max} = 0$, but particle-hole excitations $N_{max} > 0$ are possible. A compromise should be adopted between the quality of the internal wave functions and the feasibility of the cluster calculation.

Let us start with the most common applications, i.e. p-shell nuclei with Z_1 protons and N_1 neutrons ($A_1 = Z_1 + N_1$). Filling the p shell can be performed in $N_c = C_6^{Z_1-2} \times C_6^{N_1-2}$ different possibilities (C_i^j is the number of combinations of j elements among i elements). The basis therefore involves N_c Slater determinants Φ_i. For example $N_c = 6$ for ^{15}N, and $N_c = 225$ for ^{12}C.

In general, these shell-model states do not have a definite spin. Projection over the total spin I_1 is performed by diagonalization of the spin operators $\mathbf{I}^2$ and I_z which provides

$$\Phi^{I_1 K_1} = \sum_{i=1}^{N_c} d_i^{I_1 K_1} \Phi_i, \qquad (1.7.24)$$

where $d_i^{I_1K_1}$ are linear coefficients obtained from the eigenvalue problems

$$\sum_{j=1}^{N_c} d_j^{I_1K_1}\left[\langle \Phi_i \mid \mathbf{I}^2 \mid \Phi_j\rangle - I_1(I_1+1)\delta_{ij}\right] = 0$$

$$\sum_{j=1}^{N_c} d_j^{I_1K_1}\left[\langle \Phi_i \mid I_z \mid \Phi_j\rangle - K_1\delta_{ij}\right] = 0. \tag{1.7.25}$$

The fact that the eigenvalues of $\mathbf{I}^2$ and I_z provides integer or half-integer values for I_1 and K_1 is a strong test of the calculation.

In practice, further diagonalizations of the isospin $\mathbf{T}^2$, intrinsic spin $\mathbf{S}^2$ and orbital angular momentum $\mathbf{L}^2$ are performed, in order to obtain a deeper analysis of the wave functions. Basis states (1.7.24) are therefore recombined as

$$\Phi_{S_1L_1T_1c_1}^{I_1K_1} = \sum_{i=1}^{N_c} d_{S_1L_1T_1c_1,i}^{I_1K_1}\Phi_i, \tag{1.7.26}$$

where c_1 is an additional quantum number used to distinguish between states with identical values of $(I_1K_1S_1L_1T_1)$. The parity is simply obtained from the product of the parities of the individual orbitals. Finally, basis states (1.7.26) are used to diagonalize the Hamiltonian. This is necessary in collision theories, where the internal wave functions must be eigenstates of the internal Hamiltonian. Linear combinations of Slater determinants are then used in multicluster calculations.

In principle, wave functions (1.7.26) could be employed for several clusters. In practice, however, the total number of Slater determinants is given by the product of N_c values of each cluster. Consequently calculations are currently limited to systems involving a single cluster with generalized shell-model wave functions (1.7.26). In Ref. [36], we give the different sets of quantum numbers for ^{8}Li ($N_c = 120$) and ^{11}B ($N_c = 300$) described in the p shell. These wave functions are used in that reference for a microscopic calculation of the ^{8}Li$(\alpha, n)^{11}$B cross section. In principle, including excited configurations ($N_{max} > 0$) can be performed, but strongly increases the number of Slater determinants.

Let us briefly compare the use of the LS coupling (1.7.21) and of the jj coupling (1.7.23). There is an orthogonal transform between them and are equivalent as long as all orbitals of a given shell are considered. As discussed above the use of the jj coupling increases the computer times. However it allows to keep limited numbers of Slater determinants, even beyond the p shell.

A simple example is provided by ^{14}C, where both coupling modes are illustrated in Fig. 1.2. In the LS coupling mode, six neutrons fill the p shell, and four p protons can be combined in $C_6^4 = 15$ combinations. This provides two 0^+ states, one 1^+ state and two 2^+ states. Alternatively, considering the jj coupling mode only provides one 0^+ state since the $1p_{3/2}$ subshell is filled. The corresponding spectra obtained with an oscillator parameter $b=1.6$ fm, and the V2 interaction ($M=0.6$) complemented by a spin–orbit force ($S_0 = 30$ MeV.fm^5) are displayed in Fig. 1.3, and compared

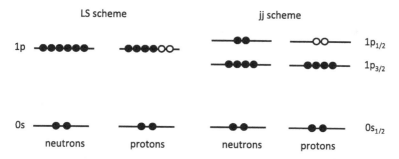

Fig. 1.2 Shell-model orbitals for ^{14}C in the LS and jj coupling modes (see text). Full and open circles represent occupied and unoccupied orbitals, respectively

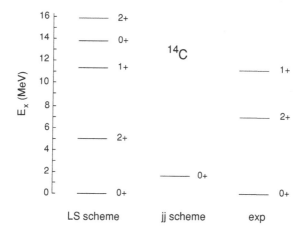

Fig. 1.3 Energy spectrum of ^{14}C in the LS and jj coupling modes. For the sake of clarity, the experimental energy of the ground state has been shifted to the LS value

with experiment. Our goal here is not to optimize the interaction, but to illustrate the problem with a typical nucleon–nucleon force. As expected the jj coupling mode, limited to the $1p_{3/2}$ subshell, does not provide excited states and the ground state is less bound (by 1.9 MeV) than in the LS coupling mode. The advantage of the former conditions is that the number of Slater determinants is limited to one.

This problem is still more apparent when going to sd-shell nuclei. Let us consider ^{17}N, which is illustrated in Fig. 1.4. In the LS coupling mode we have $N_c = 6 \times 66 = 396$ Slater determinants. This gives $1/2^-(12), 3/2^-(19), 5/2^-(18)$ states, $7/2^-(13), 9/2^-(6)$, and $11/2^-(2)$ states. The energy spectrum (limited to the 2 first levels for each angular momentum) is shown in Fig. 1.5. Using this set of basis functions for two or three-cluster calculations is highly time and memory consuming. Considering the jj coupling mode (illustrated in Fig. 1.4) provides a much smaller number of Slater determinants $N_c = 30$. Of course the number of ^{17}N states is reduced, and the binding energies are lower that in the LS coupling mode, but

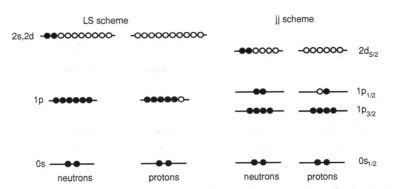

Fig. 1.4 See caption to Fig. 1.2 for ^{17}N

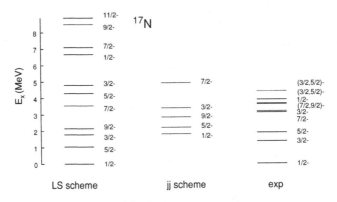

Fig. 1.5 See caption to Fig. 1.3 for ^{17}N

a multicluster calculation (e.g. ^{17}N + n) keeps the computer requirements within acceptable limits. Notice that the calculations have been performed with standard parameters in the nucleon–nucleon interaction. The comparison with experiment can be improved by slightly tuning the parameters M and S_0.

As a general statement, this problem gets more and more important when the nucleon number is far from a closed shell. For example, ^{18}O and ^{30}Ne need $N_c = C_{12}^2 = 66$ in the sd shell; in contrast ^{24}Mg would need $N_c = C_{12}^4 \times C_{12}^4 = 245025$, as 4 neutrons and 4 protons are distributed among the 12 sd orbitals. If this large number does not raise significant problems in shell-model calculations, it makes cluster approaches impossible owing to the additional clusters, and to the global angular-momentum projection. The use of the jj coupling mode, in that case, is necessary.

1.7.3 Multicluster Angular-Momentum Projection

Cluster wave functions based on Slater determinants are not defined with a definite spin value. In order to restore the spin, an angular-momentum projection is needed. We start by assuming that the individual clusters have a spin zero. A projected wave function of the system is obtained from (see Ref. [4])

$$\Phi_K^{LM}(\mathbf{S}_{\{N\}}) = \int \mathscr{D}_{MK}^{L*}(\Omega)\mathscr{R}^L(\Omega)\Phi(\mathbf{S}_{\{N\}})d\Omega, \qquad (1.7.27)$$

where $\mathscr{D}_{MK}^L(\Omega)$ is a Wigner function depending on the Euler angles $\Omega = (\alpha, \beta, \gamma)$, and $\Phi(\mathbf{S}_{\{N\}})$ is a N-cluster Slater determinant (1.5.2). In (1.7.27), K is the projection over the intrinsic axis, and the rotation operator $\mathscr{R}^L(\Omega)$ is defined as

$$\mathscr{R}^L(\Omega) = e^{i\alpha L_z}e^{i\beta L_y}e^{i\gamma L_z}, \qquad (1.7.28)$$

and performs a rotation of the wave function, or an inverse rotation [107] of the space coordinates $\mathbf{r}_i$ of the individual orbitals. Since the orbitals are defined in the harmonic-oscillator model, a rotation of the quantal coordinates $\mathbf{r}_i$ is equivalent to an inverse rotation of the generator coordinate. Consequently, we have

$$\mathscr{R}^L(\Omega)\Phi(\mathbf{S}_{\{N\}}, \mathbf{r}_1, \dots, \mathbf{r}_A) = \Phi(\mathbf{S}_{\{N\}}, \mathscr{R}^L(\Omega^{-1})\mathbf{r}_1, \dots, \mathscr{R}^L(\Omega^{-1})\mathbf{r}_A)$$
$$= \Phi(\mathscr{R}^L(\Omega)\mathbf{S}_{\{N\}}, \mathbf{r}_1, \dots, \mathbf{r}_A). \qquad (1.7.29)$$

The effect of the rotation operator is therefore equivalent to a rotation of the generator coordinates. This property is typical of harmonic oscillator functions, and greatly simplifies the calculations.

Let us consider a rotation-invariant operator O, such that

$$O = \mathscr{R}^L(\Omega^{-1})O\mathscr{R}^L(\Omega). \qquad (1.7.30)$$

A matrix element of O between projected functions (1.7.27) reads

$$\langle \Phi_K^{LM}(\mathbf{S}_{\{N\}})|O|\Phi_{K'}^{L'M'}(\mathbf{S}'_{\{N\}})\rangle$$
$$= \frac{8\pi^2}{2L+1}\delta_{LL'}\delta_{MM'}\int \mathscr{D}_{KK'}^{L\star}(\omega)\langle\Phi(\mathbf{S}_{\{N\}})|O\mathscr{R}^L(\omega)|\Phi(\mathbf{S}'_{\{N\}})\rangle d\omega, \qquad (1.7.31)$$

and therefore reduces to a three-dimensional integral over the Euler angles. If $O_{\lambda\mu}$ is an irreducible operator of rank λ, (1.7.30) is generalized as

$$\mathscr{R}^L(\Omega)O_{\lambda\mu}\mathscr{R}^L(\Omega^{-1}) = \sum_{\mu}\mathscr{D}_{\mu'\mu}^\lambda(\Omega)O_{\lambda\mu'}, \qquad (1.7.32)$$

and the matrix element (1.7.31) must be extended [35].

Let us now consider clusters with spin. An N-cluster basis function (which can be a linear combination, as in Sect. 1.7.2) therefore involves quantum numbers associated with the spin projection K_i and is denoted as $\Phi_{K_1...K_N}$. In that case, the rotation operator corresponding to the total angular momentum J is factorized as

$$\mathcal{R}^J(\Omega) = \mathcal{R}^L(\Omega)\mathcal{R}^S(\Omega), \tag{1.7.33}$$

where $\mathcal{R}^L(\Omega)$ rotates the space coordinates, and $\mathcal{R}^S(\Omega)$ the spin coordinates. Since, by definition, the cluster states have good spin, the spin rotation provides

$$\mathcal{R}^S(\Omega)\Phi_{K_1...K_N} = \sum_{K_1'...K_N'} \mathcal{D}^{I_1}_{K_1'K_1}(\Omega)\ldots\mathcal{D}^{I_N}_{K_N'K_N}(\Omega)\Phi_{K_1'...K_N'}, \tag{1.7.34}$$

where I_i are the spins of the N clusters (for the sake of clarity the generator coordinates are implied). A projected basis state (1.7.27) is then generalized to

$$\Phi^{JM}_{K,K_1...K_N} = \sum_{K_1'...K_N'} \int \mathcal{D}^{J*}_{KM}(\Omega)\mathcal{D}^{I_1}_{K_1'K_1}(\Omega)\ldots\mathcal{D}^{I_N}_{K_N'K_N}(\Omega)\mathcal{R}^L(\Omega)\Phi_{K_1'...K_N'}d\Omega, \tag{1.7.35}$$

and, as in (1.7.27), only the space rotation should be explicitly performed. Matrix elements between functions (1.7.35) are directly obtained from an extension of (1.7.31). Specific applications to 3 and 4-cluster systems can be found in Refs. [35, 108]. An important application concerns two-cluster systems, and is explained in more detail in the next subsection.

Finally the parity projection is performed with the operator

$$\Phi^{JM\pi}_{K,K_1...K_N} = \frac{1}{2}(1 + \pi P)\Phi^{JM}_{K,K_1...K_N}, \tag{1.7.36}$$

where $\pi = \pm1$ is the parity of the state, and where P reverses all nucleon coordinates as $\mathbf{r}_i \to -\mathbf{r}_i$. In some specific cases, this operator can be replaced by an equivalent rotation operator (see an example in Ref. [35]). This allows to combine angular-momentum and parity projection in a single rotation operator.

1.7.4 Multichannel Two-Cluster Systems

Until now the presentation was limited to single-channel two-cluster models. We briefly show here how to extend the formalism to multichannel calculations, and/or with spins different from zero. Although the notations are more complicated, the principles of the GCM (Sect. 1.4), as well as matrix elements between basis states (Sect. 1.5) remain unchanged.

Let us consider a channel c composed of two clusters with spins I_1 and I_2 (parities π_1 and π_2 are implied). The internal wave functions $\Phi_c^{I_1K_1}$ and $\Phi_c^{I_2K_2}$ are defined in (1.4.1), and are in general combinations of Slater determinants [see (1.7.24)]. A channel c is characterized by the properties of the clusters: masses and charges, spins, levels of excitation, etc. In transfer and inelastic reactions, the introduction of excited channels is of course necessary. However, even in spectroscopic calculations, additional channels may improve the total wave function of the system, according to the variational principle. As shown in Sect. 1.7.2, calculations involving p-shell or sd-shell clusters may contain a large number of channels.

From the internal cluster wave functions, we extend definition (1.4.7) to

$$
\begin{aligned}
\Phi_c^{IK}(\mathbf{R}) &= \frac{1}{\sqrt{N_0}} \sum_{K_1 K_2} \langle I_1 K_1 I_2 K_2 | IK \rangle \mathscr{A} \Phi_c^{I_1 K_1}\left(-\frac{A_2}{A}\mathbf{R}\right) \Phi_c^{I_2 K_2}\left(\frac{A_1}{A}\mathbf{R}\right) \\
&= \frac{1}{\sqrt{N_0}} \mathscr{A}\left[\Phi_c^{I_1}\left(-\frac{A_2}{A}\mathbf{R}\right) \otimes \Phi_c^{I_2}\left(\frac{A_1}{A}\mathbf{R}\right)\right]^{IK},
\end{aligned}
\tag{1.7.37}
$$

where I is the channel spin, and results from the coupling of I_1 and I_2. This quantum number plays an important role in reactions. In (1.7.37), we assume that the origin is at the c.m., and we have rewritten the angular-momentum coupling in the standard, compact, notation.

According to (1.7.34), projection of basis functions (1.7.37) provides

$$
\Phi_{cIK}^{JM}(R) = \int \mathscr{D}_{MK}^{J*}(\Omega)\mathscr{R}^J(\Omega)\Phi_c^{IK}(\mathbf{R})d\Omega.
\tag{1.7.38}
$$

This definition is well adapted to spectroscopy. However we define an equivalent basis as

$$
\Phi_{c\ell I}^{JM\pi}(R) = \left(\frac{2\ell+1}{256\pi^5}\right)^{1/2} \sum_K \langle IK\ell 0 | JK \rangle \Phi_{cIK}^{JM}(R),
\tag{1.7.39}
$$

which makes use of the relative angular momentum ℓ. The normalization factor allows to simplify the RGM wave function (see below). This factor, however, can be chosen arbitrarily as long as it is consistent in all matrix elements. Notice that the projection over ℓ directly provides the projection on parity which is related to the individual parities of the clusters π_1 and π_2 as

$$
\pi = \pi_1 \pi_2 (-1)^\ell.
\tag{1.7.40}
$$

Using Eqs. (1.7.27), (1.7.33) and (1.7.34), we have

$$
\Phi_{c\ell I}^{JM\pi}(R) = \frac{1}{4\pi} \int d\Omega_R \left[\Phi_c^I(R, \Omega_R) \otimes Y_\ell(\Omega_R)\right]^{JM}.
\tag{1.7.41}
$$

with

$$
\Phi_c^{IK}(R, \Omega_R) = \mathscr{R}^L(\Omega_R)\Phi_c^{IK}(\mathbf{R}).
\tag{1.7.42}
$$

The overlap between two projected basis functions (1.7.41) is obtained from a generalization of the single-channel result (1.7.31) as

$$\langle \Phi_{c\ell I}^{J\pi}(R) \mid \Phi_{c'\ell'I'}^{J\pi}(R') \rangle = \frac{\sqrt{(2\ell+1)(2\ell'+1)}}{8\pi(2J+1)}$$

$$\times \sum_{K,K'} \langle I\,K\,\ell\,0 \mid J\,K \rangle \langle I'\,K'\,\ell'\,K-K' \mid J\,K \rangle$$

$$\times \int_0^\pi d_{K-K',0}^{\ell'}(\beta) \langle \Phi_c^{IK}(R,0) \mid \Phi_{c'}^{I'K'}(R',\beta) \rangle d\cos\beta,$$

$$(1.7.43)$$

where $\Phi_c^{IK}(R,\beta)$ is a Slater determinant with the generator coordinate in the xz plane, and making an angle β with the z axis. To derive (1.7.43), we have used the symmetry of the unprojected matrix element around the z axis. For two-cluster calculations these matrix elements are obtained from one-dimensional integrals. This definition is valid for any rotation–invariant operator. The extension to more general operators can be found in Ref. [85]. Notice that the projected matrix elements (1.7.43) must be symmetric. This is not trivial since the generator coordinates R and R' are not treated in the same way (R is chosen along the z axis). The symmetry of the final result is a severe test of the calculation.

A calculation analog to those developed in Sect. 1.4 provides the equivalence between the GCM and RGM for a multichannel system. The extension of (1.4.27) is directly obtained from

$$\Phi_{c\ell I}^{JM\pi}(R) = \frac{1}{\sqrt{N_0}} \Phi_{cm} \mathscr{A} \Gamma_\ell(\rho,R) \varphi_{c\ell I}^{JM\pi}(\xi_1,\xi_2,\Omega_\rho), \qquad (1.7.44)$$

where the channel wave function reads

$$\varphi_{c\ell I}^{JM\pi}(\xi_1,\xi_2,\Omega_\rho) = \left[\left[\phi_c^{I_1}(\xi_1) \otimes \phi_c^{I_2}(\xi_2)\right]^I \otimes Y_\ell(\Omega_\rho)\right]^{JM}. \qquad (1.7.45)$$

In this definition, $\phi_c^{I_1}$ and $\phi_c^{I_2}$ are the translation-invariant internal wave functions depending on the sets ξ_1 and ξ_2 of internal coordinates. In multichannel problems, the total wave function of the system is given by

$$\Psi^{JM\pi} = \sum_{c\ell I} \Psi_{c\ell I}^{JM\pi}, \qquad (1.7.46)$$

where the contribution of each channel is defined as

$$\Psi_{c\ell I}^{JM\pi} = \int f_{c\ell I}^{J\pi}(R) \Phi_{c\ell I}^{JM\pi}(R) dR \qquad \text{(in the GCM)}$$

$$= \mathscr{A} g_{c\ell I}^{J\pi} \varphi_{c\ell I}^{JM\pi} \qquad \text{(in the RGM)}. \qquad (1.7.47)$$

As for single-channel calculations, the RGM radial function is deduced from the generator function as

$$g_{c\ell I}^{J\pi}(\rho) = \int f_{c\ell I}^{J\pi}(R) \Gamma_\ell(\rho,R) dR. \qquad (1.7.48)$$

Fig. 1.6 Multicluster configurations for two (**a**), three (**b, d**), and five (**c**) -cluster description. In model (**c**) the basis of the tetrahedron is assumed to be defined by three α particles in an equilateral configuration

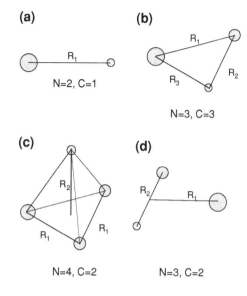

1.7.5 Multicluster Models

1.7.5.1 General Discussion

Although multicluster theories have been also developed in the RGM [33], we focus here on GCM calculations. Let us assume N clusters with internal wave functions Φ_N and centered at $\mathbf{S}_i$. As in Sect. 1.5, the set of cluster locations is denoted as $\mathbf{S}_{\{N\}} = (\mathbf{S}_1, \ldots, \mathbf{S}_N)$. From these locations, we define a set of generator coordinates $\mathbf{R}_{\{C\}} = (\mathbf{R}_1, \ldots, \mathbf{R}_C)$ where C represents the number of independent coordinates required to define the system (see Fig. 1.6). For example, two-cluster systems are characterized by one generator coordinate, the distance between the clusters (Fig. 1.6a). In a three-body model, the clusters are located at the vertices of a triangle (Fig. 1.6b). Depending on the geometry of the triangle, the number of generator coordinates can be $C = 1, 2,$ or 3 ($C = 1$ for an equilateral triangle).

A GCM basis function is defined by a multicluster generalization of Eq. (1.4.7) as

$$\Phi(\mathbf{R}_{\{C\}}) = \sqrt{\frac{A_1! \cdots A_N!}{A!}} \, \Phi_{cm}^{-1} \mathscr{A} \Phi_1(\mathbf{S}_1) \ldots \Phi_N(\mathbf{S}_N), \tag{1.7.49}$$

where the c.m. component has been removed. Again, for the sake of clarity, we do not explicitly mention the spin orientations of the clusters. Equation (1.7.49) is the starting point of all multicluster models. Matrix elements between these Slater determinants are obtained as for two-cluster calculations (see Sect. 1.5). When $N > 2$, there are, however, various applications of multicluster models, which differ by the projection technique:

(i) *Systems with a fixed geometry* (Fig. 1.6b for $N=3$ and Fig. 1.6c for $N=4$). Wave function (1.7.49) is projected with (1.7.35) and (1.7.36) on spin J and parity π, respectively. Examples for three-cluster systems are ^{11}Li or ^{24}Mg described by triangular ^{9}Li+n+n or ^{16}O+α+α configurations [109, 110]. Some four-cluster systems have been described by a tetrahedral configuration with an equilateral triangle for the three α particles, and an additional s cluster [35]. The use of a symmetric structure for the three α particles allows a reduction of the computer times for the projected matrix elements.

(ii) *Two-body systems involving a cluster nucleus* (Fig. 1.6d). This approach is essentially used to describe nucleus–nucleus collisions, where one of the colliding nuclei presents a cluster structure. Typical examples are ^{7}Be + p, with ^{7}Be = α + ^{3}He [111] and ^{12}C + α with ^{12}C = 3α [108]. In that case, the angular momentum of the cluster subsystem must be restored. The projection over angular momentum is therefore multiple.

(iii) *Multicluster hyperspherical formalism.* This development is recent [45] and is currently limited to three-cluster systems. The relative motion is described in hyperspherical coordinates [23]. This framework was recently extended to a microscopic description of three-body scattering states [95].

1.7.5.2 Fixed Geometry

In option (i), the projections over angular momentum and parity are performed with (1.7.35) and (1.7.36), respectively. As mentioned before, in some specific cases, the parity P operator can be replaced by a rotation (for example, operator P in an equilateral triangle involving three identical clusters is equivalent to a rotation by π). When the clusters have a spin zero, a projected matrix element of a rotation-invariant operator O between projected basis functions (1.7.27) is obtained from (1.7.31). The integrals are in general performed numerically (see Ref. [35] for further detail).

Finally, the total wave function of the system is obtained from a superposition of projected functions (1.7.27) as

$$
\begin{aligned}
\Psi^{JM\pi} &= \sum_K \int f_K^{J\pi}(R_{\{C\}}) \Phi_K^{JM\pi}(R_{\{C\}}) dR_1 \cdots dR_C, \\
&\approx \sum_K \sum_{R_{\{C\}}} f_K^{J\pi}(R_{\{C\}}) \Phi_K^{JM\pi}(R_{\{C\}}).
\end{aligned}
\tag{1.7.50}
$$

Coefficients $f_K^{J\pi}(R_{\{C\}})$ are obtained from the Hill–Wheeler involving the Hamiltonian and overlap kernels

$$
\sum_{R'_{\{C\}}, K'} \left[H_{KK'}^{J\pi}(R_{\{C\}}, R'_{\{C\}}) - E_\omega^{J\pi} N_{KK'}^{J\pi}(R_{\{C\}}, R'_{\{C\}}) \right] f_{K'}^{J\pi}(R'_{\{C\}}) = 0.
\tag{1.7.51}
$$

When the cluster spins are different from zero, additional quantum numbers, corresponding to the spin orientations, must be introduced. This model is well adapted to three-body halo nuclei, such as ^{6}He [112] or ^{11}Li [109]. The 3α and 4α descriptions of ^{12}C and ^{16}O are also known to significantly improve the binding energies [35].

1.7.5.3 Systems Involving a Cluster Nucleus

Multicluster models mentioned in the previous subsection are well adapted to nuclear spectroscopy. To extend these models to nucleus–nucleus reactions, a multiple angular momentum is necessary to restore, not only the spin of the total system, but also the spins of the colliding nuclei. Although a situation where both colliding nuclei present a cluster structure is possible, practical applications are currently limited to a s-shell particle with a multicluster nucleus. We therefore consider systems built from $N+1$ clusters.

Let us define nucleus 1 by N clusters with a set of generator coordinates $R_{\{C\}}$. The internal wave functions with spins I_1 and parity π_1 are therefore taken as in Eq. (1.7.50) and read

$$\Phi_\omega^{I_1 K_1 \pi_1} = \sum_{K, R_{\{C\}}} F_{K,\omega}^{I_1 \pi_1}(R_{\{C\}}) \Phi_K^{I_1 K_1 \pi_1}(R_{\{C\}}). \tag{1.7.52}$$

In this definition, index ω corresponds to the level of excitation. Wave functions with different ω values are orthogonal to each other. States with $E_\omega^{I_1 \pi_1} < 0$ correspond to bound states, whereas $E_\omega^{I_1 \pi_1} > 0$ represent pseudostates. They can be interpreted as square-integrable approximations of scattering states, and simulate the distortion of the nucleus.

Let us now consider the total $(N+1)$-cluster system. The relative motion with nucleus 2 (assumed to be described by a single cluster) requires the additional relative coordinate R. An unprojected wave function is written as

$$\Phi(\mathbf{R}_{\{C\}}, \mathbf{R}) = \frac{1}{\sqrt{N_0}} \mathscr{A} \, \Phi_1\left(\mathbf{R}_{\{C\}}, -\frac{A_2}{A}\mathbf{R}\right) \Phi_2\left(\frac{A_1}{A}\mathbf{R}\right), \tag{1.7.53}$$

where $\Phi_1(\mathbf{R}_{\{C\}}, -\frac{A_2}{A}\mathbf{R})$ is a Slater determinant (1.7.49) centred at $-\frac{A_2}{A}\mathbf{R}$. After projection over the angular momentum of nucleus 1, and summation over $R_{\{C\}}$, a basis state is defined as

$$\Phi_\omega^{IK}(\mathbf{R}) = \frac{1}{\sqrt{N_0}} \mathscr{A} [\Phi_\omega^{I_1 \pi_1}(-\frac{A_2}{A}\mathbf{R}) \otimes \Phi^{I_2 \pi_2}(\frac{A_1}{A}\mathbf{R})]^{IK}, \tag{1.7.54}$$

where (1.7.52) has been used for nucleus 1. The multichannel theory presented in Sect. 1.7.4 can therefore be applied. In particular the matrix elements (1.7.43) are still valid, after an additional projection on $I_1 \pi_1$. In general, these matrix elements involve 7-dimensional integrals [80, 113] (3 dimensions for the Euler angles in the bra and in the ket, and one additional integral for the relative motion). When the cluster nucleus involves two clusters, this integral is reduced to 5 dimensions [110].

Fig. 1.7 Three-cluster
configuration

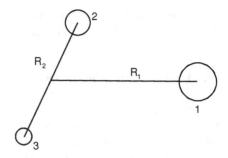

1.7.5.4 Hyperspherical Formalism

Let us consider the three-cluster system displayed in Fig. 1.7. The center of mass of
each cluster is defined as

$$
\mathbf{R}_{cm,1} = \frac{1}{A_1} \sum_{i=1}^{A_1} \mathbf{r}_i,
$$

$$
\mathbf{R}_{cm,2} = \frac{1}{A_2} \sum_{i=A_1+1}^{A_1+A_2} \mathbf{r}_i,
$$

$$
\mathbf{R}_{cm,3} = \frac{1}{A_3} \sum_{i=A_1+A_2+1}^{A} \mathbf{r}_i. \tag{1.7.55}
$$

In the hyperspherical formalism [23], the scaled Jacobi coordinates are given by

$$
\mathbf{x} = \sqrt{\mu_{23}}(\mathbf{R}_{cm,2} - \mathbf{R}_{cm,3}),
$$

$$
\mathbf{y} = \sqrt{\mu} \left[\mathbf{R}_{cm,1} - \frac{A_2 \mathbf{R}_{cm,2} + A_3 \mathbf{R}_{cm,3}}{A_{23}} \right], \tag{1.7.56}
$$

where $A_{23} = A_2 + A_3$, and where the reduced masses are

$$
\mu_{23} = \frac{A_2 A_3}{A_{23}}
$$

$$
\mu = \frac{A_1 A_{23}}{A}. \tag{1.7.57}
$$

These coordinates provide the hyperradius and hyperangle

$$
\rho^2 = \mathbf{x}^2 + \mathbf{y}^2
$$

$$
\alpha_\rho = \arctan(y/x). \tag{1.7.58}
$$

The hyperspherical formalism is well known in non-microscopic three-body
systems [30, 114], where the structure of the nuclei is neglected. In that case, the
kinetic energy can be written as

$$T = -\frac{\hbar^2}{2m_N} \left(\frac{\partial^2}{\partial \rho^2} + \frac{5}{\rho} \frac{\partial}{\partial \rho} - \frac{\mathbf{K}^2(\Omega^5)}{\rho^2} \right), \tag{1.7.59}$$

where $\Omega^5 = (\Omega_x, \Omega_y, \alpha)$ is defined from the hyperangle α, and by the directions of the Jacobi coordinates Ω_x and Ω_y. The hypermomentum operator $\mathbf{K}^2$ generalizes the concept of angular momentum in two-body systems and can be diagonalized as

$$\mathbf{K}^2 \mathscr{Y}_{KLM}^{\ell_x \ell_y}(\Omega^5) = K(K+4) \mathscr{Y}_{KLM}^{\ell_x \ell_y}(\Omega^5), \tag{1.7.60}$$

where $\mathbf{K}$ is the hypermomentum, and ℓ_x and ℓ_y are the orbital momenta associated with $\mathbf{x}$ and $\mathbf{y}$. The hyperspherical functions are [115]

$$\mathscr{Y}_{KLM}^{\ell_x \ell_y}(\Omega^5) = \phi_K^{\ell_x \ell_y}(\alpha) \left[Y_{\ell_x}(\Omega_x) \otimes Y_{\ell_y}(\Omega_y) \right]^{LM}, \tag{1.7.61}$$

with

$$\phi_K^{\ell_x \ell_y}(\alpha) = \mathscr{N}_K^{\ell_x \ell_y}(\cos \alpha)^{\ell_x}(\sin \alpha)^{\ell_y} P_n^{\ell_y + 1/2, \ell_x + 1/2}(\cos 2\alpha). \tag{1.7.62}$$

In these definition, $P_n^{\alpha\beta}(x)$ is a Jacobi polynomial, $\mathscr{N}_K^{l_x l_y}$ is a normalization factor, and $n = (K - \ell_x - \ell_y)/2$ is a positive integer. The total wave function is then expanded over the basis (1.7.61), which provides a system of coupled differential equations (see Ref. [114] for detail). There are three different choices for the Jacobi coordinates (1.7.56). However, these choices are equivalent since the corresponding hyperspherical functions (1.7.61) are related to each other by a unitary transform involving Raynal–Revai coefficients [115]. Many applications have been performed in the spectroscopy of light nuclei [114, 116] and, more recently, for three-body continuum states [98, 117].

The extension of the hyperspherical theory to microscopic three-cluster systems is recent [45]. Let us consider a three-cluster Slater determinant (1.7.49) or (1.7.53) defined by two generator coordinates $\mathbf{R}_1$ and $\mathbf{R}_2$ (see Fig. 1.7). According to (1.7.58), we define the hyperradius and hyperangle from the scaled generator coordinates

$$\begin{aligned} \mathbf{X} &= \sqrt{\mu_{23}}\,\mathbf{R}_2 \\ \mathbf{Y} &= \sqrt{\mu}\,\mathbf{R}_1 \end{aligned} \tag{1.7.63}$$

as

$$\begin{aligned} R^2 &= \mathbf{X}^2 + \mathbf{Y}^2 \\ \alpha_R &= \arctan(Y/X). \end{aligned} \tag{1.7.64}$$

This Slater determinant can be factorized as in (1.4.8) for two-cluster systems. We have

$$\Phi(\mathbf{X}, \mathbf{Y}) = \Phi_{cm} \mathscr{A} \phi_1 \phi_2 \phi_3 \exp\left(-\frac{(\mathbf{x} - \mathbf{X})^2}{2b^2} \right) \exp\left(-\frac{(\mathbf{y} - \mathbf{Y})^2}{2b^2} \right), \tag{1.7.65}$$

where Φ_{cm} is defined as in (1.4.19), and ϕ_1, ϕ_2, ϕ_3 are the translation-invariant internal wave functions of the three clusters. We take here the origin of the coordinates at the center of mass. We assume that the clusters have a spin zero, but the theory can be generalized by additional angular momentum couplings (see Ref. [45] for detail).

To develop (1.7.65), we use the expansion [115]

$$\exp\left(-(\mathbf{x}.\mathbf{X} + \mathbf{y}.\mathbf{Y})\right) = \frac{(2\pi)^3}{(\rho R)^2} \sum_{\ell_x \ell_y LMK} I_{K+2}(R\rho) \mathscr{Y}_{KLM}^{\ell_x \ell_y}(\Omega_\rho^5) \mathscr{Y}_{KLM}^{\ell_x \ell_y *}(\Omega_R^5),$$

(1.7.66)

where $I_{K+2}(x)$ is a modified Bessel function. The Slater determinant (1.7.65) can therefore be written as

$$\Phi(\mathbf{X}, \mathbf{Y}) = \sum_{\ell_x \ell_y LMK} \Phi_{\ell_x \ell_y K}^{LM}(R) \mathscr{Y}_{KLM}^{\ell_x \ell_y}(\Omega_R^5),$$

(1.7.67)

where the projected basis function reads

$$\Phi_{\ell_x \ell_y K}^{LM}(R) = \int d\Omega_R^5 \mathscr{Y}_{KLM}^{\ell_x \ell_y \star}(\Omega_R^5) \Phi(\mathbf{X}, \mathbf{Y})$$

$$= \Phi_{cm} \mathscr{A} \phi_1 \phi_2 \phi_3 G_K(\rho, R) \mathscr{Y}_{KLM}^{\ell_x \ell_y}(\Omega_\rho^5),$$

(1.7.68)

with

$$d\Omega_R^5 = R^5 \cos^2 \alpha_R \sin^2 \alpha_R d\alpha_R d\Omega_X d\Omega_Y,$$

(1.7.69)

and

$$G_K(\rho, R) = \left(\frac{b^2}{\rho R}\right)^2 \left(\frac{4\pi}{b^2}\right)^{3/2} \exp\left(-\frac{\rho^2 + R^2}{2b^2}\right) I_{K+2}\left(\frac{\rho R}{b^2}\right).$$

(1.7.70)

Definition (1.7.68) is a direct extension of (1.4.27), obtained for two-cluster systems. It only depends on a single generator coordinate, the hyperradius R. As in (1.4.27), ρ is the quantal coordinate, and R a parameter which is not affected by antisymmetrization. The total wave function is given, as in (1.4.28) for two-cluster systems, as

$$\Psi^{LM} = \sum_{\ell_x \ell_y K} \int f_{\ell_x \ell_y K}^L(R) \Phi_{\ell_x \ell_y K}^{LM}(R) dR,$$

(1.7.71)

where the generator functions $f_{\ell_x \ell_y K}^L(R)$ are obtained from a three-body Hill–Wheeler equation (1.4.32). The matrix elements involving GCM projected functions are computed as in Sect. 1.7.5.3, with a further integral over the hyperangle. This gives, for the overlap,

$$\langle \Phi^{LM}_{\ell_x \ell_y K}(R) | \Phi^{LM}_{\ell'_x \ell'_y K'}(R') \rangle = \int d\alpha_R d\alpha_{R'} \cos^2 \alpha_R \sin^2 \alpha_R \cos^2 \alpha_{R'} \sin^2 \alpha_{R'}$$

$$\times \phi_K^{\ell_x \ell_y}(\alpha_R) \phi_{K'}^{\ell'_x \ell'_y}(\alpha_{R'}) \langle \Phi^L_{\ell_x \ell_y}(X, Y) | \Phi^L_{\ell'_x \ell'_y}(X', Y') \rangle,$$

$$(1.7.72)$$

where the matrix elements in the integrand are obtained from five dimensional integrals. Matrix elements in the hyperspherical framework therefore involve 7-dimensional integrals. The advantage with respect to the fixed geometry is that this basis involves one generator coordinate only. It has been essentially applied to systems involving s clusters, such as ^{6}He [45], ^{6}Li [45], ^{9}Be [96] or ^{12}C [96]. In that case, a semi-analytic treatment of the matrix elements (1.7.72) can be used. Very recently an ^{16}O+p+p calculation was performed [118] to investigate the diproton radioactivity of ^{18}Ne. The model has also been extended to a microscopic description of three-body continuum states [95].

1.8 Scattering States With the GCM

1.8.1 Introduction

The treatment of scattering states in microscopic models is a delicate problem. Exact solutions of the Schrödinger equation (1.1.1) for positive energies must take account of the asymptotic boundary conditions. Extensions to scattering and resonant states represent however a wide range of applications: elastic and inelastic scattering, transfer, capture, etc. The two latter processes are important in nuclear astrophysics, where the low-energy cross sections relevant to stellar models are in general too small to be measured in laboratory. The need for a precise treatment of unbound states is also crucial in the study of light exotic nuclei, where the ground state is close to the particle-emission threshold, or is unbound. In that case the determination of resonance properties require an extension to scattering states. At large distances, the microscopic Hamiltonian tends to

$$H \to H_1 + H_2 - \frac{\hbar^2}{2\mu} \Delta_\rho + \frac{Z_1 Z_2 e^2}{\rho}. \qquad (1.8.1)$$

Consequently, the relative wave function (1.1.4) of a scattering state at energy E tends to, in partial wave ℓm,

$$\Psi^{\ell m}(\rho) \xrightarrow[\rho \to \infty]{} \phi_1 \phi_2 Y_\ell^m(\Omega) g_{ext}^\ell(\rho),$$

$$g_{ext}^\ell(\rho) = I_\ell(k\rho) - U_\ell O_\ell(k\rho), \qquad (1.8.2)$$

where k is the wave number, and $I_\ell(x) = O_\ell^*(x)$ are incoming and outgoing Coulomb functions [119]. At large distances, antisymmetrization between the colliding nuclei is negligible. In Eq. (1.8.2), U_ℓ is the collision matrix, which determines the scattering cross sections. In single-channel calculations, U_ℓ is parametrized as

$$U_\ell = \exp(2i\delta_\ell), \tag{1.8.3}$$

where δ_ℓ is the phase shift. It is real in microscopic theories, since nucleon–nucleon interactions are real. In optical models, it can be complex owing to complex optical potentials. In a multichannel problem [120], U_ℓ is a symmetric and unitary matrix whose size is equal to the number of open channels. Here we restrict the presentations to single-channel calculations. An extension can be found in Refs. [85, 121].

In the RGM approach, solutions of (1.3.17) (or its angular-momentum extension) can be derived at positive energies by using finite-difference methods, or by the Lagrange-mesh approach [122, 123]. The GCM variant, however, cannot be directly adapted to scattering states since any finite combination of (projected) Slater determinants (1.4.27) presents a Gaussian behaviour. This problem is addressed by using the microscopic R-matrix method [119, 124], which is briefly described in Sect. 1.8.3.

1.8.2 Cross Sections

The collision matrices U_ℓ provide the elastic cross section. As in previous sections, we only consider systems with spinless particles (see Ref. [124] for a generalization). At the scattering angle $\Omega = (\theta, \phi)$ the elastic cross section is given by

$$\frac{d\sigma_{\text{el.}}}{d\Omega} = |f_C(\Omega) + f_N(\Omega)|^2, \tag{1.8.4}$$

where the Coulomb and nuclear amplitudes read

$$f_C(\Omega) = -\frac{\eta}{2k\sin^2\theta/2} e^{2i(\sigma_0 - \eta \ln \sin \theta/2)},$$

$$f_N(\Omega) = \frac{i}{2k} \sum_\ell (2l+1) e^{2i\sigma_l} (1 - U_\ell) P_\ell(\cos\theta). \tag{1.8.5}$$

In these definitions, $\sigma_\ell = \arg \Gamma(\ell + 1 + i\eta)$ is the Coulomb phase shift, and $\eta = Z_1 Z_2 e^2 / \hbar v$ is the Sommerfeld parameter (v is the relative velocity). As the Coulomb amplitude diverges at small angles, the integrated elastic cross section is not defined for charged-particle scattering.

Radiative capture is an electromagnetic transition from a scattering state to a bound state. The electromagnetic aspects of this process can be studied at the first order of the perturbation theory [119], with a scattering state at positive energy E and a bound state in partial wave $J_f \pi_f$ as final state at negative energy E_f. The definition of capture cross sections can be found in Refs. [124, 125] for example.

1.8.3 The Microscopic R-Matrix Method

As mentioned before, GCM basis functions have a Gaussian asymptotic behaviour, and cannot directly describe scattering states. This problem is typical of variational calculations, where the basis functions can only reproduce the short-range part of the wave functions. The R-matrix method provides an efficient way to use a finite basis for the determination of scattering properties. In this approach the configuration space is divided in two regions, separated by the channel radius a. In the internal region, the wave function is given by the GCM expansion (1.4.28)

$$\Psi_{\text{int}}^{\ell m} = \sum_n f_\ell(R_n) \Phi^{\ell m}(R_n)$$
$$= \mathscr{A}\phi_1\phi_2 g_{\text{int}}^\ell(\rho) Y_\ell^m(\Omega_\rho). \tag{1.8.6}$$

The channel radius is chosen large enough to make the nuclear force as well as antisymmetrization between the clusters negligible. Consequently, the external wave function is defined as

$$\Psi_{\text{ext}}^{\ell m} = \phi_1\phi_2 g_{\text{ext}}^\ell(\rho) Y_\ell^m(\Omega_\rho), \tag{1.8.7}$$

where the radial function $g_{\text{ext}}^\ell(\rho)$ is given by (1.8.2).

The quantities to be determined are the collision matrix U_ℓ and the coefficients $f_\ell(R_n)$. The principle of the R-matrix theory is to solve the Schrödinger equation in the internal region, and to use the continuity condition

$$g_{\text{int}}^\ell(a) = g_{\text{ext}}^\ell(a). \tag{1.8.8}$$

However, as the kinetic-energy operator is not Hermitian over a finite interval, the Schrödinger equation is replaced by the Bloch-Schrödinger equation

$$(H + \mathscr{L} - E)\Psi_{\text{int}}^{\ell m} = \mathscr{L}\Psi_{\text{ext}}^{\ell m}, \tag{1.8.9}$$

where the Bloch operator [126] acts at $\rho = a$ and is defined as

$$\mathscr{L} = \frac{\hbar^2}{2\mu}\delta(\rho - a)\frac{d}{d\rho}. \tag{1.8.10}$$

Using expansion (1.8.6) in (1.8.9) gives the linear system

$$\sum_{n'} C_{nn'}^\ell f_\ell(R_{n'}) = \langle \Phi^\ell(R_n) \mid \mathscr{L} \mid \Psi_{\text{ext}}^\ell \rangle, \tag{1.8.11}$$

where matrix $\mathbf{C}^\ell$ is defined at energy E by

$$C_{nn'}^\ell = \langle \Phi^\ell(R_n)|H + \mathscr{L} - E|\Phi^\ell(R_{n'})\rangle_{\text{int}}. \tag{1.8.12}$$

These matrix elements are defined over the internal region. This is achieved by subtracting the external contributions [121]. By definition of the channel radius a, antisymmetrization effects and the nuclear interaction are negligible in the external region. The relevant matrix elements are therefore given by

$$\langle \Phi^\ell(R_n)|\Phi^\ell(R_{n'})\rangle_{\text{int}} = \langle \Phi^\ell(R_n)|\Phi^\ell(R_{n'})\rangle - \int_a^\infty \Gamma_\ell(\rho, R_n)\Gamma_\ell(\rho, R_{n'})\rho^2 d\rho,$$

$$\langle \Phi^\ell(R_n)|H|\Phi^\ell(R_{n'})\rangle_{\text{int}} = \langle \Phi^\ell(R_n)|H|\Phi^\ell(R_{n'})\rangle$$

$$- \int_a^\infty \Gamma_\ell(\rho, R_n)(T_\rho + V_C(\rho) + E_1 + E_2)\Gamma_\ell(\rho, R_{n'})\rho^2 d\rho,$$

(1.8.13)

where the first terms in the r.h.s. are matrix elements over the whole space, involving Slater determinants. The second terms represent the external contributions of the basis functions and are computed numerically.

From matrix $\mathbf{C}^\ell$, one defines the R matrix

$$R^\ell = \frac{\hbar^2 a}{2\mu} \sum_{nn'} \Gamma_\ell(a, R_n)(\mathbf{C}^\ell)_{nn'}^{-1} \Gamma_\ell(a, R_{n'}) \qquad (1.8.14)$$

which provides the collision matrix as

$$U_\ell = \frac{I_\ell(ka) - ka I_\ell'(ka)R^\ell}{O_\ell(ka) - ka O_\ell'(ka)R^\ell}. \qquad (1.8.15)$$

For single-channel calculations the R-matrix and the collision matrix are of dimension one and, strictly speaking, are therefore not matrices. However the tradition is to keep the terminology "matrix", even for single-channel calculations. When the collision matrix is known, coefficients $f_\ell(R_n)$ can be determined from the system (1.8.11). Notice that the channel radius a is not a parameter. In practice, it stems from a compromise: it should be large enough to satisfy the R-matrix conditions, but should be kept as small as possible to limit the number of basis states in the internal region. The stability of the collision matrix and of the wave function with respect to the channel radius is a strong test of the calculation. Further detail concerning the R-matrix method, and its application to microscopic calculations can be found in Refs. [85, 121, 124].

1.9 Applications of the GCM

1.9.1 The 2α and 3α Systems

1.9.1.1 Conditions of the Calculations

The $\alpha + \alpha$ system has been well known for many years, and was one of the first appli-cations of microscopic cluster models. Owing to the large binding energy of the α particle, ^{8}Be is an ideal example of nuclear cluster structure. Two-alpha calculations are rather simple; the matrix elements can be computed from the analytical expres-sions of Tables 1.3, 1.4 and 1.5. The phase shifts are well known experimentally [127] and can be accurately reproduced by microscopic cluster models associated with the R-matrix method (see, for example, Ref. [128]).

On the other hand, more complicated 3α calculations have also been performed in various three-body models. Here we present a simultaneous study of both systems, as well as a comparison between different 3α descriptions of ^{12}C. The calculations are performed within the same conditions: an oscillator parameter $b = 1.36$ fm, and the Minnesota interaction with an admixture parameter $u = 0.94687$, as adopted in Ref. [93]. This u value provides a good description of the $\alpha + \alpha$ phase shifts up to 20 MeV, i.e. below the proton threshold. In these conditions, the binding energy of the α particle (independent of u) is $E_\alpha = -24.28$ MeV, and the r.m.s. radius is $\sqrt{< r^2 >_\alpha} = \sqrt{9/8} b = 1.44$ fm.

1.9.1.2 The $\alpha + \alpha$ System

The generator coordinates R are taken from 0.8 to 8 fm by step of 0.8 fm. We first present the energy curves, defined as the energy of the system for a fixed generator coordinate R

$$E^\ell(R) = \frac{H_\ell(R, R)}{N_\ell(R, R)}, \qquad (1.9.16)$$

and involve the Hamiltonian and overlap kernels.

Using the asymptotic behaviour (1.8.1) of the Hamiltonian provides, at large R values

$$E^\ell(R) \to 2E_\alpha + \frac{\hbar^2}{2\mu} \frac{\ell(\ell+1)}{R^2} + \frac{Z_1 Z_2 e^2}{R} + \frac{1}{4}\hbar\omega, \qquad (1.9.17)$$

where the last contribution comes from a residual kinetic-energy term. The energy curves cannot be considered as nucleus–nucleus potentials, as they do not include forbidden states (see Sect. 1.6). However they provide qualitative properties of the system. In particular the existence of a minimum suggests bound states or resonances, and the location of this minimum provides an estimate of clustering effects.

Fig. 1.8 $\alpha + \alpha$ energy curves (1.9.16). Horizontal lines represent threshold energies (see Eq. 1.9.17)

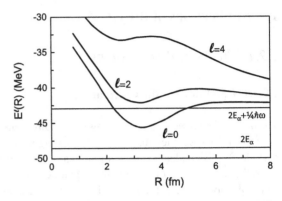

Fig. 1.9 $\alpha + \alpha$ phase shifts for $\ell = 0, 2, 4$. Experimental data are taken from Ref. [127]

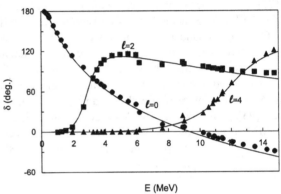

The energy curves for $\ell = 0, 2, 4$ are presented in Fig. 1.8. As it is well known, the minimum for $\ell = 0$ is located at fairly large distances ($R \approx 3.2$ fm), which is a strong support for $\alpha + \alpha$ clustering in ^{8}Be. When ℓ increases this minimum moves to smaller distances. It almost disappears for $\ell = 4$, where only a broad resonance can be expected.

The $\alpha + \alpha$ phase shifts are presented in Fig. 1.9 with the experimental data of Ref. [127]. The 0^+ ground state is found at $E = 0.098$ MeV, in fair agreement with experiment ($E = 0.092$ MeV). The broad 2^+ and 4^+ resonances are also well reproduced by the $\alpha + \alpha$ model. Further developments in the $\alpha + \alpha$ system, including monopole distortion of the α particle (i.e. the α wave function is defined by a combination of several b values) can be found, for example, in Refs. [37, 60]. A discussion of the sensitivity to the channel radius is presented in Ref. [124].

1.9.1.3 The 3α System

The ^{12}C nucleus described by a 3α cluster structure has been studied in various microscopic approaches: with a frozen geometry (see, e.g., Refs. [5, 35, 129]), with an $\alpha + ^8$Be model [33, 130], and with the hyperspherical formalism [93]. Here we aim at comparing the different approaches within the same conditions. We also complement the hyperspherical calculation presented in Ref. [93].

In multichannel (or multicluster) calculations, the energy curves are defined by a generalization of (1.9.16). For a given generator coordinate R, the Hamiltonian matrix is diagonalized as

$$\sum_j \left(H_{ij}^{J\pi}(R, R) - E^{J\pi}(R) N_{ij}^{J\pi}(R, R) \right) c_j^{J\pi} = 0, \qquad (1.9.18)$$

where i, j represent the channels (or additional generator coordinates in multicluster problems). We only consider the lowest eigenvalue.

The calculations are performed as follows:

(a) For the frozen geometry, we take an equilateral structure (see Fig. 1.10), with $R_C = 1$–7 fm (by step of 1 fm). A minimum is found near $R_C \approx 2.1$ fm, which is smaller than with the Volkov force [5, 35]. The Volkov force is known to give rise to stronger clustering effects.

(b) For the $\alpha + ^8$Be model (Fig. 1.11), ^{8}Be is described by $I = 0, 2, 4$ and by 4 generator coordinates $R_2 = 1.4, 2.6, 3.8, 5.0$ fm. These values cover the minima observed in the energy curves (Fig. 1.8). For the $\alpha + ^8$Be motion, we take $R_1 = 1.5$–12.3 fm by step of 1.2 fm. Matrix elements are determined as explained in Ref. [110]. Figure 1.11 displays the $\alpha + ^8$Be energy curves as a function of the generator coordinate R_1. The 0^+ and 2^+ partial waves present a minimum near $R_1 \approx 2$ fm, whereas the 3^- energy curve is typical of a stronger deformation. We illustrate the influence of $\alpha + ^8$Be* excited channels by keeping only the $\alpha + ^8$Be(0^+) configurations (dotted lines). The energy surface (generalization of Eq. (9.18) for two generator coordinates) is presented in Fig. 1.12 for $J = 0^+$. The minimum, corresponding to the ground state of ^{12}C, is obtained for rather small values of R_1 and R_2. At large R_1 values, the dependence on R_2 follows the $\alpha + \alpha$ energy curves of Fig. 1.8.

(c) For the hyperspherical description of the 3α system, we take the generator coordinates $R = 1.5$–15 fm by step of 1.5 fm, and K values up to $K_{\max} = 12$. The energy curves (Fig. 1.13) present a minimum near $R \approx 4$ fm for $J = 0^+, 2^+$ and at larger distance for $J = 3^-$. In Fig. 1.14 we analyze the convergence with respect to $K_{\max}$. This convergence is rather fast, much faster than in non-microscopic models [116].

The ^{12}C energies, obtained by the diagonalization of the full basis, i.e. including all generator coordinates, are shown in Table 1.6 for the $0_1^+, 0_2^+, 2^+$ and 3^- states. Except for the 0_2^+ resonance, the differences between calculations (b) and (c) are of the order of 0.02 MeV, which shows that both bases are equivalent. In contrast, the much simpler model (a) gives a significant underbinding (~ 0.7 MeV). In option (b), considering only $I = 0$ in ^{8}Be provides a non-negligible difference. For the 0_2^+ state which plays a key role in He burning, the $\alpha + ^8$Be description is slightly better than the hyperspherical approach (larger K and R values would be necessary to reach

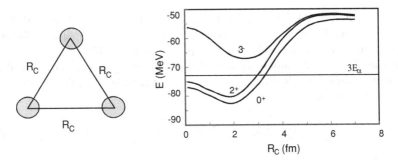

Fig. 1.10 3α structure in model (a) (*left*) and energies (*right*). The α particle is represented by a gray circle

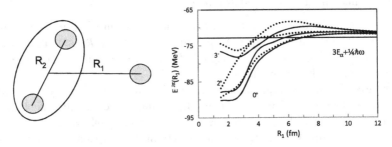

Fig. 1.11 3α structure in model (b) (*left*) and energy curves for different J values (*right*). The dotted lines are obtained with the $\alpha + {}^{8}\text{Be}(0^{+})$ channel only

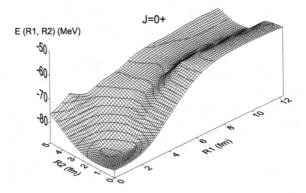

Fig. 1.12 ${}^{12}\text{C}$ energy surface for $J = 0^{+}$ as a function of R_1 and R_2

full convergence). For this state the frozen equilateral triangle configuration is not adapted.

In general the 2^{+} excitation energy is underestimated. This result is due to the lack of spin–orbit force, whose matrix elements vanish in an α model. Introducing α

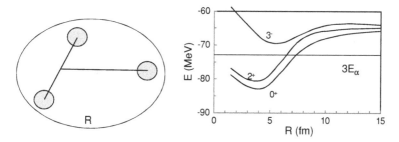

Fig. 1.13 3α structure in hyperspherical model (c) (*left*) and energies (*right*)

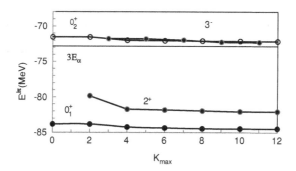

Fig. 1.14 3α binding energies as a function of K_{max}

Table 1.6 Binding energies (in MeV) for various ^{12}C states in models (a), (b) and (c). For model (b), the bracketed values are obtained with $I=0$ only

State	(a)	(b)	(c)
0_1^+	-83.69	$-84.44(-84.10)$	-84.46
0_2^+	-63.86	$-72.23(-72.11)$	-72.14
2^+	-81.29	$-82.04(-81.41)$	-82.06
3^-	-69.03	$-72.30(-71.81)$	-72.28

breakup configurations increases the 2^+ excitation energy [131], in agreement with experiment. In that case, however, the simplicity of the α cluster model is lost.

1.9.2 Other Applications of the Multicluster Model

In this subsection, we aim to illustrate the multicluster approach with typical results obtained with a five-cluster model [35, 80, 132]. It allows the description of reactions between a nucleus denoted as 1 and a nucleus denoted as 2 and/or to describe spectroscopic properties of the unified nucleus $(1+2)$.

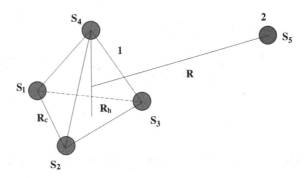

Fig. 1.15 Schematic representation of the five-center model. S_i are the corresponding vertices

Table 1.7 Binding energies (in MeV) of one-center and four-center wave functions.

	Ground-state	One center	Four centers
^{12}C	0^+	-76.3	-88.0
^{13}C	$1/2^-$	-83.7	-91.7
^{14}C	0^+	-96.1	-103.4
^{15}N	$1/2^-$	-120.7	-126.8
^{16}O	0^+	-140.4	-148.8

A schematic representation is given in Fig. 1.15. Nucleus 1 is described by a tetrahedral structure with three alpha clusters located at the vertices of an equilateral triangle and an additional s-cluster. Typical examples are ^{12}C $= 3\alpha$, ^{13}C $= 3\alpha + n$, and ^{15}N $= 3\alpha + t$. Nucleus 2 is described by a s-cluster and corresponds to an α particle or a nucleon. The set of generator coordinates defining nucleus 1 is defined as $R_{\{C\}} = (R_c, R_h)$ (see Fig. 1.15).

To test the cluster description of nucleus 1, we first analyze properties of some p-shell nuclei described by a tetrahedral structure [35]. The calculations with the four-cluster model are performed with a mixing of (R_c, R_h) configurations, whereas one-center results are obtained with $R_c = R_h = 0$. The oscillator parameter is optimized to minimize the binding energy. The nucleon–nucleon interaction is the Volkov force V2 (with the standard value $M=0.6$), and the spin–orbit amplitude is chosen as $S_0 = 30$ MeV.fm^5. The spin–orbit force does not contribute to ^{12}C, ^{14}C and ^{16}O since the clusters have an intrinsic spin zero.

Table 1.7 compares the ground-state binding energies in a one-center model and in a four-centers model. The binding energies obtained within the four-center model are always lower with a quite substantial difference.

Table 1.8 shows typical $E2$ transition probabilities and the ^{12}C quadrupole moment. Clearly the introduction of clustering effects in these nuclei improve the wave functions. The four-cluster results are in good agreement with experiment, whereas the no-cluster approximation underestimates the $E2$ properties.

We illustrate five-cluster calculations with the $\alpha + ^{16}$O system, described by five α clusters [80]. It is well-known that several ^{20}Ne states present a marked $\alpha + ^{16}$O

Table 1.8 ^{12}C quadrupole moment (in $e.fm^2$) and reduced transition probabilities (in W.u.). Experimental data are taken from Refs. [133, 134].

		One-center	Four-center	Experiment
^{12}C	$Q(2^+)$	3.0	5.4	6 ± 3
	$B(E2, 2^+ \rightarrow 0^+)$	2.0	4.5	4.65 ± 0.26
^{13}C	$B(E2, 3/2^- \rightarrow 1/2^-)$	1.5	3.4	3.5 ± 0.8
	$B(E2, 5/2^- \rightarrow 1/2^-)$	1.0	3.1	3.1 ± 0.2

Table 1.9 α-width (in keV) of some ^{20}Ne states in the $K = 0^-$ band. Experimental data are taken from Ref. [135]

J^π	Two centers	Five centers (0^+)	Five-centers $(0^+, 3^-, 1^-)$	Experiment
1^-	0.042	0.032	0.031	0.028 ± 0.003
3^-	13.0	10.8	10.6	8.2 ± 0.3
5^-	200	173	169	145 ± 40
7^-	700	570	530	310 ± 30

structure. The ^{20}Ne nucleus is described by $\alpha + {}^{16}$O channels where the 0^+ ground state and the 1^- and 3^- excited states are considered. For computer-time reasons, only one set of generator coordinates ($R_c = 1.8$ fm, $R_h = 2.5$ fm) is selected to describe the ^{16}O nucleus. These values minimize the ^{16}O binding energy.

We focus here on α widths of some states in the $K = 0^-$ band. They are obtained within the R-matrix formalism. Results are gathered in Table 1.9 and obtained in three different ways: the standard two-cluster model where ^{16}O is described by a closed p-shell structure, the multicluster approach with only the ^{16}O ground state, and with some excited channels. The α widths are overestimated in the two-center approach, but are significantly reduced when clustering effects are included in ^{16}O. Excited channels still improve the comparison with experiment.

In Ref. [80] the spectroscopy of ^{20}Ne was complemented by calculations of the $\alpha + {}^{16}$O phase shifts and of the ^{16}O$(\alpha, \gamma)^{20}$Ne radiative-capture cross section. As mentioned in Sect. 1.8, the cluster model can be extended to scattering states with the R-matrix method, which provides scattering properties (such as resonance widths) and cross sections.

1.9.3 Multichannel Study of the $^{17}F(p, \gamma)^{18}Ne$ Reaction

The knowledge of the ^{17}F(p, γ)^{18}Ne reaction rate is important for the understanding of novae and X-ray bursts [136]. The energy range characteristic of such astrophysical events can be evaluated by the calculation of the Gamow energy E_G and the width of the Gamow peak Δ_0 [137]. For a typical temperature T=0.5 GK, these values

Fig. 1.16 ^{18}Ne and ^{18}O
energy spectra (taken from
Ref. [136]) with respect to
the nucleon threshold (*dotted
line*)

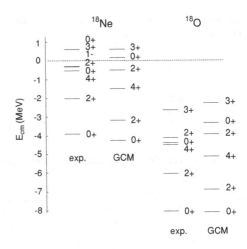

are $E_G = 0.32$ MeV and $\Delta_0 = 0.28$ MeV. Until now, a direct measurement of the
^{17}F(p, γ)^{18}Ne cross section down to these energies has not been performed.

It is now well established that the 3_1^+ ($\ell = 0$) resonance at $E_{cm} = 0.64$ MeV
dominates the ^{17}F(p, γ)^{18}Ne reaction rate at stellar temperatures. The energy and
the proton width have been measured by Bardayan et al. [138]. However, the gamma
width which determines the reaction rate is experimentally unknown and is estimated
from theoretical calculations.

The predictive power of the GCM is of particular interest in such a context. Indeed,
the small number of parameters allows reliable calculations in the astrophysical
energy range. The ^{18}Ne wave functions are defined as a combination of ^{17}F + p and
^{14}O + α channel functions. The ^{17}F internal wave functions are defined from all
possible Slater determinants with one proton in the sd shell, the s and p shells being
filled. This provides the well known shell-model states with spin $I_1 = 5/2^+, 1/2^+$,
and $3/2^+$. Similarly, the ^{14}O internal wave functions are defined from four neutrons
in the p shell, the s shell being filled for the neutron part, and the s and p shells
being filled for the proton part. This provides two states with $I_1 = 0^+$, one state with
$I_1 = 1^+$ and two states with $I_1 = 2^+$ (see Sect. 7.2). The nucleon–nucleon interaction
is fitted to the 3^+ energy. Further detail about the conditions of the calculation is given
in Ref. [136].

The ^{18}Ne spectrum is shown in Fig. 1.16 along with the ^{18}O mirror nucleus. We
find a good overall agreement with experiment. The state ordering below the ^{17}F + p
threshold is well reproduced, except for the 0_2^+ state, slightly unbound in the GCM.
However the difference with the experimental value is only of 0.49 MeV. We can also
notice the good description of the 2_2^+ state.

The important 3_1^+ resonance is known to have a single-particle structure, and is
well described in a ^{17}F + p model. The energy is adjusted by the nucleon–nucleon
force, but the proton width $\Gamma_p = 21.1$ keV is obtained without any fitting procedure.
The GCM value is in very good agreement with experiment ($\Gamma_p = 18.0 \pm 2_{\text{stat}} \pm$

Fig. 1.17 ^{17}F(p, γ)^{18}Ne astrophysical S-factor with the contribution of different multipolarities (taken from Ref. [136])

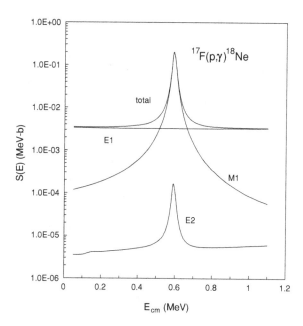

1_{sys} keV). The predicted gamma width $\Gamma_\gamma = 33$ meV is similar to values deduced from the shell model, and used in astrophysics ($\Gamma_\gamma = 30 \pm 20$ meV) [138].

The total and partial S-factors are displayed in Fig. 1.17. The astrophysical S-factor is related to the cross section $\sigma(E)$ as

$$S(E) = E\sigma(E)\exp(2\pi\eta), \tag{1.9.19}$$

where η is the Sommerfeld parameter defined in Sect. 1.8.2. The calculations are performed for the $E1$, $E2$, and $M1$ multipolarities, and the 0_1^+, 2_1^+, 2_2^+ and 4_1^+ bound states are considered. As expected, the non-resonant $E1$ transitions give the dominant contribution below the 3_1^+ state. At zero energy, the S-factor is entirely determined by the $E1$ term. The calculation gives $S(0)=3.5$ keV-b. On the contrary, the $M1$ contribution is dominant near the resonance (the $E2$ term is negligible). More detailed calculations [136] show that transitions to the 2_1^+ and 2_2^+ bound states represent the main parts of the S-factor.

The ^{17}F(p, γ)^{18}Ne reaction rate is usually calculated as a sum of a direct component and of a resonant term taking into account the 1_1^-, 3_1^+ and 0_3^+ contributions in ^{18}Ne [139]. One of the main advantages of our method is to perform calculations without separation between resonant and non-resonant contributions. However, in the present case our model is unable to reproduce the 1_1^- and 0_3^+ low-energy resonances. Their contribution to the rate can be treated separately taking energies and total widths from experiment [139].

All these results illustrate the adequacy and limitations of the present framework for reactions of astrophysical interest. Here, two states located in the astrophysical energy range are not reproduced by the GCM: the 1_1^- and 0_3^+ states. This means that the multichannel basis is not sufficient, and that other configurations should be introduced. Owing to the fact that the main contribution comes here from the 3_1^+ resonance, this problem is however minor.

Another drawback comes from the lack of degrees of freedom in the Volkov interaction. The present parameter choice leads to a strong overestimate of the $(^{14}O + \alpha) - (^{17}F + p)$ threshold: 15.43 MeV to be compared with the experimental value 1.19 MeV [140]. This can be explained by the fact that the model gives a better description of ^{17}F than of ^{14}O. This problem prevents the simultaneous study of the $^{14}O(\alpha, p)^{17}F$ transfer reaction. For these reasons, we have developed the EVI [141] (see Sect. 1.2 for more detail), a new interaction with an additional parameter which allows to fit two important properties of the system such as a resonance energy and a threshold value. This interaction is well suited to transfer reactions where the reproduction of the Q value is crucial. Applications in nuclear astrophysics can be found, for example, in Ref. [142].

1.9.4 ^{12}Be as an Example of a Light Exotic Nucleus

Due to technical difficulties, experimental informations related to the study of exotic light nuclei are in general limited. From a theoretical point of view, cluster models appear to be particularly well adapted to study such nuclei. In particular, the exact treatment of the asymptotic behavior of the wave functions through the MRM (see Sect. 1.8) allows the description of unbound states. A cluster model is also well suited to molecular states, which present a strong deformation.

We illustrate applications of the GCM with a recent study performed on the ^{12}Be nucleus [14], by focusing on molecular states. Above the $^6He+^6He$ threshold, 4^+, 6^+ and 8^+ states have been identified in the breakup of ^{12}Be into the $^6He+^6He$ and $^8He + \alpha$ channels by Freer et al. [143]. They are believed to be members of a molecular band.

A multi-channel wave function is given by a superposition of $^8He + \alpha$ and $^6He+^6He$ components. The 0^+ internal wave functions of ^6He and ^8He are built in a one-center harmonic oscillator model with $(p_{3/2})^2$ and $(p_{3/2})^4$ configurations, respectively and the 2^+ states, with $(p_{3/2})(p_{1/2})$ and $(p_{3/2})^3(p_{1/2})^1$ configurations, respectively. Details on the conditions of the calculations are given in Ref. [14]. The present calculation updates an earlier study [13], where only ground-state configurations were included.

The calculation (see Fig. 1.18) supports the existence of a molecular band, as proposed by Freer et al. [143] and by Saito et al. [144]. The large reduced widths support the molecular structure of this band. The analysis of the dimensionless reduced widths shows that the 0^+ wave function is dominated by the $^6He(0^+) +^6He(0^+)$ channel. The theoretical 2^+ and 4^+ energies are in good agreement with the results of Saito et al. [144] and of Freer et al. [143], respectively.

Fig. 1.18 Positive-parity ^{12}Be states predicted by the GCM (full symbols) and experimental candidates (open symbols [143–145])

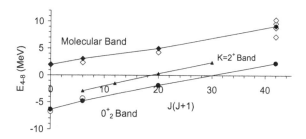

The wave functions are dominated by the ^{6}He(0^+) $+^6$He(0^+) and ^{8}He(0^+) $+ \alpha$ ground-state configurations.

The 0^+_2 and 2^+_2 states are well reproduced by the GCM. Indeed, the energy difference with experiment is less than 0.5 MeV for both states (see Fig. 1.18). According to Refs. [146–148], we confirm that these states belong to a same 0^+_2 band and we propose other band members. The calculation shows that the wave functions of the 0^+_2-band members are dominated by the ^{8}He $+ \alpha$ channels.

With this example, we have illustrated the ability of the GCM to describe molecular states. Indeed, we have reproduced many known states of ^{12}Be, in particular those belonging to a molecular band. We have also predicted new ^{12}Be bands which could be searched for in future experimental studies.

1.10 Conclusions

In this work, we have reviewed various aspects of microscopic cluster models. With respect to non-microscopic variants, microscopic theories offer several advantages: in particular, they only depend on a nucleon–nucleon interaction, and excited configurations can be introduced without further parameters. The cluster approximation makes them tractable, even for fairly large nucleon numbers. Of course, cluster models use effective interactions.

Microscopic cluster models are applied in many topics; using the microscopic R-matrix method, they can be consistently extended to scattering states [124]. This property opens many perspectives in low-energy nuclear physics. Not only cross sections can be studied, but spectroscopic applications can be extended to unbound states, even with a broad width [149]. The microscopic treatment of two-cluster scattering states is well known, but this formalism has been recently extended to three-cluster scattering states [95].

We have illustrated the formalism with some typical examples, both in spectroscopy and in reactions. In the literature, the GCM has been applied to many fields, ranging from the spectroscopy of exotic nuclei to reactions of astrophysical interest. Microscopic cluster models represent a efficient tool for the investigation of nuclei located near or beyond the driplines. These nuclei, such as ^{16}B or ^{18}B for example [150], are now actively studied in large-scale facilities. They combine several difficulties: they are unbound even in their ground state, excited states of the core are expected to be important, and core-neutron interactions (such as ^{15}B + n or ^{17}B+n) are not available. Investigations of Bose–Einstein condensation [8] in nuclear physics also benefit from cluster models. In the future, scattering theories could be developed with microscopic cluster wave functions of the projectile, in particular for the Continuum Discretized Coupled Channel (CDCC) method [99], or for the eikonal method [151]. The merging of precise scattering models with microscopic descriptions of the projectile represents a challenge for the upcoming years.

Acknowledgments This text presents research results of the IAP program P6/23 initiated by the Belgian-state Federal Services for Scientific, Technical and Cultural Affairs.

References

1. Brink, D.M.: J. Phys. Conf. Ser. **111**, 012001 (2008)
2. Wheeler, J.A.: Phys. Rev. **52**, 1083 (1937)
3. Margenau, H.: Phys. Rev. **59**, 3 (1941)
4. Brink, D.: Proc. Int. School "Enrico Fermi" 36, Varenna 1965, p. 247. Academic Press, New-York (1966)
5. Fujiwara, Y., Horiuchi, H., Ikeda, K., Kamimura, M., Katō, K., Suzuki, Y., Uegaki, E.: Prog. Theor. Phys. Suppl. **68**, 29 (1980)
6. Hoyle, F.: Astrophys. J. Suppl. **1**, 121 (1954)
7. Ikeda, K., Takigawa, N., Horiuchi, H.: Prog. Theor. Phys. Suppl., Extra Number p. 464 (1968)
8. Tohsaki, A., Horiuchi, H., Schuck, P., Röpke, G.: Phys. Rev. Lett. **87**, 192501 (2001)
9. von Oertzen, W., Freer, M., Kanada-En'yo, Y.: Phys. Rep. **432**, 43 (2006)
10. Freer, M.: Rep. Prog. Phys. **70**, 2149 (2007)
11. Fujiwara, Y., Tang, Y.C.: Phys. Rev. C **31**, 342 (1985)
12. Freer, M., Angélique, J.C., Axelsson, L., Benoit, B., Bergmann, U., Catford, W.N., Chappell, S.P.G., Clarke, N.M., Curtis, N., D'Arrigo, A., DeGóes Brennand, E., Dorvaux, O., Fulton, B.R., Giardina, G., Gregori, C., Grevy, S., Hanappe, F., Kelly, G., Labiche, M., Brun, C.L., Leenhardt, S., Lewitowicz, M., Markenroth, K., Marqués, F.M., Motta, M., Murgatroyd, J.T., Nilsson, T., Ninane, A., Orr, N.A., Piqueras, I., Laurent, M.G.S., Singer, S.M., Sorlin, O., Stuttgé, L., Watson, D.L.: Phys. Rev. Lett. **82**, 1383 (1999)
13. Descouvemont, P., Baye, D.: Phys. Lett. **505B**, 71 (2001)
14. Dufour, M., Descouvemont, P., Nowacki, F.: Nucl. Phys. A **836**, 242 (2010)
15. Wildermuth, K., Kanellopoulos, T.: Nucl. Phys. **9**, 449 (1958)
16. Tang, Y.C.: in Topics in Nuclear Physics II Lecture Notes in Physics. Vol. 145, p. 572 Springer, Berlin (1981)
17. Johnson, E.D., Rogachev, G.V., Goldberg, V.Z., Brown, S., Robson, D., Crisp, A.M., Cottle, P.D., Fu, C., Giles, J., Green, B.W., Kemper, K.W., Lee, K., Roede, B.T., Tribble, R.E.: Eur. Phys. J. A **42**, 135 (2009)
18. Neudatchin, V.G., Kukulin, V.I., Korotkikh, V.L., Korennoy, V.P.: Phys. Lett. **34B**, 581 (1971)
19. Buck, B., Dover, C.B., Vary, J.P.: Phys. Rev. C **11**, 1803 (1975)
20. Wildermuth, K., Tang, Y.C.: A Unified Theory of the Nucleus. Vieweg, Braunschweig (1977)
21. Suzuki, Y., Varga, K.: Stochastic variational approach to quantum-mechanical few-body problems. Lecture Notes in Physics Vol. m54 (1998)
22. Baye, D.: Phys. Rev. Lett. **58**, 2738 (1987)
23. Lin, C.D.: Phys. Rep. **257**, 1 (1995)

24. Faddeev, L., Merkuriev, S.: Quantum Scattering Theory for Several Particle Systems. Kluwer Academic Publishers, Dordrecht (1993)
25. Tohsaki, A.: Phys. Rev. C **49**, 1814 (1994)
26. Caurier, E., Martínez-Pinedo, G., Nowacki, F., Poves, A., Zuker, A.P.: Rev. Mod. Phys. **77**, 427 (2005)
27. Navrátil, P., Vary, J.P., Barrett, B.R.: Phys. Rev. Lett. **84**, 5728 (2000)
28. Kanada-En'yo, Y., Horiuchi, H., Ono, A.: Phys. Rev. C **52**, 628 (1995)
29. Neff, T., Feldmeier, H.: Eur. Phys. J. Spec. Top. **156**, 69 (2008)
30. Kievsky, A., Rosati, S., Viviani, M., Marcucci, L.E., Girlanda, L.: J. Phys. G **35**, 063101 (2008)
31. Saito, S.: Prog. Theor. Phys. Suppl. **62**, 11 (1977)
32. Horiuchi, H.: Prog. Theor. Phys. Suppl. **62**, 90 (1977)
33. Kamimura, M.: Nucl. Phys. A **351**, 456 (1981)
34. Varga, K., Suzuki, Y., Tanihata, I.: Phys. Rev. C **52**, 3013 (1995)
35. Dufour, M., Descouvemont, P.: Nucl. Phys. A **605**, 160 (1996)
36. Descouvemont, P.: Nucl. Phys. A **596**, 285 (1996)
37. Baye, D., Kruglanski, M.: Phys. Rev. C **45**, 1321 (1992)
38. Descouvemont, P., Baye, D.: Phys. Lett. B **169**, 143 (1986)
39. Wada, T., Horiuchi, H.: Phys. Rev. C **38**, 2063 (1988)
40. Dufour, M., Descouvemont, P.: Nucl. Phys. A **726**, 53 (2003)
41. Tanihata, I., Hamagaki, H., Hashimoto, O., Shida, Y., Yoshikawa, N., Sugimoto, K., Yamakawa, O., Kobayashi, T., Takahashi, N.: Phys. Rev. Lett. **55**, 2676 (1985)
42. Jonson, B.: Phys. Rep. **389**, 1 (2004)
43. Varga, K., Suzuki, Y., Ohbayasi, Y.: Phys. Rev. C **50**, 189 (1994)
44. Descouvemont, P.: Nucl. Phys. A **584**, 532 (1995)
45. Korennov, S., Descouvemont, P.: Nucl. Phys. A **740**, 249 (2004)
46. Adahchour, A., Descouvemont, P.: Nucl. Phys. A **813**, 252 (2008)
47. Baye, D., Descouvemont, P.: Phys. Rev. C **38**, 2463 (1988)
48. Tursunov, E.M., Baye, D., Descouvemont, P.: Phys. Rev. C **73**, 014303 (2006)
49. Liu, Q.K.K., Kanada, H., Tang, Y.C.: Phys. Rev. C **23**, 645 (1981)
50. Dufour, M., Descouvemont, P.: Phys. Rev. C **78**, 015808 (2008)
51. Langanke, K.: Adv. Nucl. Phys. **21**, 85 (1994)
52. Descouvemont, P.: Phys. Rev. C **70**, 065802 (2004)
53. Iliadis, C.: Nuclear Physics of Stars. Wiley, Weinheim (2007)
54. Baye, D., Descouvemont, P., Kruglanski, M.: Nucl. Phys. A **550**, 250 (1992)
55. Tang, Y.C., LeMere, M., Thompsom, D.R.: Phys. Rep. **47**, 167 (1978)
56. Wildermuth, K., Kanellopoulos, E.J.: Rep. Prog. Phys. **42**, 1719 (1979)
57. Friedrich, H.: Phys. Rep. **74C**, 209 (1981)
58. Bertsch, G., Borysowicz, J., Mcmanus, H., Love, W.G.: Nucl. Phys. A **284**, 399 (1977)
59. Volkov, A.B.: Nucl. Phys. **74**, 33 (1965)
60. Thompson, D.R., LeMere, M., Tang, Y.C.: Nucl. Phys. A **286**, 53 (1977)
61. Baye, D., Pecher, N.: Bull. Cl. Sci. Acad. Roy. Belg. **67**, 835 (1981)
62. Navrátil, P., Quaglioni, S., Stetcu, I., Barrett, B.R.: J. Phys. G **36**, 083101 (2009)
63. Navrátil, P., Kamuntavičius, G.P., Barrett, B.R.: Phys. Rev. C **61**, 044001 (2000)
64. Wiringa, R.B., Stoks, V.G.J., Schiavilla, R.: Phys. Rev. C **51**, 38 (1995)
65. Machleidt, R.: Phys. Rev. C **63**, 024001 (2001)
66. Navrátil, P., Gueorguiev, V.G., Vary, J.P., Ormand, W.E., Nogga, A.: Phys. Rev. Lett. **99**, 042501 (2007)
67. Tohsaki-Suzuki, A.: Prog. Theor. Phys. Suppl. **62**, 191 (1977)
68. Quaglioni, S., Navrátil, P.: Phys. Rev. C **79**, 044606 (2009)
69. Hesse, M., Sparenberg, J.M., Van Raemdonck, F., Baye, D.: Nucl. Phys. A **640**, 37 (1998)
70. Thompson, D.R., Tang, Y.C.: Phys. Rev. C **4**, 306 (1971)
71. Theeten, M., Baye, D., Descouvemont, P.: Phys. Rev. C **74**, 044304 (2006)

72. Suzuki, Y.: Nucl. Phys. A **405**, 40 (1983)
73. J. Raynal, Computing as a language of physics, Trieste 1971, p. 281. IAEA, Vienna (1972)
74. Baye, D.: Phys. Stat. Sol. (b) **243**, 1095 (2006)
75. Horiuchi, H.: Prog. Theor. Phys. **43**, 375 (1970)
76. Timofeyuk, N.K., Descouvemont, P.: Phys. Rev. C **71**, 064305 (2005)
77. Dufour, M., Descouvemont, P.: Nucl. Phys. A **785**, 381 (2007)
78. Lawson, R.D.: Theory of The Nuclear Shell Model. Clarendon, Oxford (1980)
79. Bethe, H.A., Rose, M.E.: Phys. Rev. **51**, 283 (1937)
80. Dufour, M., Descouvemont, P., Baye, D.: Phys. Rev. C **50**, 795 (1994)
81. Aoki, K., Horiuchi, H.: Prog. Theor. Phys. **68**, 2028 (1982)
82. Descouvemont, P., Baye, D.: Nucl. Phys. A **517**, 143 (1990)
83. Gartenhaus, S., Schwartz, C.: Phys. Rev. **108**, 482 (1957)
84. Rose, H.J., Brink, D.M.: Rev. Mod. Phys. **39**, 306 (1967)
85. Baye, D., Descouvemont, P.: Nucl. Phys. A **407**, 77 (1983)
86. Tohsaki-Suzuki, A.: Prog. Theor. Phys. **59**, 1261 (1978)
87. Hill, D.L., Wheeler, J.A.: Phys. Rev. **89**, 1102 (1953)
88. Friedrich, H.: Nucl. Phys. A **224**, 537 (1974)
89. Abramowitz, M., Stegun, I.A.: Handbook of Mathematical Functions. Dover, London (1972)
90. Baye, D., Salmon, Y.: Nucl. Phys. A **331**, 254 (1979)
91. Varga, K., Lovas, R.G.: Phys. Rev. C **37**, 2906 (1988)
92. Saito, S.: Prog. Theor. Phys. **41**, 705 (1969)
93. Suzuki, Y., Matsumura, H., Orabi, M., Fujiwara, Y., Descouvemont, P., Theeten, M., Baye, D.: Phys. Lett. B **659**, 160 (2008)
94. Baye, D., Descouvemont, P.: Ann. Phys. **165**, 115 (1985)
95. Damman, A., Descouvemont, P.: Phys. Rev. C **80**, 044310 (2009)
96. Theeten, M., Matsumura, H., Orabi, M., Baye, D., Descouvemont, P., Fujiwara, Y., Suzuki, Y.: Phys. Rev. C **76**, 054003 (2007)
97. Fujiwara, Y., Suzuki, Y., Miyagawa, K., Kohno, M., Nemura, H.: Prog. Theor. Phys. **107**, 993 (2002)
98. Descouvemont, P., Tursunov, E.M., Baye, D.: Nucl. Phys. A **765**, 370 (2006)
99. Austern, N., Iseri, Y., Kamimura, M., Kawai, M., Rawitscher, G., Yahiro, M.: Phys. Rep. **154**, 125 (1987)
100. Fujiwara, Y., Suzuki, Y., Kohno, M., Miyagawa, K.: Phys. Rev. C **77**, 027001 (2008)
101. Kanada, H., Kaneko, T., Nagata, S., Nomoto, M.: Prog. Theor. Phys. **61**, 1327 (1979)
102. Buck, B., Friedrich, H., Wheatley, C.: Nucl. Phys. A **275**, 246 (1977)
103. Hutson, J.M.: Comput. Phys. Commun. **84**, 1 (1994)
104. Buck, B., Baldock, R.A., Rubio, J.A.: J. Phys. G **11**, L11 (1985)
105. Descouvemont, P.: Nucl. Phys. A **655**, 440 (1999)
106. Descouvemont, P.: Astrophys. J. **543**, 425 (2000)
107. Messiah, A.: Quantum Mechanics. Dover Publications, New York (1999)
108. Descouvemont, P.: Phys. Rev. C **47**, 210 (1993)
109. Descouvemont, P.: Nucl. Phys. A **615**, 261 (1997)
110. Descouvemont, P., Baye, D.: Nucl. Phys. A **463**, 629 (1987)
111. Descouvemont, P., Baye, D.: Nucl. Phys. A **573**, 28 (1994)
112. Baye, D., Suzuki, Y., Descouvemont, P.: Prog. Theor. Phys. **91**, 271 (1994)
113. Descouvemont, P.: Phys. Rev. C **44**, 306 (1991)
114. Zhukov, M.V., Danilin, B.V., Fedorov, D.V., Bang, J.M., Thompson, I.J., Vaagen, J.S.: Phys. Rep. **231**, 151 (1993)
115. Raynal, J., Revai, J.: Nuovo Cim. A **39**, 612 (1970)
116. Descouvemont, P., Daniel, C., Baye, D.: Phys. Rev. C **67**, 044309 (2003)
117. Thompson, I.J., Danilin, B.V., Efros, V.D., Vaagen, J.S., Bang, J.M., Zhukov, M.V.: Phys. Rev. C **61**, 024318 (2000)
118. Adahchour, A., Descouvemont, P.: J. Phys. G **37**, 045102 (2010)

119. Lane, A.M., Thomas, R.G.: Rev. Mod. Phys. **30**, 257 (1958)
120. Thompson, I.J.: Comput. Phys. Rep. **7**, 167 (1988)
121. Baye, D., Heenen, P.-H., Libert-Heinemann, M.: Nucl. Phys. A **291**, 230 (1977)
122. Hesse, M., Roland, J., Baye, D.: Nucl. Phys. A **709**, 184 (2002)
123. Quaglioni, S., Navrátil, P.: Phys. Rev. Lett. **101**, 092501 (2008)
124. Descouvemont, P., Baye, D.: Rep. Prog. Phys. **73**, 036301 (2010)
125. Descouvemont, P.: Theoretical Models for Nuclear Astrophysics. Nova Science, New York (2003)
126. Bloch, C.: Nucl. Phys. **4**, 503 (1957)
127. Afzal, S.A., Ahmad, A.A.Z., Ali, S.: Rev. Mod. Phys. **41**, 247 (1969)
128. Baye, D., Heenen, P.-H.: Nucl. Phys. A **233**, 304 (1974)
129. Uegaki, E., Okabe, S., Abe, Y., Tanaka, H.: Prog. Theor. Phys. **57**, 1262 (1977)
130. Descouvemont, P., Baye, D.: Phys. Rev. C **36**, 54 (1987)
131. Itagaki, N., Aoyama, S., Okabe, S., Ikeda, K.: Phys. Rev. C **70**, 054307 (2004)
132. Dufour, M., Descouvemont, P.: Phys. Rev. C **56**, 1831 (1997)
133. Ajzenberg-Selove, F.: Nucl. Phys. A **506**, 1 (1990)
134. Ajzenberg-Selove, F.: Nucl. Phys. A **523**, 1 (1991)
135. Ajzenberg-Selove, F.: Nucl. Phys. A **475**, 1 (1987)
136. Dufour, M., Descouvemont, P.: Nucl. Phys. A **730**, 316 (2004)
137. Clayton, D.D.: Principles of Stellar Evolution and Nucleosynthesis. The University of Chicago Press, Chicago (1983)
138. Bardayan, D.W., Blackmon, J.C., Brune, C.R., Champagne, A.E., Chen, A.A., Cox, J.M., Davinson, T., Hansper, V.Y., Hofstee, M.A., Johnson, B.A., Kozub, R.L., Ma, Z., Parker, P.D., Pierce, D.E., Rabban, M.T., Shotter, A.C., Smith, M.S., Swartz, K.B., Visser, D.W., Woods, P.J.: Phys. Rev. C **62**, 055804 (2000)
139. García, A., Adelberger, E.G., Magnus, P.V., Markoff, D.M., Swartz, K.B., Smith, M.S., Hahn, K.I., Bateman, N., Parker, P.D.: Phys. Rev. C **43**, 2012 (1991)
140. Tilley, D.R., Weller, H.R., Cheves, C.M., Chasteler, R.M.: Nucl. Phys. A **595**, 1 (1995)
141. Dufour, M., Descouvemont, P.: Nucl. Phys. A **750**, 218 (2005)
142. Dufour, M., Descouvemont, P.: Phys. Rev. C **72**, 015801 (2005)
143. Freer, M., Angélique, J.C., Axelsson, L., Benoit, B., Bergmann, U., Catford, W.N., Chappell, S.P.G., Clarke, N.M., Curtis, N., D'Arrigo, A., DeGóes Brennand, E., Dorvaux, O., Fulton, B.R., Giardina, G., Gregori, C., Grevy, S., Hanappe, F., Kelly, G., Labiche, M., Le Brun, C., Leenhardt, S., Lewitowicz, M., Markenroth, K., Marqués, F.M., Murgatroyd, J.T., Nilsson, T., Ninane, A., Orr, N.A., Piqueras, I., Laurent, M.G.S., Singer, S.M., Sorlin, O., Stuttgé, L., Watson, D.L.: Phys. Rev. C **63**, 034301 (2001)
144. Saito, A., Shimoura, S., Takeuchi, S., Motobayashi, T., Minemura, T., Matsuyama, Y., Baba, H., Akiyoshi, H., Ando, Y., Aoi, N., Fülöp, Z., Gomi, T., Higurashi, Y., Hirai, M., Ieki, K., Imai, N., Iwasa, N., Iwasaki, H., Iwata, Y., Kanno, S., Kobayashi, H., Kubono, S., Kunibu, M., Kurokawa, M., Liu, Z., Michimasa, S., Nakamura, T., Ozawa, S., Sakurai, H., Serata, M., Takeshita, E., Teranishi, T., Ue, K., Yamada, K., Yanagisawa, Y., Ishihara, M.: Nucl. Phys. A **738**, 337 (2004)
145. Shimoura, S., Ota, S., Demichi, K., Aoi, N., Baba, H., Elekes, Z., Fukuchi, T., Gomi, T., Hasegawa, K., Ideguchi, E., Ishihara, M., Iwasa, N., Iwasaki, H., Kanno, S., Kubono, S., Kurita, K., Kurokawa, M., Matsuyama, Y., Michimasa, S., Miller, K., Minemura, T., Motobayashi, T., Murakami, T., Notani, M., Odahara, A., Saito, A., Sakurai, H., Takeshita, E., Takeuchi, S., Tamaki, M., Teranishi, T., Yamada, K., Yanagisawa, Y., Hamamoto, I.: Phys. Lett. B **654**, 87 (2007)
146. Kanada-En'yo, Y., Horiuchi, H.: Phys. Rev. C **68**, 014319 (2003)
147. Ito, M., Itagaki, N., Sakurai, H., Ikeda, K.: Phys. Rev. Lett. **100**, 182502 (2008)
148. Bohlen, H.G., von Oertzen, W., Kokalova, T., Schulz, C., Kalpakchieva, R., Massey, T.N., Milin, M.: Int. J. Mod. Phys. E **17**, 2067 (2008)
149. Baye, D., Descouvemont, P., Leo, F.: Phys. Rev. C **72**, 024309 (2005)

150. Lecouey, J.L., Orr, N., Marqués, F., Achouri, N., Angélique, J.C., Brown, B., Carstoiu, F., Catford, W., Clarke, N., Freer, M., Fulton, B., Grévy, S., Hanappe, F., Jones, K., Labiche, M., Lemmon, R., Ninane, A., Sauvan, E., Spohr, K., Stuttgé, L.: Phys. Lett. B **672**, 6 (2009)
151. Suzuki, Y., Lovas, R.G., Yabana, K., Varga, K.: Structure and Reactions of Light Exotic Nuclei. Taylor & Francis, London (2003)

Chapter 2
Neutron Halo and Breakup Reactions

T. Nakamura and Y. Kondo

Abstract Characteristic features of neutron halos in the context of breakup experiments at intermediate/high energies are discussed. Neutron halos have been found for light neutron rich nuclei along the neutron drip line, as intense radioactive nuclear beams have become available in recent years. A neutron halo nucleus is composed of a tightly bound core surrounded by one or two neutrons which extend outside of the mean field potential due to quantum tunneling. Coulomb breakup of halo nuclei shows extremely enhanced cross sections for such systems, originating from the characteristic electric dipole response of halo nuclei at low excitation energies, called soft $E1$ excitation. Such features are shown for one-neutron halo nuclei by Coulomb breakup experiments of ^{11}Be on Pb at about 70 MeV/nucleon, performed at RIKEN, where it was found that the direct breakup mechanism is responsible for this excitation. We then show how the $E1$ excitation spectrum can be related to properties of the halo distribution, and hence that the method of Coulomb breakup is a powerful spectroscopic tool. As such, applications of the Coulomb breakup of ^{15}C and ^{19}C are shown. The ^{19}C case is valuable for extracting the microscopic structure of ^{19}C, which has only recently been clarified. The ^{15}C case can be used to extract the radiative capture cross section of the ^{14}C$(n, \gamma)^{15}$C reaction. We then demonstrate recent applications of the "inclusive" Coulomb breakup method to a new-region of loosely bound nuclei near the island of inversion (N $\sim$ 20). There, evidence obtained, of the $1n$ halo structure in ^{31}Ne, is presented. In the $2n$ halo case the Coulomb breakup of ^{11}Li at 70 MeV/nucleon, measured at RIKEN, is shown. This reaction has provided evidence of dineutron-like structure, revealed by the strong enhancement of the soft $E1$ excitation. For nuclear dominated breakup, where a light target is used, the momentum distribution of the core fragment has key information on the halo

T. Nakamura (✉)· Y. Kondo
Department of Physics, Tokyo Institute of Technology,
2-12-1 O-Okayama, Meguro, Tokyo 152-8551, Japan
e-mail: nakamura@phys.titech.ac.jp

Y. Kondo
e-mail: kondo@phys.titech.ac.jp

C. Beck (ed.), *Clusters in Nuclei, Vol.2*, Lecture Notes in Physics 848,
DOI: 10.1007/978-3-642-24707-1_2, © Springer-Verlag Berlin Heidelberg 2012

distribution, and the single-particle properties of the valence neutron. Here we show an example of a spectroscopic study of ^{13}Be, populated by removing one neutron from the two neutron halo nucleus ^{14}Be. These results show that the breakup reactions play significant roles in elucidating the structures along the neutron drip line. This feature will be very important for further investigations of the drip-line nuclei towards the heavier region, as will be produced using the new-generation RI (Rare-Isotope) beam facilities, as has just been completed in RIKEN (RIBF) in Japan. Such enhanced RI-beam facilities are soon to be commissioned in Europe (SPIRAL2, FAIR etc.), in Asia (KoRIA), and in the US (FRIB).

2.1 Introduction

In 1985, the anomalously large matter radius of the most neutron-rich Li isotope, ^{11}Li, significantly beyond the conventional $r_0 A^{1/3}$ law, was discovered from a measurement of its interaction cross section with a carbon target at 790 MeV/nucleon [1]. Soon after this epoch-changing finding, an extremely narrow transverse momentum distribution of ^{9}Li, following the breakup of ^{11}Li on a carbon target, was observed [2]. This was later confirmed by the measurement of the corresponding longitudinal component [3]. Combined with the large matter radius, the narrow momentum width of the core fragment was interpreted as a manifestation of a "halo", an extended density distribution of two valence neutrons spilling out of the densely-packed ^{9}Li core. This basic concept was discussed in Ref. [4]. The third important finding of this nucleus was made by observing its anomalously large Coulomb breakup (CB, or electromagnetic dissociation EMD) cross section [5], which was interpreted as evidence of an enhanced electric dipole response at low excitation energies, known as "soft $E1$ excitation". These features are all related to the cluster-like structure of ^{11}Li, i.e., a three-body ^{9}Li$+n+n$ structure appearing near the breakup threshold.

This chapter focuses on "breakup reactions" of halo nuclei. Breakup reactions have played significant roles since the early days of halo physics. After about two decades from the first experimental confirmation of halo structure in ^{11}Li, many experimental methods and techniques related to breakup reactions of halo nuclei have made very significant progress.

It should also be noted that several RI-beam facilities have been upgraded, commissioned, or are expected to soon be commissioned, to improve the quality and intensity of the secondary beams. In 2007, the new-generation RI-beam facility RIBF (RI beam Factory) at RIKEN in Japan has been commissioned [6, 7]. This facility enhances capabilities of drip-line physics thanks to its extraordinary improvement in intensities of exotic nuclei, by more than three orders of magnitude for most nuclei, compared to those available at the pre-existing facilities. The power of the RIBF facility was demonstrated recently by the discovery of about 50 new neutron-rich isotopes near the predicted astrophysical r-process path [8, 9]. Later in this chapter, we show results of breakup measurement of ^{31}Ne performed in this facility [10]. Other new-generation facilities include FAIR at GSI in Germany [11], SPIRAL2 at

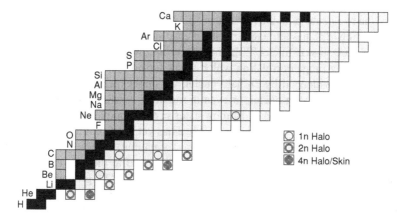

Fig. 2.1 Nuclear chart up to $Z = 20$ (Ca), whose drip-lines are based on the identified bound nuclei. The known neutron halo nuclei are indicated. Evidence for halo structures of ^{22}C and ^{31}Ne have recently been obtained experimentally [10, 14]

GANIL in France [12], FRIB at MSU in the USA [13], which will be in operation in 3–10 years.

As in ^{11}Li, neutron halo nuclei have been discovered at the limit of nuclear stability on the neutron-rich side (neutron drip line) in the lower mass part of the nuclear chart, as shown in Fig. 2.1. A neutron halo nucleus is composed of a tightly bound core surrounded by a low-density neutron cloud, called a halo, which contains one or two neutrons. Halo phenomena occur basically due to the quantum tunneling effect for weakly bound neutron(s) which extend outside of the surface of nuclear mean-field potential due to the core. Since the halo neutron(s) are spatially decoupled from the core, a halo nucleus has a cluster-like property, and a $1n$ halo nucleus is regarded as a two-body system composed of a core+$1n$, and a $2n$ halo nucleus as a three-body system composed of a core+n+n .

As discussed in the review articles [15–19], and references therein, the main issues in halo physics are summarized as follows:

1. Microscopic structure of $1n$ halo nuclei: A key issue is the single-particle configuration of the valence neutron relative to the core. The condition of $1n$-halo formation is that a nucleus should have small S_n ($1n$ separation energy), typically $S_n < 1$ MeV. Another important requirement is that the valence neutron has low (or no) orbital angular momentum to avoid the hindrance of the tunneling effect due to the centrifugal barrier ($\ell(\ell + 1)\hbar^2/(2\mu r^2)$). It is suggested that the orbital angular momentum should be $\ell = 0$ or 1 for halo formation. In fact, only when $\ell = 0, 1$ can one obtain a divergent r.m.s. radius ($\sqrt{\langle r^2 \rangle}$) of the tail density distribution in the limit that $S_n \to 0$ [18, 20, 21].
2. Microscopic structure of $2n$ halo nuclei: A $2n$-halo nucleus has a three-body structure, called Borromean, where the two-body subsystems are weakly unbound but the three-body system is bound. Recently, a possible strong dineutron correlation,

where the two neutrons at the low-density nuclear surface take a spatially compact configuration [22], has attracted special interest since this may happen for $2n$ halo nuclei or neutron-skin nuclei [23]. The dineutron correlation is different from the BCS-type long range correlation which arises for normal nuclei [24]. If such a correlation is evidenced by $2n$ halo nuclei, we may expect similar correlations in heavier neutron-skin nuclei, or even in the inner crust of a neutron star. The possibility of the revelation of Efimov states in three-body halo nuclei has also been discussed [25–27], but whose evidence has yet to be provided.

3. Interplay between shell evolution and halo formation: In most halo nuclei the lowering of a low-ℓ single-particle orbital, compared to the conventional shell order is suggested. A famous example is the intruder $1/2^+$ ground state below the $1/2^-$ state in ^{11}Be (parity inversion), as was identified 50 years ago [28] well before the discovery of its $1n$ halo structure. Such a shell evolution enhances the halo formation due to the appearance of a low-ℓ orbital. In turn, we should note that low kinetic energy of halo neutron(s) may cause the lowering of ℓ.

4. Interplay between deformation and the halo: Deformation can influence the lowering of a low-ℓ orbital in some cases, which could cause formation of a halo structure [29, 30].

5. Other many-body correlations as a cause of halo formation: The possibility that tensor correlations enhance s-p degeneracy of the two valence neutrons in ^{11}Li is discussed [31, 32].

6. Characteristic nuclear responses of halo nuclei: "Soft $E1$ excitation" is a unique response of halo nuclei to an electric dipole operator. Such exotic excitation modes may appear in different responses, such as $E0$, $E2$ [33], and spin-isospin excitations [34].

7. Halo formation in heavier nuclei: So far, halo nuclei have been found in the mass region less than $A \sim 20$, except for the newly-found $1n$ halo nucleus ^{31}Ne (see Fig. 2.1)[10]. We thus raise the question: can we observe many halo cases in heavier nuclei?; how? In what form can halo states appear in heavier nuclei? The anti-halo effect predicted theoretically [35, 36], where neutron halo formation is suppressed due to the pairing correlation, may limit the number of $2n$ halo cases. It is also suggested that the dominance of larger ℓ orbits for heavier nuclei may restrict the cases of neutron halo to limited regions where s or p shells are located near the Fermi surface. On the other hand, deformation and shell evolution effects may increase the possibility of halo cases along the neutron drip line. In heavier regions, we may expect more complex halo structure due to a mixture of different shell configurations. For example, Meng and Ring have predicted the existence of giant halo structure, involving about 6 neutrons for neutron-rich Zr isotopes [37].

8. Understanding of possible $4n$ halo systems (among other multi-neutron halos): ^{8}He and ^{19}B are the $4n$ halo candidates, whose structures are not well known. These systems could provide the first step to understand multi-neutron halos expected in heavier neutron rich nuclei.

9. Finally, we would like to raise the following questions:

    a. Where is the neutron drip line located in heavier nuclei?
    b. Does the existence of halo states change the location of the drip line? (*i.e.*, will
       the existence of halo structures change the nuclear stability?)

Such questions are fundamental to understand many-body nucleonic systems and their stability at the limit of extreme isospin asymmetry.

Breakup reactions of halo nuclei play important roles in solving many of the above issues. Breakup reactions are categorized into two processes, (1) Coulomb Breakup, and (2) Nuclear Breakup, depending on the primary interaction involved. The breakup on a heavy target, such as Pb or U, is dominated by Coulomb breakup, while the breakup with a light target, such as proton, Be, and C, is dominated by nuclear breakup. Both reaction processes are important in the spectroscopy of halo nuclei since the observables in breakup reactions are strongly correlated to the microscopic structures of the ground and continuum states of halo nuclei; as described later.

In what follows, we show the characteristic features of breakup reactions by illustrating recent experimental results. Section 2.2 describes the Coulomb breakup method in general. Section 2.3 describes the characteristic features of Coulomb breakup and soft $E1$ excitation of "$1n$" halo nuclei. There, we explain the mechanism of soft $E1$ excitation and how the single-particle configuration of a $1n$ halo nucleus can be studied. The examples shown there are for exclusive Coulomb breakup of ^{11}Be [38–40], ^{19}C [41], ^{15}C [42–44], and recent inclusive Coulomb breakup of ^{31}Ne measured at RIBF [10]. Section 2.4 describes examples for "$2n$" halo nuclei, showing ^{11}Li [45–50]. There we discuss the possible application to probe the dineutron correlation in $2n$ halo nuclei. In Sect. 2.5, we discuss the use of nuclear breakup experiments on ^{14}Be to study the spectroscopic information on the unbound nucleus ^{13}Be [51]. In Sect. 2.6, we summarize and provide an outlook of the potential of breakup reactions on halo nuclei for the future.

2.2 Coulomb Breakup at Intermediate/High Energies

Coulomb excitation has been a powerful spectroscopic tool in nuclear physics for many years. In earlier days, heavy ion beams of low-energy, below the Coulomb barrier, were primarily used. Recently, Coulomb excitation has been applied successfully to neutron-rich/proton-rich nuclei at intermediate/high energies. When Coulomb excitation is applied to drip line nuclei, such as halo nuclei, the Coulomb excitation leads to a breakup channel, associated with 1–2 nucleon removals, called Coulomb breakup or equivalently electromagnetic dissociation, due to their weak binding.

In Coulomb breakup, a projectile is excited by an impulse due to the Coulomb field of a high-Z target as shown schematically in Fig. 2.2. Here Lorentz-contracted electric field acts on the projectile as it passes the high-Z target with a relativistic velocity at an impact parameter b. For relativistic speeds, the magnetic field also becomes effective, so the more general term electromagnetic dissociation (EMD) is also used. In fact, for a relativistic incident velocity $\beta = v/c$ of a projectile, the

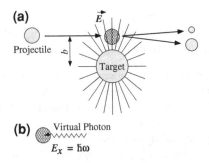

Fig. 2.2 a The Coulomb breakup process is shown schematically. When a relativistic projectile passes a high-Z target at an impact parameter b, it experiences an impulse due to the Lorentz-contracted electromagnetic field. When the intermediate excited state is above the particle decay threshold, as is often the case for halo nuclei, the projectile breaks up into the core fragment and nucleon(s). **b** The Coulomb breakup process can be regarded as the absorption of a virtual photon

electric field $\mathbf{E}$ and magnetic field $\mathbf{B}$ are almost perpendicular to each other, related by $|\mathbf{B}| = \beta|\mathbf{E}|$ [52]. This implies that the *EM* field approximates a real electromagnetic wave as $\beta \to 1$.

Hence, Coulomb breakup can be expressed as a photo-absorption process induced by a virtual photon, as shown schematically in Fig. 2.2b. Such a treatment is called the equivalent-photon [52, 53] or Weizsäcker-Williams method of virtual quanta. There, the Coulomb excitation cross section at an excitation energy E_x is expressed simply as a product of the photo-absorption cross section $\sigma_\gamma^{E\lambda}(E_x)$ and the virtual photon number $N_{E\lambda}(E_x)$. The latter is obtained by integrating $N_{E\lambda}(E_x, b)$ (the photon flux at an impact parameter b) from the cutoff impact parameter b_0 to infinity or, equivalently, by integrating the angle differential photon number with respect to the scattering angle from 0 to the grazing angle θ_{gr},

$$\frac{d\sigma(E\lambda)}{dE_x} = \int_{b_0}^{\infty} 2\pi b\, db\, N_{E\lambda}(E_x, b) \frac{\sigma_\gamma^{E\lambda}(E_x)}{E_x} \tag{2.1}$$

$$= \int_0^{\theta_{gr}} d\Omega \frac{dN_{E\lambda}(E_x, \theta)}{d\Omega} \frac{\sigma_\gamma^{E\lambda(E_x)}}{E_x} \tag{2.2}$$

$$= N_{E\lambda}(E_x) \frac{\sigma_\gamma^{E\lambda}(E_x)}{E_x}. \tag{2.3}$$

Here, λ represents a multipolarity of the transition, and E identifies an electric transition (M is used for the magnetic transitions). The photo-absorption cross section is related to the reduced transition probability $B(E\lambda)$ as in,

$$\sigma_\gamma^{E\lambda}(E_x) = \frac{(2\pi)^3(\lambda+1)}{\lambda[(2\lambda+1)!!]^2} \left(\frac{E_x}{\hbar c}\right)^{2\lambda-1} \frac{dB(E\lambda)}{dE_x}. \tag{2.4}$$

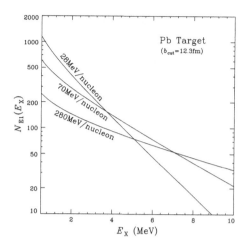

Fig. 2.3 $E1$ virtual photon spectra as a function of the energy of a virtual photon ($E_\gamma = E_x$). Three cases, for incident energies of 28, 70, and 280 MeV/nucleon on a Pb target are shown. These incident energies are those for the Coulomb breakup experiments of ^{11}Li in Refs. [45, 46], [50], and [49], respectively

Since $N_{E\lambda}(E_x)$ is provided in an analytical form, a measurement of $d\sigma/dE_x$ can be used to extract $dB(E\lambda)/dE_x$ directly.

For neutron halo nuclei, the electric dipole ($E1$) transition is dominant. For $E1$ excitation, Eqs. (2.3) and (2.4) can be combined as,

$$\frac{d\sigma(E1)}{dE_x} = \frac{16\pi^3}{9\hbar c} N_{E1}(E_x) \frac{dB(E1)}{dE_x}, \tag{2.5}$$

where $N_{E1}(E_x)$ is the number of $E1$ virtual photons with photon energy E_x. When we measure the scattering angle, as in the example of the ^{11}Be Coulomb breakup experiment shown later [40], the formula for the corresponding double differential cross section is also useful.

$$\frac{d^2\sigma}{d\Omega dE_x} = \frac{16\pi^3}{9\hbar c} \frac{dN_{E1}(E_x, \theta)}{d\Omega} \frac{dB(E1)}{dE_x}. \tag{2.6}$$

Note that in the semiclassical approach the center of mass (c.m.) of the projectile and that of the outgoing particles after the breakup approximately follows a Rutherford trajectory. Hence, there is a one-to-one correspondence between b and θ, stated as $b = a \cot(\theta/2) \simeq 2a/\theta$, where a is half the distance of the closest approach in the classical Rutherford scattering.

The typical $E1$ virtual photon spectra (photon flux) as a function of γ-ray energy ($E_\gamma = E_x$) are shown in Fig. 2.3. The electric field experienced by the projectile, as a function of time, is bell-shaped, with a width of $b/(\beta\gamma)$. Hence, the energy spectrum, which is the Fourier transform of the time spectrum, extends to higher excitation energy (higher frequency) for projectiles with higher incident energy, as detailed in Ref. [52].

2.3 Coulomb Breakup and Soft $E1$ Excitation of $1n$ Halo Nuclei

A large EMD cross section (0.89 ± 0.10 b) was observed for ^{11}Li at 0.8 GeV/nucleon, which was associated with the large enhancement of the $E1$ strength at low excitation energies [5]. As the virtual-photon flux falls rapidly with energy, as shown in Fig. 2.3, such a large cross section can only be explained by the concentration of significant $E1$ strength at low excitation energies of 1–2 MeV; which we now call soft $E1$ excitation. Such a picture is contrary to the conventional picture of the $E1$ response of nuclei where most of the $E1$ strength is exhausted by the giant dipole resonance (GDR), located at $E_x \sim 80A^{-1/3}$ MeV or $31.2A^{-1/3} + 20.6A^{-1/6}$ MeV [54].

Ikeda predicted that such a low-energy $E1$ transition is caused by the low-frequency motion of the charged core nucleus against the low-density neutron halo, which is called a soft dipole resonance (SDR) [32, 54]. On the other hand, it has been known that the enhancement occurs due to a direct breakup mechanism, where the halo nucleus breaks up without forming any intermediate resonances, and that the enhancement occurs due to the spatially-decoupled nature of the halo, whose detail is shown later.

Three earlier exclusive Coulomb breakup experiments on ^{11}Li, in the 1990s [45–49], failed to clarify the mechanism of the soft $E1$ excitation, since the $B(E1)$ strength distributions obtained from the three experiments were inconsistent (see Sect. 2.4). In addition, theoretical interpretation of the soft $E1$ excitation of ^{11}Li is difficult due to the correlations of the two neutrons. Hence, the Coulomb breakup of the $1n$ halo nucleus ^{11}Be was of great importance for understanding the underlying mechanism of soft $E1$ excitation due to its simpler structure. This section is devoted to the Coulomb breakup of a $1n$ halo nucleus, to clarify the mechanism and characteristic features of the soft $E1$ response of a $1n$ halo nucleus.

2.3.1 Coulomb Breakup of ^{11}Be and Characteristic Feature of Soft E1 Excitation of One-Neutron Halo Nuclei

The $1n$ halo structure of ^{11}Be was first suggested by the enhancement of its reaction cross section at a high incident energy of 790 MeV/nucleon at LBL [56], and at a lower incident energy of 33 MeV/nucleon at RIKEN [57]. A narrow momentum distribution of ^{10}Be following the breakup of ^{11}Be on a light target was observed [58], and later investigated further with γ ray coincidences with the ^{10}Be fragment [59] at MSU. These measurements show clearly the one neutron halo structure of ^{11}Be. The first exclusive Coulomb breakup experiment was studied at 72 MeV/nucleon at RIKEN [38]. Later, Coulomb breakup was studied at 520 MeV/nucleon at GSI [39], and was investigated in more detail at 69 MeV/nucleon at RIKEN [40]. In this section, we show the characteristic features of Coulomb breakup of $1n$ halo nuclei by showing these experimental results for ^{11}Be.

The wave function of the ground state of ^{11}Be($J^\pi = 1/2^+$) can be written as,

$$|^{11}\text{Be}(1/2^+; \text{gs})\rangle = \alpha|^{10}\text{Be}(0^+)\rangle \otimes 2s_{1/2} + \beta|^{10}\text{Be}(2^+)\rangle \otimes 1d_{5/2} + ... \quad (2.7)$$

Here, the main issue is the occupancy of the halo configuration in the ground state, namely, the spectroscopic factor for the first term involving the s-wave neutron. Already in the 1970s, although the halo structure was not known, there were some spectroscopic studies of ^{11}Be by using the transfer reaction, ^{10}Be(d, p)^{11}Be, from which the dominance of $2s_{1/2}$ neutron in the ground state ($\alpha^2(= C^2S) = 0.73$ [60], 0.77 [61]) was extracted. The one-neutron separation energy is known very precisely as 504 ± 5 keV. Since its ground state property is better established, Coulomb breakup of ^{11}Be is better suited for investigating the mechanism of soft $E1$ excitation, compared to the $2n$ halo nucleus ^{11}Li.

2.3.1.1 Typical Experimental Setup and the Invariant Mass Method

The three exclusive Coulomb breakup experiments used to map the $B(E1)$ distribution of ^{11}Be [38–40] were largely consistent with each other. Hence, hereafter, we mainly show the result from Ref. [40].

To extract the $B(E1)$ distribution, an exclusive (kinematically complete) measurement is required, where momentum vectors of all the outgoing particles, in this case ^{10}Be and the neutron, are measured. The experimental setup used in the experiment of Ref. [40] is shown in Fig. 2.4. In this experiment, the ^{11}Be secondary beam, produced by fragmentation of an ^{18}O beam at 100 MeV/nucleon at RIPS at RIKEN, bombarded a Pb target at an average energy of 68.7 MeV/nucleon. The momentum vectors of the beam ($\mathbf{P}(^{11}\text{Be})$) as well as those of the outgoing ^{10}Be fragment ($\mathbf{P}(^{10}\text{Be})$) and the neutron ($\mathbf{P}(n)$) were measured in coincidence.

The excitation energy of the projectile can be extracted by reconstructing the invariant mass $M(^{11}\text{Be}^*)$ of the intermediate excited state of ^{11}Be as follows.

$$M(^{11}\text{Be}^*) = \sqrt{\left(\sum_i E_i\right)^2 - \left(\sum_i P_i\right)^2} \quad (2.8)$$

$$= \sqrt{(E(^{10}\text{Be}) + E(n))^2 - (\mathbf{P}(^{10}\text{Be}) + \mathbf{P}(n))^2}, \quad (2.9)$$

where $E(^{10}\text{Be})$ and $E(n)$ are the total energy of the ^{10}Be fragment and that of the neutron, respectively. The relative energy E_{rel} between ^{10}Be and the neutron is then related to $M(^{11}\text{Be}^*)$ as,

$$E_{\text{rel}} = M(^{11}\text{Be}^*) - M(^{10}\text{Be}) - m_n. \quad (2.10)$$

Here $M(^{10}\text{Be})$ and m_n denote the mass of ^{10}Be and of the neutron, respectively. The excitation energy E_x is related to the relative energy by $E_x = E_{\text{rel}} + S_n$.

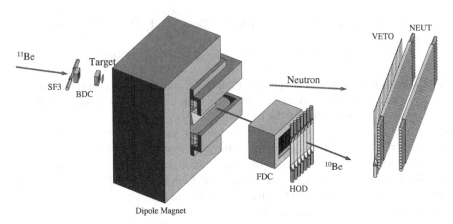

Fig. 2.4 Experimental setup used in the exclusive Coulomb breakup experiment of ^{11}Be at RIPS at RIKEN. The figure is from Ref. [40]

It is useful to note approximate forms of E_{rel}. In the non-relativistic limit, E_{rel} can be written as,

$$E_{rel} = \frac{\mu}{2} \left[\mathbf{v}(^{10}Be) - \mathbf{v}(n) \right]^2, \tag{2.11}$$

where μ stands for the reduced mass of ^{10}Be and the neutron, and $\mathbf{v}(^{10}Be)$ and $\mathbf{v}(n)$ the velocities of the outgoing particles. When these velocities are transformed to the projectile rest frame, this result is a good approximation. This can be also approximately written as

$$E_{rel} = \frac{\mu}{2} \left[\left(v(^{10}Be) - v(n) \right)^2 + (\bar{v}\theta_{12})^2 \right], \tag{2.12}$$

where $\bar{v}$ and θ_{12} represent the mean velocity and the opening angle between the two outgoing particles. This approximate form implies that the measurement of the relative energy is primarily determined by the difference of the velocities (first term) and the opening angle (second term) of the two outgoing particles. An approximate formula of the energy resolution can then be given as,

$$\Delta E_{rel} \cong \sqrt{2 \cdot \frac{E}{A} \frac{A_1 A_2}{A_1 + A_2} \cdot E_{rel}} \cdot \sqrt{\left(\frac{\Delta v_1}{v_1}\right)^2 + \left(\frac{\Delta v_2}{v_2}\right)^2 + \Delta\theta_{12}^2}, \tag{2.13}$$

where E and A denote the kinetic energy and mass number of the projectile. In the case of 70 MeV/u ^{11}Be, decaying into ^{10}Be and a neutron with a relative energy of 1 MeV, a resolution of 190 keV(r.m.s.) is achieved for the condition of $\Delta v_1/v_1 = \Delta v_2/v_2 = 1\%$, and $\Delta\theta_{12} = 10$ mrad. In the experiment of Ref. [40], the same resolution value was obtained by a detailed Monte Carlo analysis including the known velocity and angular resolutions. Note that the energy resolution is proportional to $\sqrt{E_{rel}}$ and $\sqrt{E/A}$. Because of this dependence, one can achieve better energy resolution in the invariant mass method compared to the missing mass method where the resolution is determined by the order of E instead of $\sqrt{E/A}$. It is also noted that in the invariant mass method, energy resolution is independent of the angular and energy spreads of the beam, the source of one of the major uncertainties for missing mass spectroscopy.

2.3.1.2 Breakup Cross Sections and the $B(E1)$ Spectrum

The breakup cross sections of ^{11}Be on lead and carbon targets are shown in Figs. 2.5a and b, respectively. The breakup data on the carbon target, where nuclear breakup is expected to be dominant, is used to investigate the characteristic features of target dependence. The cross sections are plotted for the different scattering angular ranges shown. Here, the scattering angle θ is that in the c.m. frame of the target+projectile. This was extracted from the difference of the incoming, $\mathbf{P}(^{11}\text{Be})$, and the sum of the outgoing, $\mathbf{P}(^{10}\text{Be}) + \mathbf{P}(n)$, momentum vectors. The angular ranges, $0 \leq \theta \leq 6°$ for the Pb target and $0 \leq \theta \leq 12°$ for the C target, are almost identical in the laboratory frame, corresponding to the acceptance range in this experiment. We call these angular ranges the "whole acceptance". All the spectra shown in these figures are corrected for the acceptance.

One can see clearly the distinctive difference between the spectra for the two targets. The total breakup cross section into ^{10}Be $+ n$ for the Pb target (whole acceptance, $E_{rel} \leq 5\,\text{MeV}$) is 1790 ± 20 (stat.) ± 110 (syst.) mb, which is larger by a factor of about 20 than that for the C target (93.3 ± 0.8 (stat.)$^{+5.6}_{-10.3}$ (syst.) mb). Such a big difference is due to the fact that the reaction on the Pb target is dominated by Coulomb breakup, which is of the order of one barn for the soft $E1$ excitation. On the other hand, reaction with the C target is dominated by nuclear breakup. Note that if only nuclear breakup were to occur on both for C and Pb targets, the ratio of the cross sections for these two targets would be about 2, roughly equal to the ratio of the target+projectile radii [40].

The second remarkable difference is the spectral shape. The spectra for the Pb target, both for the whole acceptance and the angular-selected one, have an asymmetric peak at very low E_{rel}, while that for the carbon target for the whole acceptance shows two resonance peaks ($E_x = 1.78$ MeV, $5/2^+$, and $E_x = 3.41$ MeV, $3/2^+$) superimposed on the continuum. It is interesting to note that the angular-selected spectrum for the C target has a similar asymmetric peak as for the Pb target.

This asymmetric shape of the peaks is in good agreement of the direct breakup mechanism shown by the solid curves. Namely, the direct breakup mechanism explains how the soft $E1$ excitation occurs for a $1n$ halo nucleus. The detailed explanation of the direct breakup mechanism, and its characteristic features, is described in the next subsection. We find that the agreement with the direct breakup mechanism is perfect for the angular selected data for Pb, and approximately so for the whole acceptance data. The selection of the most forward angles, well within the grazing angle ($\theta_{gr} = 3.8°$), has been found useful to extract almost purely the "Coulomb"

Fig. 2.5 Breakup cross
sections as a function of E_{rel}
of ^{10}Be and the neutron in
the breakup of ^{11}Be+Pb at
69 MeV/nucleon (**a**) and
^{11}Be + C at 67 MeV/nucleon
(**b**). See text and Ref. [40]
for details. (The figure is
from Fukuda et al. [40])

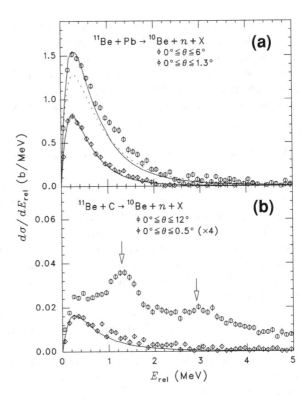

component of the breakup as was demonstrated in this experiment and later by
detailed reaction theories [62–64]. For the whole acceptance data, we see a slight
deviation which is attributed to nuclear breakup and higher order Coulomb breakup
effects. Even for the C target data, when we select sufficiently forward angles, we
find the direct Coulomb breakup dominates. These features were also seen in the
angular distributions shown in Ref. [40].

The $B(E1)$ distribution was extracted from the angular-selected data. In this case,
it is useful to adopt the double differential cross section of Eq. (2.6), integrated over
the selected angular range. The resultant $B(E1)$ spectrum is shown in Fig. 2.6, which
is again explained by the direct breakup mechanism shown in the solid curve. The
integrated $B(E1)$ obtained from the data selected for the forward angles amounts to
$1.05 \pm 0.06 e^2 fm^2$ corresponding to 3.29 ± 0.19 W.u. (Weisskopf unit) for $E_x \leq 4$
MeV, which is huge, considering that the $E1$ strength in this energy region for normal
nuclei is negligible (below 0.1 W.u.).

2.3.1.3 Direct Breakup Mechanism

We now explain the direct breakup mechanism. As demonstrated by the comparison
with the Coulomb breakup data of ^{11}Be, the direct breakup mechanism explains well
the soft $E1$ excitation of $1n$ halo nuclei. In the direct breakup mechanism, the $1n$-halo

Fig. 2.6 $B(E1)$ strength distribution for ^{11}Be as a function of E_x, obtained from the angle-selected Coulomb breakup data on a Pb target ($\theta < 1.3$ degrees). The solid curve is the result of a calculation with the direct breakup mechanism with $\alpha^2 = 0.72$

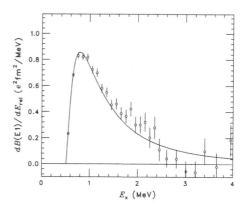

nucleus, in this case ^{11}Be, breaks up into ^{10}Be and a neutron without forming any intermediate resonances. The possibility of the soft dipole resonance (SDR) as a main mechanism of this large $E1$ strength is now rejected. Hence, it is more appropriate to call this phenomenon soft $E1$ "excitation".

The direct breakup mechanism is explained simply by the following matrix element,

$$\frac{dB(E1)}{dE_{\rm rel}} =| \langle \Phi_f(\mathbf{r}, \mathbf{q}) \mid e_{\rm eff}^{E1} \hat{T}(E1) \mid \Phi_i(\mathbf{r}) \rangle |^2, \qquad (2.14)$$

where $\Phi_i(\mathbf{r})$ and $\Phi_f(\mathbf{r}, \mathbf{q})$ represent the wave function of the ground state and the final state in the continuum of the neutron relative to the core, respectively. $\mathbf{r}$ is the relative coordinate of the valence neutron relative to the c.m. of the core. $\Phi_f(\mathbf{r}, \mathbf{q})$ is also a function of the relative momentum $q = \sqrt{2\mu E_{\rm rel}}/\hbar$. The electric dipole operator is $\hat{T}(E1)(= r Y^{(1)}(\Omega))$. For the case of ^{11}Be, Φ_i is $|^{11}{\rm Be}(1/2^+; {\rm g.s.})\rangle$.

For simplicity, we consider a pure single particle state and a spinless core, as in $|^{10}{\rm Be}(0^+) \otimes v2s_{1/2}\rangle$. The Eq. (2.14) can then be expressed in a simpler form as

$$\frac{dB(E1)}{dE_{\rm rel}} = \frac{3}{4\pi} (e_{\rm eff}^{E1})^2 \langle \ell_i 010 \mid \ell_f 0\rangle^2 \left| \int dr r^2 \phi_f(r, q) \, r\phi_i(r) \right|^2, \qquad (2.15)$$

where ℓ_i, ℓ_f are the orbital angular momenta of the valence neutron in the initial state, and the scattered neutron in the final state, respectively [65–67]. ϕ_i is the radial wave function involved in Φ_i, and ϕ_f is the radial component of the final scattering state of the neutron. The $E1$ effective charge $e_{\rm eff}^{E1}$ for the one-neutron halo nucleus is Ze/A. Although the actual analysis is performed with a distorted wave for ϕ_f, it is instructive to consider the plane wave approximation

$$\phi_f(r, q) = \sqrt{\frac{2\mu q}{\hbar^2 \pi}} j_{\ell_f}(qr), \qquad (2.16)$$

where $j_{\ell_f}(qr)$ is the spherical Bessel function. With this approximation, it is found that the above equations (Eqs. (2.14) and (2.15)) have the form of the Fourier transform of $r\phi_i(r)$. Hence, the $B(E1)$ distribution at low q (low E_{rel}) is the amplified Fourier image of the density distribution of the valence neutron (or halo). It should be noted that this image is amplified due to the factor "r" in the $E1$ operator. It is shown in Ref. [64] that due to this amplification, $B(E1)$ at low E_{rel} is almost solely determined by the density distribution outside of the range of the mean-field potential. That is, in the integration of Eq. (2.14), most of the contribution comes from the radial region outside of the range of the neutron-core binding potential.

This interpretation in terms of a Fourier transform also implies that the spatial decoupling of the halo and the core is important. The spatially decoupled halo state overlaps very well with the $E1$ operator which further emphasizes large separations of the valence neutron and the core. Such an intermediate state has a large overlap with the final scattering state where the outgoing neutron is by definition well separated from the core fragment. Conversely, since neutrons in ordinary nuclei are not spatially decoupled from the core, only a high energy $E1$ photon can separate the whole neutron fluid from the proton fluid as arises in the GDR.

Another instructive formula is obtained by further approximating Eq. (2.15), by replacing the ground state radial wave function by its asymptotic form at large distance. In general, $\phi_i(r)$ can be written

$$\phi_i(r) = N h_{\ell_i}^{(1)}(i\eta r), \tag{2.17}$$

where N is the normalization factor, and $h_{\ell_i}^{(1)}$ is the spherical Hankel function with $\eta = \sqrt{2\mu S_n}/\hbar$. In particular, the s-wave state as in $|^{10}\mathrm{Be}(0^+) \otimes \nu 2s_{1/2}\rangle$ is expressed by the Yukawa function,

$$\phi_i(r) = -N \frac{\exp(-\eta r)}{\eta r} \tag{2.18}$$

Then Eq. (2.15) leads to the following analytical form:

$$\frac{dB(E1)}{dE_{rel}} = N^2 \frac{3\hbar^2}{\pi^2 \mu} \left(\frac{Ze}{A}\right)^2 \frac{\sqrt{S_n} E_{rel}^{3/2}}{(E_{rel} + S_n)^4}. \tag{2.19}$$

This formula shows the essence of the direct breakup of an s-wave halo state. We find that the peak of this analytical form appears at $E_{rel} = (3/5)S_n$ (i.e., $E_x = (8/5)S_n$), and the integrated strength is inversely proportional to S_n. Hence, the lower the separation energy, the stronger the soft $E1$ excitation is. Thus, the observable is dictated by S_n or the threshold energy. This is why the soft $E1$ excitation is considered as a threshold effect. However, we should note that S_n is simply the reflection of the halo distribution. Hence, the spatially decoupled nature of the halo nucleus is considered to be more essential. Such analytical forms, for different combinations of the initial and final states, were described in Ref. [66, 67] and their features are also discussed in Sect. 2.3.2

2.3.1.4 Spectroscopic Factor

The $B(E1)$ spectrum can be used to extract the spectroscopic information of the halo structure. Namely, for the two-component configuration, Eq. (2.14) can be rewritten as,

$$
\begin{aligned}
\frac{dB(E1)}{dE_{\text{rel}}} &= \alpha^2 \mid \langle \Phi_f \mid \frac{Ze}{A} \hat{T}(E1) \mid^{10}\text{Be}(0^+) \otimes \nu 2s_{1/2}\rangle \mid^2 + \\
&\quad \beta^2 \mid \langle \Phi_f^* \mid \frac{Ze}{A} \hat{T}(E1) \mid^{10}\text{Be}(2^+) \otimes \nu 1d_{5/2}\rangle \mid^2 \\
&\simeq \alpha^2 \mid \langle \Phi_f \mid \frac{Ze}{A} \hat{T}(E1) \mid^{10}\text{Be}(0^+) \otimes \nu 2s_{1/2}\rangle \mid^2,
\end{aligned}
\tag{2.20}
$$

where Φ_f^* represents the scattering state between the excited state of $^{10}\text{Be}(2_1^+)$ and the neutron. Note that the interference term vanishes since the final state of ^{10}Be is definite.

As was mentioned the $B(E1)$ strength at low E_{rel} is only sensitive to the tail of the density distribution. Accordingly, only the amplitude for the $\mid^{10}\text{Be}(0^+) \otimes \nu 2s_{1/2}\rangle$ component, the halo configuration, survives as shown in Eq. (2.20). The spectroscopic factor α^2 of $\mid^{10}\text{Be}(0^+) \otimes \nu 2s_{1/2}\rangle$ can thus be extracted from the amplitude of the $B(E1)$ spectrum. In the experiment of Ref. [40], the spectroscopic factor for the halo configuration $\alpha^2 = 0.72 \pm 0.04$ was extracted. The spectroscopic factors from other experiments range from 0.6 to 0.9 (see Table II of Ref. [40]), which are over all consistent with the current result.

2.3.1.5 Sum Rules

There are two significant sum rules for the $E1$ response of a $1n$ halo nucleus; the non-energy weighted cluster sum rule [68–70] and the energy weighted cluster sum rule [71]. The non-energy weighted cluster sum rule is described as,

$$
B(E1) = \int_{-\infty}^{+\infty} \frac{dB(E1)}{dE} dE = \frac{3}{4\pi} \left(\frac{Ze}{A}\right)^2 \langle r^2\rangle,
\tag{2.21}
$$

which shows that the integrated $B(E1)$ is proportional to $\langle r^2\rangle$, where r represents the distance of the halo neutron from the c.m. of the core. For ^{11}Be, Ref. [40] estimated $\sqrt{\langle r^2\rangle} = 5.77 \pm 0.16$ fm by summing the $E1$ strength up to $E_x = 4$ MeV, which is $1.05 \pm 0.06\,e^2\text{fm}^2$. Note, however, the sum should be taken to infinity, and also take account of bound final states [68, 69]. The extrapolation to infinity amounts to $0.12e^2\text{fm}^2$, while for bound states, we should also add the $E1$ strength of $0.100(15)$ $e^2\text{fm}^2$ due to excitations of the 1st excited state at $E_x = 320$ keV ($J^\pi = 1/2^-$) [72] as well as other contributions ($1p_{3/2}$) calculated as $0.037e^2\text{fm}^2$ [70]. Then the sum of $E1$ strength amounts to $1.31 \pm 0.07e^2\text{fm}^2$, where the uncertainty is only statistical.

Considering the systematic errors arising from the model [69], the estimated $\sqrt{\langle r^2 \rangle}$ is 6.4 ± 0.5 fm, about 10% larger than the original estimate.

It is useful to investigate the other dependence on the integrated $B(E1)$ strength. Combined with the result of the previous section, for the case of the s-wave $1n$ halo, one can extract the following relations:

$$B(E1) \propto \alpha^2 / S_n \tag{2.22}$$

$$\propto \langle r^2 \rangle. \tag{2.23}$$

It is now evident that $B(E1)$ of the $1n$ halo is determined by and scales with S_n, α^2, and is related to the size, $\langle r \rangle^2$, according to the non-energy weighted cluster sum rule.

Another important sum rule is the energy-weighted cluster sum rule [71], which is described as,

$$\int \sigma_\gamma(E1) dE_x = \int \frac{16\pi^3}{9\hbar c} E_x \frac{dB(E1)}{dE_x} dE_x = \frac{60NZ}{A} - \frac{60N_c Z_c}{A_c} \text{MeVmb}, \tag{2.24}$$

where A_c, N_c, and Z_c represent the mass-, neutron- and atomic number of the core fragment. For ^{11}Be, this means that the cluster sum rule exhausts 5.7% of the TRK (Thomas-Reich-Kuhn) sum rule. Ref. [40] extracted the value 4.0(5)%, which is about 70 % of the cluster sum. It is interesting to note that this value is consistent with the spectroscopic factor α^2 for the halo configuration.

2.3.2 Spectroscopy Using Coulomb Breakup of 1n Halo Nuclei - Application to ^{19}C

As was demonstrated for ^{11}Be, Coulomb breakup of a $1n$ halo nucleus occurs primarily by the direct breakup mechanism, as expressed by the matrix element shown in Eqs. (2.14) or (2.15). The $B(E1)$ distribution has a characteristic shape depending on S_n, ℓ (denoted by ℓ_i in this section) of the valence neutron, and the spectroscopic factor of the configuration. Here, the characteristic shapes and amplitude of $B(E1)$ spectra are discussed, whose characteristic differences can be used to clarify the shell configuration and spin-parity of the loosely bound state, in this case, the ground state of ^{19}C. The structure of ^{19}C was unknown for many years before the exclusive Coulomb breakup experiment [41] was performed. We also demonstrate that the angular distribution of the Coulomb breakup reaction is useful in the determination of S_n. These features provide the basis for the Coulomb breakup method offering a useful spectroscopic tool for a loosely-bound $1n$ nuclear system.

2.3.2.1 Issues with ^{19}C

The structure of ^{19}C has not been well known partially because of a large uncertainty in the four direct mass measurements: The Time-Of-Flight Isochronous (TOFI) at Los Alamos measured S_n (^{19}C) to be 700 ± 240 keV [73], and later 230 ± 120 keV [74], while at GANIL, the high-resolution spectrograph (SPEG) combined with the TOF measured S_n (^{19}C) to be 50 ± 420 keV [75], and later -70 ± 240 keV [76]. The mass evaluation in 1993 based on these direct mass measurements was 0.16 ± 0.11 MeV [77, 78]. Such a small S_n value, even smaller than that of ^{11}Be, suggests a significant halo state in ^{19}C, which has drawn much attention. However, we should note that the extracted S_n value ranged from almost 0 to 700 keV.

Since the 13th neutron in ^{19}C most-likely occupies an orbital in the sd shell, the following two possibilities for the ground state configuration should be examined:

$$|^{19}C(1/2^+)\rangle = \alpha|^{18}C(0^+) \otimes \nu 2s_{1/2}\rangle + \beta|^{18}C(2^+) \otimes \nu 1d_{5/2}\rangle, \qquad (2.25)$$

$$|^{19}C(5/2^+)\rangle = \gamma|^{18}C(2^+) \otimes \nu 2s_{1/2}\rangle + \delta|^{18}C(0^+) \otimes \nu 1d_{5/2}\rangle, \qquad (2.26)$$

where α, β, γ, and δ denote the spectroscopic amplitude for each configuration. The shell model calculations in Ref. [79, 80] also suggest such configurations, where these two states are almost degenerate in energy.

Before the exclusive Coulomb breakup measurement that we present here was performed, inclusive measurements of either ^{18}C [79–81] or the neutron [82] momentum components following ^{19}C breakup were performed. The momentum distribution of ^{18}C measured at 77 MeV/nucleon at MSU [79, 80] exhibited a narrow width of 42 ± 4 MeV/c, while that measured at 914 MeV/nucleon at GSI [81] had a broader width of 69 ± 3 MeV/c. Later, the semi-exclusive nuclear breakup on the ^{9}Be target was performed at MSU, where the γ-coincidence information was also incorporated. There, the inclusive ^{18}C momentum distribution is closer to the high energy result of Ref. [81].

Because of such ambiguities in the early inclusive measurements, the interpretation of the ground state of ^{19}C had long been controversial. Reference [79] suggested a $J^\pi = 1/2^+$ assignment for the ^{19}C ground state, and indicated halo formation arising from a large component of the $|^{18}C(0^+) \otimes \nu 2s_{1/2}\rangle$ configuration. However, the revised analysis [80] suggested the $J^\pi = 5/2^+$ assignment with a large portion of a $|^{18}C(2^+) \otimes \nu 2s_{1/2}\rangle$ configuration. A similar possibility was also suggested in the GSI paper [81]. The halo formation should be strongly suppressed for this configuration because of the effective increase of S_n by 1.62 MeV corresponding to the cost of the core excitation energy of the ^{18}C (2^+) state. As is later shown, this would affect the spectral shape of the Coulomb breakup spectrum as well.

2.3.2.2 Characteristic Features of Spectral Shape of Direct Breakup

Since the density distribution outside of the potential will vary strongly, depending on the orbital angular momentum of the valence neutron ℓ_i and the S_n value, the $B(E1)$

Table 2.1 Characteristics of the spectral shape of $dB(E1)/dE_{rel}$ depending on the orbital angular momentum of the valence neutron (ℓ_i) in the initial state, and that of the scattered neutron (ℓ_f) in the final state [21, 67]

$\ell_i \rightarrow \ell_f$	$dB(E1)/dE_{rel} \propto E_{rel}^{\ell_f+1/2}$ (for small E_{rel})	Peak of $dB(E1)/dE_{rel}$
$s \rightarrow p$	$\propto E_{rel}^{3/2}$	$E_{rel} = \frac{3}{5}S_n$
$p \rightarrow s$	$\propto E_{rel}^{1/2}$	$E_{rel} \simeq 0.18 S_n$
$p \rightarrow d$	$\propto E_{rel}^{5/2}$	$E_{rel} = \frac{5}{3}S_n$
$d \rightarrow p$	$\propto E_{rel}^{3/2}$	$E_{rel} = \frac{5}{3}S_n$

spectrum is also characterized by these values. There is also a dependence on ℓ_f, the orbital angular momentum of the dissociated neutron. In the realistic calculation of $dB(E1)/dE_{rel}$ based on the direct breakup mechanism, Eq. (2.14) or (2.15) are used. However, it is instructive to investigate the analytical forms when using a spherical Hankel function for the ground state, and a spherical Bessel function (plane wave) for the final state, as in Eq. (2.19) for the s-wave halo. One obtains a general form, for instance, in Eq. (2) of Ref. [67]. In this reference, the explicit analytical forms are presented for $p \rightarrow s$ and $p \rightarrow d$, in addition to that for $s \rightarrow p$ (Eq. (2.19)). It can also be shown that, for a very small E_{rel} value,

$$\frac{dB(E1)}{dE_{rel}} \propto E_{rel}^{\ell_f+1/2}, \tag{2.27}$$

and $dB(E1)/dE_{rel}$ reaches maximum at the value shown in Table 2.1 , depending on ℓ_i and ℓ_f [21]. This table illustrates that the distinctive difference of spectral shape of $dB(E1)/dE_{rel}$ allows us to determine the ℓ_i value as well as to estimate S_n. For some cases, J^π of the ground state can be extracted. Therefore, the exclusive Coulomb breakup of a $1n+$ core system is a very powerful tool for investigating microscopic structure of such a weakly-bound nucleus.

2.3.2.3 Energy Spectrum and the Ground State Configuration of ^{19}C

Figure 2.7 shows the Coulomb breakup cross section as a function of E_{rel}, which was obtained by using the ^{19}C beam with an average energy of 67 MeV/nucleon at RIKEN [41]. Here, the nuclear breakup component was subtracted using the breakup spectra for a C target. The method of selecting only very forward angles, as was used in more recent ^{11}Be data [40], was not adopted in these data due to lower statistics. The Coulomb breakup spectrum in Fig. 2.7 shows a typical asymmetric shape with a large cross section of 1.19 ± 0.11 barn, a typical feature for a neutron halo nucleus.

Calculations were performed for the direct breakup mechanism for 4 different assumed ground state configurations, depending on the configuration in Eqs. (2.25) and (2.26). First of all, we have found that for $J^\pi = 5/2^+$, any mixture of the two components, $|^{18}C(2^+) \otimes \nu 2s_{1/2}\rangle$ and $|^{18}C(0^+) \otimes \nu 1d_{5/2}\rangle$ failed to reproduce the

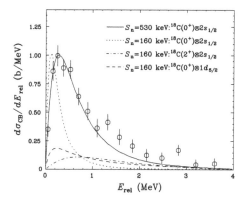

Fig. 2.7 Coulomb breakup cross sections for ^{19}C measured at 67 MeV/nucleon on a Pb target. The *dot-dashed* and *dashed curves* are calculations for the configurations, $|^{18}C(2^+) \otimes \nu 2s_{1/2}\rangle$ and $|^{18}C(0^+) \otimes \nu 1d_{5/2}\rangle$, respectively, of $J^\pi = 5/2^+$ assignment with unit spectroscopic factors and $S_n = 160$ keV. The *dotted* and *solid curves* are calculations for the configuration $|^{18}C(0^+) \otimes \nu 2s_{1/2}\rangle$ with $S_n = 160$ keV and $S_n = 530$ keV, respectively. The latter S_n value is from the angular distribution shown below

data with a former adopted S_n value of 160 keV. Varying S_n over a wide range (from 100 to 700 keV) was not able to reproduce the data, since the calculation could not explain the magnitude of the cross section.

For the $J^\pi = 1/2^+$ case, on the other hand, the peak height of the spectrum can be reproduced with $\alpha^2 = 0.064$ (dotted line) for $S_n = 160$ keV. However, a further tuning of the S_n value was required to obtain an overall fit to the spectrum. In fact, good agreement was obtained using a higher S_n value of 530 keV as shown by the solid line. This S_n value was obtained from an independent analysis using the angular distribution, as shown below. In this case a large spectroscopic factor of $\alpha^2 = 0.67$ was deduced. As for the $|^{18}C(2^+) \otimes \nu 1d_{5/2}\rangle$ configuration, only negligible contribution is obtained and thus this is not shown.

We concluded that the ground state of ^{19}C is $J^\pi = 1/2^+$, involving a dominant configuration of $|^{18}C(0^+) \otimes \nu 2s_{1/2}\rangle$ and the S_n value of about 500 keV. This configuration is rather close to that predicted by a shell model calculation with the WBP interaction [79, 83]. The large amplitude for an s-wave valence neutron with such a small S_n value shows that ^{19}C is a $1n$ halo nucleus. There are a number of theoretical work on dynamics and spectroscopic significance of Coulomb breakup of ^{19}C. S. Typel and R. Shyam used a time-dependent dynamical model of the breakup with a heavy target, which could explain well the energy spectrum of ^{19}C as well as that of ^{11}Be [84]. We also refer to Ref. [63] and references therein.

Fig. 2.8 Angular
distribution of the Coulomb
breakup cross section of ^{19}C
on a Pb target. (The figure is
from T. Nakamura et al. [41])

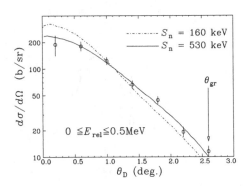

2.3.2.4 Mass from the Angular Distribution of Coulomb Breakup

We now demonstrate that the S_n value can be determined from an analysis of the
angular distribution of ^{19}C (i.e., of the ^{18}C $+ n$ c.m.). As was shown in the case
of the Coulomb breakup of ^{11}Be [40], the scattering angle of the excited ^{19}C* in
the c.m. frame of the target+projectile can be extracted using the momentum vectors of the incident ^{19}C particle, and of the outgoing ^{18}C and the neutron. As in
Eq. (2.6), the angular distribution of the cross section is proportional to that of the $E1$
virtual photon ($dN_{E1}/d\Omega$). The point is that the angular distribution of the photon is
the function of E_x, which is related to S_n by $E_x = E_{rel} + S_n$. For a given E_{rel}, the
angular distribution is then controlled by the parameter S_n, and is independent of the
choice of the final state wave function. Hence, S_n can be extracted from the measured
angular distribution. Note that the determination of S_n from the angular distribution
is less model dependent compared to the determination of S_n from the peak position
of the $B(E1)$ spectrum which may be affected by other $E1$ excitations.

Figure 2.8 shows the measured angular distribution, for $0 \leq E_{rel} \leq 0.5$ MeV, in
comparison with the calculated spectra, where an angular resolution of 8.4 mrad
(in 1σ) was incorporated. The angular distribution is best fitted with $S_n = 530 \pm 130$
keV as indicated by the solid curve, while the previously adopted value of $S_n = 160$
keV was unable to reproduce the angular distribution.

Such a higher S_n value was also extracted from the investigation of the momentum
distribution of ^{18}C in coincidence with the de-excitation γ ray from the first excited
state of ^{18}C at MSU [85]. Note that the newer mass evaluation S_n (^{19}C) $= 0.577 \pm 94$
MeV [86, 87]) now takes into account the angular distribution of this Coulomb
breakup measurement as well as the momentum distribution in the γ-ray tag method.

In this section, we have shown that Coulomb breakup is a powerful spectroscopic
tool to investigate otherwise unknown states of loosely bound nuclei. It is important
that the technique of using the angular-distribution analysis for a $1n$ halo system was
established through this work [41].

2.3.3 Application to the Radiative Capture Reaction $^{14}C(n, \gamma)^{15}C$ of Astrophysical Interest

Coulomb breakup associated with one neutron emission can be useful to extract the radiative neutron capture reaction (n, γ), the latter being the inverse of the (γ, n) reaction. As such, Coulomb breakup is also a useful tool in application to nuclear astrophysics [88, 89]. Here, we show the case of the Coulomb breakup of ^{15}C whose inverse process is ^{14}C $(n, \gamma)^{15}C$. The principle of detailed balance provides the relation between $\sigma_{\gamma n}$ (photo-absorption cross section: $\sigma_{\gamma n} = \sigma_{\gamma}$ in Eq. (2.4)) and the radiative neutron capture cross section $\sigma_{n\gamma}$ as in,

$$\sigma_{n\gamma}(E_{\text{c.m.}}) = \frac{2I_A + 1}{2I_{A-1} + 1} \frac{E_\gamma^2}{2\mu c^2 E_{\text{c.m.}}} \sigma_{\gamma n}(E_\gamma), \tag{2.28}$$

where $I_A = 1/2$, and $I_{A-1} = 0$ for the present case, and μ denotes the reduced mass of $^{14}C + n$, and $E_{\text{c.m.}}$ is the c.m. energy of $n + {}^{14}C$ which is equivalent to E_{rel}. Due to the phase-space factor, the photo-absorption cross section is larger by 2–3 orders of magnitude. For instance, for the current case at $E_{\text{c.m.}} = 0.5$ MeV, $\sigma_{\gamma n}(E_\gamma)/\sigma_{n\gamma}(E_{\text{c.m.}}) \simeq 150$. We also note that the photon numbers are further multiplied, which is of 2–3 orders of magnitude. Considering, in addition, the kinematic focusing and the possibility of using a thick target, of the order of 100 mg/cm^2, Coulomb breakup is more advantageous, even using a weak radioactive beam, compared to the neutron capture experiment. In this particular experiment on the neutron capture on ^{14}C, there is an additional difficulty in preparing the ^{14}C target, which is radioactive, while Coulomb breakup is free from such a difficulty.

2.3.3.1 The Radiative Neutron Capture Reaction $^{14}C(n, \gamma)$

The neutron capture reaction on ^{14}C has drawn much attention due to its importance in several nucleosynthesis processes. There are three cases where this reaction may be of significance.

1. The neutron induced CNO cycle, $^{14}C(n, \gamma)^{15}C(\beta^-)^{15}N(n, \gamma)^{16}N(\beta^-)^{16}O(n, \gamma)$ $^{17}O(n, \alpha)^{14}C$, which occurs in the burning zone of asymptotic giant branch (AGB) stars with 1–3 $M_\odot$ [90]. The $^{14}C(n, \gamma)^{15}C$ reaction is the slowest in this reaction chain, and thus controls the cycle.
2. Synthesis of heavy elements: Terasawa [91, 92] proposed a new-type of r-process initiating from the lightest elements, and thus contains light neutron rich nuclei in the path. This r process was inferred from the similarity of the nuclear abundance pattern in metal-deficient halo stars to that of ordinary ones. The neutron capture on ^{14}C lies in one of the critical reaction flows of this process.
3. Inhomogeneous big bang models: In such models, the neutron-rich zone induces nucleosynthesis and reaction paths involving light neutron-rich nuclei can appear.

In this scenario, the $^{14}C(n, \gamma)^{15}$ C reaction is also considered to be one of the key reactions [93].

Due to these possible stellar reactions, the ^{14}C $(n, \gamma)^{15}$C reaction has attracted much attention.

2.3.3.2 Soft $E1$ Excitation and the p-Wave Direct Neutron Capture

From the nuclear structure point of view, the ground state of ^{15}C, i.e., the final state of the capture reaction, is intriguing because it has a moderate-sized neutron halo with a neutron separation energy S_n of only 1.218 MeV. Its main configuration is $|^{14}C(0^+) \otimes \nu 2s_{1/2}\rangle$. Therefore, Coulomb breakup of ^{15} C is expected to proceed via soft $E1$ excitation due to the direct breakup mechanism.

Then, the direct capture of a p-wave neutron [65] is considered to be a dominant process. This capture process is called direct neutron capture since the neutron in the continuum is captured onto the ^{14}C nucleus, just as in the inverse process where ^{15}C breaks up into its p-wave continuum by the Coulomb breakup (direct breakup). The p-wave capture process, as a stellar reaction, is exceptional, considering that the neutron capture of a low-energy s-wave neutron that obeys the $1/v$ law is usually dominant. The dominance of p-wave neutron capture has also been discussed for stable nuclei, although the final state in this case is an excited state in an s orbital as in ^{13}C $(1/2^+)$ [94, 95].

2.3.3.3 Issues with ^{15}C Coulomb Breakup and Direct Capture Experiments on ^{14}C

There has been controversy over the previous experimental results for this neutron capture process. The pioneering experiment to extract the ^{14}C $(n, \gamma)^{15}$C reaction rate was made by a direct measurement of the neutron capture cross section on a ^{14}C radioactive target [96]. The extracted MACS (Maxwellian averaged capture cross section) of $1.72 \pm 0.43\mu$b at kT=23 keV was about a factor of 4-5 smaller than predicted by the p-wave direct neutron capture calculations [93, 97].

As for the Coulomb breakup approach, A. Horváth et al. measured ^{15}C Coulomb breakup at 35 MeV/nucleon at MSU [42]. There, the energy spectrum of the inverse capture cross section was very different from the expected p-wave behavior. Their extracted capture cross section was about twice as large as that from the above-mentioned direct measurement. More recently, Datta Pramanik et al. measured the Coulomb breakup at higher energies, 605 MeV/nucleon at GSI [98, 99]. The result showed a typical direct breakup spectrum, indicating the p-wave direct capture in the inverse process.

More recently, the direct capture measurement was performed for the second time by Reifarth et al. [100]. The result shows a significantly higher cross section compared to the first direct measurement. In this new measurement, an energy spectrum ranging from ~10 keV to 800 keV was also obtained so that a direct comparison to the Coulomb breakup energy spectrum is now possible.

Here, we primarily show the result from the most recent Coulomb breakup experiment at RIKEN using a ^{15}C secondary beam at 68 MeV/nucleon on a Pb target [44]. It should be noted that the ^{14}C$(n, \gamma)^{15}$C case is a rare one, where both directions of the reactions, neutron capture and Coulomb breakup, can be measured, so that such data can be used to make a rigorous test of the Coulomb breakup method. Therefore, precise measurements in both directions are highly desirable. Recent reaction theories [101–103] for treating Coulomb breakup reactions can also be tested using such measurements.

2.3.3.4 Coulomb Breakup and Neutron-Capture Spectra

Figure 2.9 shows the relative energy spectrum of ^{14}C $+ n$ in the breakup of ^{15}C on a Pb target at 68 MeV/nucleon. To extract the Coulomb breakup component, we adopt the same procedure as in the Coulomb breakup of ^{11}Be [40], using the angular selection at forward angles. The solid squares show the breakup cross section integrated over the scattering angular range $0° \leq \theta \leq 6.0°$ which nearly corresponds to the whole acceptance. Open circles show the cross section for a selected angular range, $0° \leq \theta \leq 2.1°$, which corresponds to the impact parameter range $b > 20$ fm.

The breakup cross sections for $E_{rel} \leq 4$ MeV are 670 $\pm$ 14 (stat.)$\pm$40 (syst.) mb and 294 $\pm$ 12 (stat.)$\pm$18 (syst.) mb for the whole acceptance and the selected angular range, respectively. Such large breakup cross sections due to Coulomb breakup on a heavy target are typical for halo nuclei. We should note, however, that the ^{15}C breakup cross section is about 1/5 $\sim$ 1/3 of those for the conventional halo nuclei such as ^{11}Be [40], and ^{19}C [41], both of which were measured at about 70 MeV/nucleon. This indicates that the size of the halo in ^{15}C is not as extended as in those more weakly-bound halo nuclei. The $B(E1)$ spectrum, extracted from the spectrum for $0° \leq \theta \leq 2.1°$, is shown in Fig. 2.10, which is in excellent agreement with the direct breakup calculation, as shown in the solid curve (dot-dashed curve shows the calculation before folding the experimental resolutions).

2.3.3.5 Neutron Capture Cross Section

Figure 2.11 shows the neutron capture cross section $\sigma_{n\gamma}$ extracted from the $B(E1)$ spectrum by applying the principle of detailed balance (Eq. 2.28). An excellent agreement is obtained with the p-wave direct radiative capture model calculation [66]. This result is consistent with the consideration that the final state of the ^{14}C capture reaction (namely ^{15}C g.s.) is a halo state with a dominant s wave component. The result is also found to be consistent with the neutron capture measurement in Ref. [100], with only a slight deviation at around $E_{c.m.} \sim 0.5$ MeV. The overall agreement of the Coulomb breakup result with the direct capture measurement suggests that Coulomb breakup can be a good alternative for obtaining the neutron

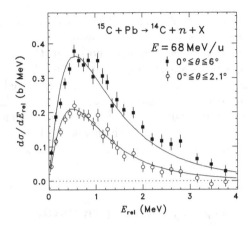

Fig. 2.9 Relative energy spectra for Coulomb breakup of ^{15}C. *Solid squares* represent the data for scattering angles up to 6°, and open circles represent the data for selected scattering angles up to 2.1°. The *solid curves* are calculations for a direct breakup model with a spectroscopic factor $\alpha^2 = 0.91$, for the halo configuration ($a = 0.5$ fm and $r_0 = 1.223$ fm) (The figure is from T. Nakamura et al. [44])

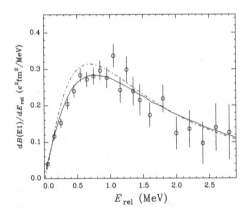

Fig. 2.10 $B(E1)$ spectrum for ^{15}C excitation. The *solid curve* corresponds to the direct breakup model calculation. The *dot-dashed curve* is the same calculation, before folding with the experimental resolution. (The figure is from T. Nakamura et al. [44])

capture cross sections involving radioactive nuclei. Note that in general the Coulomb breakup method can be applied to neutron-rich nuclei where the corresponding direct neutron capture measurement is not feasible.

As for the comparison with previous Coulomb breakup results, we find that only the data obtained from the Coulomb breakup measurement performed at GSI [98, 99] is consistent with the presented measurement. The neutron capture cross section derived from the Coulomb breakup experiment performed at MSU is significantly smaller (about 1μ b at $E_{c.m.} = 23.3$ keV, and 4μb around 0.05–0.10 MeV).

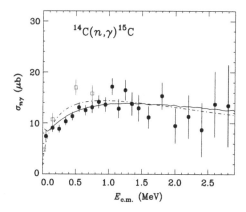

Fig. 2.11 Neutron capture cross section on ^{14}C leading to the ^{15}C ground state. *Solid circles* are the results of the current experiment, and *open squares* are those from the most recent direct capture measurement [97]. The *dot-dashed curve* is the calculation based on the direct radiative capture model, while the *solid curve* is the same one but includes experimental resolution. Both calculations were done using a spectroscopic factor $\alpha^2 = 0.91$ and with potential parameters $a = 0.5$ fm and $r_0 = 1.223$ fm (The figure is from T. Nakamura et al. [44])

Since our results are consistent with those of Reifarth et al. [100], the argument and implications in the nucleosynthesis scenario follows closely those of Ref. [100], where the role of the ^{14}C$(n, \gamma)^{15}$C in the neutron induced CNO cycles, and the impact of such a cycle on the neutron flux are discussed. The confirmation of the reciprocity between Coulomb breakup and the radiative neutron capture is encouraging for further studies of (n, γ) reactions in heavier neutron-rich nuclei. One should note, however, that in the inverse reaction of Coulomb breakup, one can restrict the reaction to the ground state. For the case of ^{15}C, there is only one excited state at 0.74 MeV, and the neutron capture to this state is found negligible from theories [93, 97]. Although, such fortunate situations may be scarce for heavier regions, we may expect lower level densities near closed shells, such as $N = 50$ and 82, where Coulomb breakup can be applicable. These nuclei may be relevant to the r-process, and thus also be of importance.

2.3.4 Inclusive Coulomb Breakup of ^{31}Ne

As was shown above, Coulomb breakup reaction is a powerful tool for investigating $1n$ halo nuclei when one can map the $dB(E1)/dE_{rel}$ function by a kinematically complete measurement (exclusive Coulomb breakup). Here, we show that "inclusive" Coulomb breakup can also be a useful tool. The case presented in this section is an inclusive Coulomb breakup measurement of ^{31}Ne [10], which was performed at the new-generation RI-beam facility, RIBF, at RIKEN as one of the day-one ^{48}Ca beam campaign experiments in Dec. 2008.

2.3.4.1 Issues with ^{31}Ne

For about a decade, the heaviest $1n$ halo nucleus experimentally known has been ^{19}C. Since the first observation of ^{31}Ne in the late 1990s [104], ^{31}Ne has been the next heavier candidate for a $1n$ halo nucleus due to the small $1n$ separation energy. According to the mass evaluation in 2003 [85, 86], S_n was theoretically estimated as 0.332 ± 1.069 MeV, while the recent direct mass measurement showed S_n to be 0.29 ± 1.64 MeV [105].

Experimental studies of ^{31}Ne are thus important as to how and in what form heavier halo nuclei appear, as raised in the questions in the Introduction. Once halo structure of ^{31}Ne is established, then we may obtain a key to understand how a single particle configuration plays a role in halo formation. ^{31}Ne ($N = 21$) could be inside "the island of inversion", where shell model calculations showed that $N = 20$ magicity is lost and significant $2p$-$2h$ configurations are mixed [106–108]. The neigboring nuclei ^{32}Mg [109], ^{32}Na [110], ^{30,32}Ne [111, 112] were found within this island, experimentally.

For the conventional shell order, where the valence neutron resides in the $1f_{7/2}$ orbital upon the ^{30}Ne core, an enhanced tail of the density distribution (halo) never develops, being blocked by the high centrifugal barrier. The halo formation of ^{31}Ne is only possible when a strong shell modification occurs, such that the $2p_{3/2}$ orbit lowers below the $1f_{7/2}$ orbit, for instance.

2.3.4.2 Inclusive Coulomb Breakup

We show that inclusive Coulomb breakup can be used to obtain a signal of the existence of a halo structure. In an inclusive breakup experiment for the $1n$ halo nucleus candidate, we measure $1n$ removal cross section on a heavy target. Namely, we measured the counts of ^{30}Ne fragments relative to the counts of ^{31}Ne projectiles which bombarded a Pb target.

As was mentioned, an exlusive (kinematically complete) measurement requires that of four momentum vectors of all the outgoing particles. In this case we need a coincidence measurement of the neutron, which requires 1-2 order larger yield for the beam. On the other hand, an inclusive measurment is feasible with beam intensity of the order of counts per second or even less, suitable for the earlier stage of the new facility. In the experiment of Ref. [10], typical ^{31}Ne beam intensity was about 5 counts per second and the data was taken only for about 10 hours.

The inclusive Coulomb breakup works in the following way. The inclusive Coulomb breakup cross section can be written as,

$$\sigma(E1) = \int_{S_n}^{\infty} \frac{16\pi^3}{9\hbar c} N_{E1}(E_x) \frac{dB(E1)}{dE_x} dE_x. \tag{2.29}$$

Namely, the product of $N_{E1}(E_x)$ and $dB(E1)/dE_x$ is contained in the cross section. As in Fig. 2.3, and in Fig. 2.12 for this particular case, the photon spectrum falls

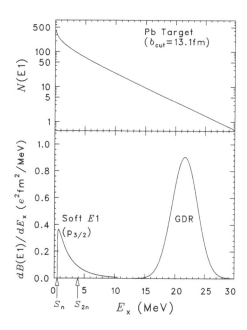

Fig. 2.12 *Top*: The $E1$ virtual photon spectrum on Pb target for the projectile at 230 MeV/nucleon at the impact parameter cut of 13.1 fm. *Bottom*: Calculated $B(E1)$ spectrum assuming the soft $E1$ excitation of the ^{31}Ne with the p-wave halo neutron bound by $S_n = 0.5$ MeV (*left*) and the giant dipole resonance (GDR) at $E_x = 21.6$ MeV with the sigma width of 2.13 MeV ($\Gamma = 5$ MeV) and with the full strength of TRK sum rule (*right*)

exponentially with E_x. Thus, $\sigma(E1)$ becomes significant only when the $B(E1)$ is concentrated at low excitation energies (soft $E1$ excitation) as in the case of halo nuclei. The bottom part of Fig. 2.12 shows the comparison of calculations of $B(E1)$ for a halo nucleus (soft $E1$ excitation) and the ordinary nucleus (GDR) for a $A = 31$ nucleus. For the GDR of the ordinary nucleus, we assume that the GDR peak is located at $31.2A^{-1/3} + 20.6A^{-1/6}$ MeV = 21.6 MeV [53] with a width of 5 MeV (Gaussian) exhausting the full TRK sum rule, namely,

$$\int \sigma_\gamma(E1)dE_x = \int \frac{16\pi^3}{9\hbar c} E_x \frac{dB(E1)}{dE_x} dE_x = \frac{60NZ}{A} \text{MeV mb}. \qquad (2.30)$$

The TRK sum for the $A = 31$ nucleus is 420 MeV mb. Then the total Coulomb breakup cross section $\sigma(E1)$ for ^{31}Ne on Pb target at 230 MeV amounts to 58 mb. [1]

On the other hand, assuming that ^{31}Ne is a halo nucleus and has soft $E1$ strength caused by the valence neutron in $p_{3/2}$ with $S_n = 0.5$ MeV, then the total Coulomb breakup cross section up to $E_x = 10$ MeV is 510 mb, which is about one order of magnitude larger than that due to the GDR. In the measurement, we obtain $1n$ removal Coulomb breakup cross section, whose integration ranges up to $E_x \sim S_{2n}$ instead of infinity. For this range, the GDR strength can yield essentially null cross section, while the soft $E1$ excitation gives rise to the Coulomb breakup cross section in the $1n$ removal channel as 470 mb. With such a distinctive difference, $1n$-removal Coulomb breakup cross section can be used as a signal showing a halo state.

---

[1] When we use the $80A^{-1/3}$ formula, we obtain even less cross section.

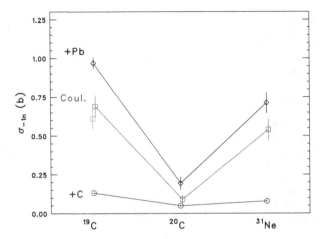

Fig. 2.13 One neutron removal cross sections for ^{19}C, ^{20}C, and ^{31}Ne on Pb (*diamonds*) and C (*circles*), and extracted Coulomb breakup cross sections (*squares*) obtained at about 230 MeV/nucleon. The large Coulomb cross section of ^{31}Ne close to that of ^{19}C (established halo nucleus) indicates the $1n$ halo structure of ^{31}Ne. The Coulomb breakup cross section for ^{19}C estimated from the exclusive one ([41], also see Sect. 3.2 is shown by the *square*, symbol just left of the current data, which shows consistency. The data for ^{19}C and ^{31}Ne are from Ref. [10]. The data for ^{20}C are preliminary

2.3.4.3 Results of ^{31}Ne Breakup

Figure 2.13 shows the $1n$ removal cross sections of ^{19,20}C and ^{31}Ne on Pb and C targets at 230–240 MeV/nucleon. It is readily seen that the cross sections of ^{19}C and ^{31}Ne are both significantly larger than that of ^{20}C. Secondly, the ratio of the cross section for Pb to that for C is 9.0 ± 1.1 for ^{31}Ne and 7.4 ± 0.4 for ^{19}C, much larger than the ratio for nuclear breakup only, which is estimated to be about 1.7–2.6. This demonstrates that the cross section for a Pb target is dominated by Coulomb breakup for ^{19}C and ^{31}Ne.

The Coulomb breakup component of the $1n$ removal cross section on Pb was deduced by subtracting the nuclear component estimated from $\sigma_{-1n}(\text{C})$. For this purpose, it was assumed that $\sigma_{-1n}(\text{C})$ arises entirely from the nuclear contribution, and that the nuclear component for a Pb target scales with the parameter Γ, as in,

$$\sigma_{-1n}(E1) = \sigma_{-1n}(\text{Pb}) - \Gamma \sigma_{-1n}(\text{C}), \tag{2.31}$$

where Γ was estimated to be ~ 1.7–2.6. The lower value is the ratio of target+projectile radii, as in Ref. [5], while the upper one is that of radii of the two targets as in the Serber model [113]. The Coulomb breakup cross section for ^{31}Ne was thus obtained to be $\sigma_{-1n}(E1) = 540 \pm 70$ mb, which takes into consideration the ambiguity arising from the choice of these two models. The dominance of the Coulomb breakup for the reaction of ^{31}Ne on Pb and the deduced $\sigma_{-1n}(E1)$ of some 0.5 b, nearly as high as the established halo nucleus ^{19}C, indicates the occurrence of

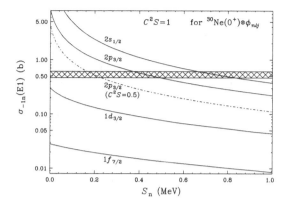

Fig. 2.14 The Coulomb breakup cross section for ^{31}Ne on Pb at 234 MeV/nucleon is compared with calculations for configurations of the valence neutron in $2s_{1/2}, 2p_{3/2}, 1d_{3/2}, 1f_{7/2}$ coupled to ^{31}Ne (g.s.) for $C^2S = 1$ as a function of S_n. An example of lower C^2S value ($C^2S = 0.5$) is shown by the *dot-dashed curve* for the $2p_{3/2}$ configuration

soft $E1$ excitation for ^{31}Ne, thereby providing evidence of $1n$ halo structure of this nucleus.

According to the direct breakup mechanism, the single-particle structure of the ground state of ^{31}Ne can be examined. Figure 2.14 compares the experimentally deduced $\sigma_{-1n}(E1)$ with calculations for possible valence-neutron configurations as a function of S_n. Since there is a large experimental uncertainty in the S_n value of ^{31}Ne, the calculations were shown as a function of S_n.

The figure deals with direct-breakup calculations for a pure single particle configuration $C^2S = 1$ for the valence neutron either in the $2s_{1/2}, 1d_{3/2}, 1f_{7/2}$, or $2p_{3/2}$ orbital, being coupled to the ground state of ^{30}Ne. More detailed analysis including possible configurations of the valence neutron coupled to the ^{30}Ne (2_1^+) state is presented in Ref. [10]. The essential point is, however, included in this figure. Namely, the comparison in Fig. 2.14 shows that the data can be reproduced only by the configuration of $|^{30}\mathrm{Ne}(0_1^+) \otimes \nu 2p_{3/2}\rangle$ ($J^\pi = 3/2^-$) or $|^{30}\mathrm{Ne}(0_1^+) \otimes \nu 2s_{1/2}\rangle$ ($J^\pi = 1/2^+$), but not by $|^{30}\mathrm{Ne}(0_1^+) \otimes \nu 1d_{3/2}\rangle$ nor $|^{30}\mathrm{Ne}(0_1^+) \otimes \nu 1f_{7/2}\rangle$. The significant contribution of such low-ℓ valence neutron is again consistent with the $1n$ halo structure in ^{31}Ne. The other important implication of this result is that the conventional shell model configuration of $|^{30}\mathrm{Ne}(0_1^+) \otimes \nu 1f_{7/2}\rangle$ for the $N=21$ nucleus does not represent a primary configuration of the ^{31}Ne ground state. Large-scale Monte-Carlo Shell Model (MCSM) calculations employing the SDPF-M effective interactions [108] support the assignment of $J^\pi = 3/2^-$ having a $|^{30}\mathrm{Ne}(0_1^+) \otimes \nu 2p_{3/2}\rangle$ contribution, which is consistent with the current findings.

The shell model calculations (MCSM) also showed the large configuration mixing, where $|^{30}\mathrm{Ne}(0_1^+) \otimes \nu 2p_{3/2}\rangle$ could be mixed with $|^{30}\mathrm{Ne}(2_1^+) \otimes \nu 2p_{3/2}\rangle$ and $|^{30}\mathrm{Ne}(2_1^+) \otimes \nu 1f_{7/2}\rangle$. Such a large configuration mixing may be better described in terms of deformation. In fact, recently, the current data was interpreted by a deformed mean-field model [114]. There, the 21st neutron ($p_{3/2}$) can be orbiting in a deformed mean-field potential, namely in the Nilsson levels $[330]1/2^-$ or $[321]3/2^-$, although the possibility of an s-wave valence neutron $([200]\ 1/2^+)$ in a strongly deformed mean field $(\beta > 0.59)$ was not fully excluded. More recently, the ground state properties of ^{31}Ne were also discussed using the particle-rotor model, which takes into account the rotational excitation of the ^{30}Ne core [115]. It is interesting to note that halo properties, such as direct breakup dynamics and soft $E1$ excitations, are kept for the wave functions obtained in a deformed mean-field potential.

Recently, there have been debates over the J^π assignment of the ground state of ^{33}Mg [116–119], either with $J^\pi = 3/2^-$ or $J^\pi = 3/2^+$. Note that ^{33}Mg has the same neutron number $N = 21$ as ^{31}Ne. The Coulomb breakup results of ^{31}Ne shown here may provide some hints to understand the structure of ^{33}Mg and other isotones nearby.

We also have a preliminary result of inclusive Coulomb breakup of ^{22}C ($2n$-halo candidate), where we have obtained evidence for $2n$ halo structure of this nucleus, whose details will be published elsewhere. For this nucleus, the large radius of ^{22}C was recently suggested in the reaction cross section measurement [14], in accordance with the $2n$ halo structure of this nucleus.

As was demonstrated, inclusive measurements are in particular important when the beam intensity is not sufficient. Such a study provides us with a first step to approach the exotic property of extremely neutron rich nuclei. However, for a full understanding of the microscopic structure of ^{31}Ne, exclusive Coulomb breakup experiments would be desired, where C^2S and S_n can be extracted for this $1n$ halo nucleus. At RIBF at RIKEN, such experiments will be realized soon by the completion of the SAMURAI (Superconducting Analyser for MUlti-particles from Radio-Isotope Beam) facility.

2.4 Coulomb Breakup and Soft $E1$ Excitation of $2n$ Halo Nuclei

The phenomena of soft $E1$ excitation for $2n$ halo nuclei have been known since the first inclusive Coulomb breakup experiment for ^{11}Li was performed [5]. However, we still lack a complete understanding of the nature of soft $E1$ excitation for $2n$ halo nuclei due to experimental and theoretical difficulties compared to $1n$ halo cases.

So far, the exclusive Coulomb breakup measurements have been performed for ^{6}He [120, 121], ^{11}Li [45–50], and ^{14}Be [122]. For ^{6}He, the two data [120, 121] are not in agreement. Theoretically, there are a number of theoretical publications for ^{6}He [123–132], although theoretical interpretations of the data [120, 121] are not fully established. For ^{14}Be, the statistics of the experiment in Ref. [118] is apparently not satisfactory to be compared with theories.

The soft $E1$ excitation of ^{11}Li was also controversial, partly due to the inconsistency of three data obtained in the 90's: one measured at MSU at 28 MeV/nucleon [45, 46], one at RIKEN at 43 MeV/nucleon [47], and one at GSI at 280 MeV/nucleon [49] (see Fig. 2.17). More recently, the new data on ^{11}Li was obtained [50] with much higher statistics and with higher sensitivity for two neutron detections. In this section, we review this data, and discuss the obtained implication of the soft $E1$ excitation and the microscopic structure of the $2n$ halo nucleus ^{11}Li.

2.4.1 Exclusive Coulomb Breakup of ^{11}Li

2.4.1.1 Experimental Setup and Treatment of Neutron Cross Talks

A kinematically complete measurement of the Coulomb breakup of ^{11}Li on Pb at 70 MeV/nucleon was performed at RIKEN [50]. The experimental setup is shown in Fig. 2.15, where the incoming ^{11}Li was excited by absorption of a virtual photon and then broke up into the ^{9}Li fragment and the two neutrons. The difference from the experiment of $1n$ halo nuclei is that we should take a special care of measurements of the fast two neutrons in coincidence. In order to veto cross talk events which can happen in the two neutron detection, the neutron detectors composed of 54 rods of plastic scintillators (214(H) ×6.1 (V) ×6.1 (D) cm^3 each) were arranged into two walls (12 × 2 rods for the front (NEUT-A) and 15 × 2 rods for the rear (NEUT-B)), separated by 1.09 m.

Since a neutron is scattered easily by the scintillator material, cross-talk events, where one single neutron leaves more than one signal, can occur. Figure 2.16 (left) schematically illustrates how such a cross-talk event occurs. There, one neutron is scattered in a scintillator rod in the first neutron-detector wall (NEUT-A), which reaches another scintillator rod in the second wall (NEUT-B) to be fired again. Such an event arising from a single neutron can be mistaken for two neutrons firing independently these two scintillator rods. When two signals are obtained from the two scintillator rods, each of which belong to a different wall (different-wall event), one can extract the apparent velocity between the two scintillator rods (v_{AB}). One can then utilize the v_{AB} information to reject the cross-talk events by imposing the condition such that the velocity to the first scintillator v_A is smaller than v_{AB}. This condition excludes almost fully the cross-talk events since the neutron has smaller velocity after the scattering. The validity of this method of vetoing the cross-talk events was confirmed by the calibration run using the ^{7}Li $(p, n)^7$Be (g.s.+0.43 MeV) reaction, which emits only one neutron at the forward angles.

We have also events where two neutrons hit two scintillator rods that belong to the same neutron-detector wall. In this case, one cannot distinguish a two-neutron event from a one-neutron event when two hits take place in a close distance. Figure 2.16 (right) shows how the cross-talk events affect the sensitivity at low E_{rel} in $^9Li+n+n$, where E_{rel} between the two neutrons is also low. There is a dip in the efficiency curve at $E_{rel} \sim 0$ MeV for the same-wall events due to the insensitivity mentioned, while the different-wall events have a smooth efficiency curve towards $E_{rel} \sim 0$, which enabled us to measure $B(E1)$ strength down to $E_{rel} = 0$ MeV. In treating the same-wall events, we confirmed that energy spectrum after the efficiency correction in the same wall events is essentially identical to that for the different wall events.

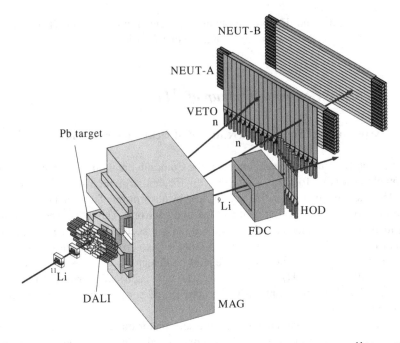

Fig. 2.15 The experimental setup for the Coulomb breakup measurement of ^{11}Li on Pb at 70 MeV/nucleon at RIKEN. The setup contains a dipole magnet (MAG), drift chamber (FDC), hodoscope (HOD), and two-walls of neutron detector arrays (NEUT-A,B). (The figure is from T. Nakamura et al. [50])

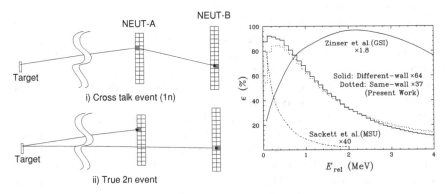

Fig. 2.16 *Left* Schematic explanation of the cross-talk event: (**i**) Cross-talk event where one neutron fires both NEUT-A and NEUT-B, and (**ii**) Two neutron event hitting independently NEUT-A and NEUT-B. The cross-talk events shown in (**i**) can be excluded by imposing the condition $v_A < v_{AB}$, where v_A denotes the neutron velocity between the target and the NEUT-A, and v_{AB} denotes the apparent velocity obtained from the timings at NEUT-A and NEUT-B. *Right* Estimated efficiencies for the different-wall events and same-wall events. The efficiency is also compared with those from the experiments at MSU [45, 46] and at GSI [49] (The latter figure is from Ref. [50]).

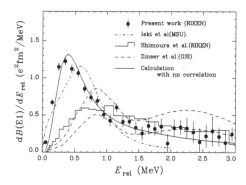

Fig. 2.17 $B(E1)$ spectrum obtained in the Coulomb breakup of ^{11}Li at 70 MeV/nucleon Ref. [50] compared with the previous data obtained at MSU at 28 MeV/nucleon (*dot-dashed-line*) [45, 46], RIKEN at 45 MeV/nucleon (*solid histogram*) [47, 48], and at GSI at 280 MeV/nucleon (*zone between dashed lines*) [49]. The data is also compared with the three-body calculation shown in the *solid curve* [68, 69]

2.4.1.2 $B(E1)$ Spectrum and Two Neutron Correlation

Figure 2.17 shows the obtained $B(E1)$ distribution, which is compared with the previous three data sets. The result reveals substantial $E1$ strength that peaks at very low E_{rel} around 0.3 MeV. This feature is in contrast to the previous data, which showed much reduced strength towards low relative energies which may be due to the low neutron detection efficiency near $E_{rel} \sim 0$ MeV in those experiments.

The spectrum amounts to a large energy-integrated $B(E1)$ strength of $1.42 \pm 0.18 \, e^2 \text{fm}^2$ (4.5(6) W.u.), for $E_{rel} \leq 3$ MeV, which is the largest soft $E1$ strength ever observed. In fact, this $E1$ strengths is about 40% larger than that for the one-neutron halo nucleus ^{11}Be [39, 40].

In extracting this spectrum, the $E1$ virtual photon spectrum assumed $S_{2n} = 0.3$ MeV (2003 Mass Evaluation) [85, 86]. Recently, the mass values with higher precisions have been experimentally obtained at ISOLDE(CERN) [133] and at TRIUMF [134]. The former used the mass spectrometer MISTRAL, which deduced the S_{2n} value to be 0.378(5) MeV, while the latter used the Penning trap technique which enabled an extremely precise mass evaluation of ^{11}Li as $S_{2n} = 0.36915(65)$ MeV. These results provided consistently higher values by about 20% compared to the 2003 mass evaluation, which affects the evaluation of the photon number that is dependent on $E_x = S_{2n} + E_{rel}$. For the latter S_{2n} value of 0.369 MeV, the $E1$ strength for ^{11}Li re-evaluated is 1.49(19) (e^2fm^2) for $E_{rel} \leq 3$ MeV, or about 5% larger. Such a slight change within the experimental uncertainty does not essentially change the conclusion.

Figure 2.17 also compares the present $B(E1)$ distribution with a calculation using the three-body model proposed by Esbensen and Bertsch [67, 68], which includes the two-neutron correlations both in the initial and final states in the $E1$ excitation. Note that this calculation reproduces the data very well with no adjustment of normalization. Recently, a revised 3-body calculation was made by Esbensen et al. [135], where the recoil effect was taken into account, in addition. Although an overall agreement with the data has been obtained, some deviation occurs around $E_{rel} \sim 0.5$ MeV as shown in Fig. 2 of Ref. [135]. This may be an indication that the other effects, such as core excitation, may be important. The recent calculation by Myo et al. [31, 32], which incorporated the effect of the core polarization using the tensor optimized shell model, reproduced better the peak region. These overall agreements of both the spectral shape and absolute strength in these calculations indicate the presence of a strong two neutron correlation in ^{11}Li, both in the ground and final states.

2.4.1.3 Non Energy Weighted Cluster Sum Rule for 2n Halo Nuclei and Dineutron Correlation

The spatial two-neutron correlation in the ground state (initial state) of two-neutron halo nuclei can be quantitatively estimated by the non-energy weighted $E1$ cluster sum rule [68, 69]. Such a sum rule was discussed for $1n$ halo nuclei in Sect. 2.3.1.5. For the case of a two-neutron halo nucleus, the $E1$ operator $\mathbf{r}$ is replaced by $\mathbf{r}_1 + \mathbf{r}_2$, where $\mathbf{r}_1$ and $\mathbf{r}_2$ are the position vectors of the two valence neutrons relative to the c.m. of the core. Namely, the sum rule is written as,

$$B(E1) = \frac{3}{4\pi} \left(\frac{Ze}{A}\right)^2 \langle r_1^2 + r_2^2 + 2\mathbf{r}_1 \cdot \mathbf{r}_2 \rangle = \frac{3}{\pi} \left(\frac{Ze}{A}\right)^2 \langle r_{c,2n}^2 \rangle. \qquad (2.32)$$

Here, $\mathbf{r}_1 + \mathbf{r}_2$ can be related to $r_{c,2n}$, which is the distance between the c.m. of the core and that of the two halo neutrons. Importantly, the term of $(\mathbf{r}_1 \cdot \mathbf{r}_2)$ involves the opening angle θ_{12} between the two position vectors of the two valence neutrons. The value of $\langle r_{c,2n} \rangle$, and hence $B(E1)$, becomes larger for the smaller spatial separation of the two neutrons, when θ_{12} approaches $0°$. The integrated $B(E1)$ thus provides a good measure of the two-neutron spatial correlation.

The integral of the $E1$-response calculation up to $E_{rel} = 3$ MeV (solid curve in Fig. 2.17) accounts for about 80% of the total $E1$ cluster sum-rule strength above S_{2n}. Assuming that the $B(E1)$ distribution follows this solid curve, the observed $B(E1)$ strength is then translated into $1.78 \pm 0.22 e^2 fm^2$ for $E_x \geq S_{2n}$, which corresponds to the value of $\sqrt{\langle r_{c,2n}^2 \rangle} = 5.01 \pm 0.32$ fm. Adopting the sum rule value, $1.07 e^2 fm^2$, for two non-correlated neutrons calculated in Ref. [68], the value $1.78 \pm 0.22 e^2 fm^2$ corresponds to $\langle \theta_{12} \rangle = 48^{+14}_{-18°}$. This angle is significantly smaller than the mean opening angle of $90°$ expected for the two non-correlated neutrons. Hence, an appreciable two-neutron spatial correlation is suggested for the two halo neutrons. Reference [135] estimated $\sqrt{\langle r_{c,2n}^2 \rangle} = 5.22$ fm, while Ref. [31] calculated $\sqrt{\langle r_{c,2n}^2 \rangle}$ to be 5.69 fm. These predicted values are close to the experimental estimation.

The important point for the soft $E1$ excitation of $2n$ halo nucleus is that the larger polarization of the charge due to the stronger dineutron-like correlation enhances the soft $E1$ excitation. The charge radius measurement of ^{11}Li also extracted the distance $\sqrt{\langle r_{c,2n}^2 \rangle} = 5.97\ (22)$ [135–137], which is higher than the value from the Coulomb breakup. This discrepancy could be related to the possible core polarization effect [135].

It can be shown that spatial nn correlation is related to a mixture of different parity states in the valence neutrons of ^{11}Li. In a simple shell model picture where the inert ^{9}Li core is assumed, the ^{11}Li could be described as,

$$\Psi(^{11}\text{Li(g.s.)}) = \alpha|^9\text{Li(g.s.)} \otimes \nu(2s_{1/2})^2\rangle + \beta|^9\text{Li(g.s.)} \otimes \nu(1p_{1/2})^2\rangle. \quad (2.33)$$

Here, we are interested in the mean opening angle between $\mathbf{r}_1$ and $\mathbf{r}_2$. The expectation value of $\cos\theta_{12}$ can be written as,

$$\langle\cos\theta_{12}\rangle = \alpha^2\langle(2s)^2|\cos\theta_{12}|(2s)^2\rangle + \beta^2\langle(1p)^2|\cos\theta_{12}|(1p)^2\rangle$$
$$+ 2\alpha\beta\langle(1p)^2|\cos\theta_{12}|(2s)^2\rangle \quad (2.34)$$

$$= 2\alpha\beta\langle(1p)^2|\cos\theta_{12}|(2s)^2\rangle \quad (2.35)$$

The terms for α^2 and β^2 are null since these are odd functions of $\cos\theta_{12}$. Hence, a non-zero expectation value implies a mixture of these different parity configurations. Namely, no mixture of different parity states (either $\alpha^2 = 0$ or $\beta^2 = 0$) gives rise to $\langle\theta_{12}\rangle = 90°$ and no $2n$ correlation, while the mixture leads to the $2n$ spatial correlation.

Other than shown above, there are extensive theoretical studies on the three-body nature and the breakup dynamics of ^{11}Li [23, 68, 131, 135, 138–146]. Our Coulomb breakup result, in combination with these theoretical studies, should provide fruitful information on the crucial properties of this intriguing Borromean system.

To summarize the soft $E1$ excitation of $2n$ halo nuclei, as in the case of $1n$ halo, the strong low-energy $E1$ transitions certainly occurs for $2n$ halo nuclei. For ^{11}Li, the integrated soft $E1$ strength is about 40% larger than for ^{11}Be. Such an enhancement is also related to the nn spatial correlation (dineutron-like correlation), which can be understood by the $\mathbf{r}_1 \cdot \mathbf{r}_2$ term in the non-energy weighted cluster sum rule. Experimentally, it is of great importance to detect two neutrons in coincidence unambiguously. A novel detection scheme, in combination with an analysis procedure using the kinematical condition for excluding cross-talk events, was introduced in Ref. [50]. Such a method would be evolved further for multi-neutron detections for a breakup of $4n$ halo nuclei in the near future.

2.5 Spectroscopy of Unbound States via the Nuclear Breakup

We now focus on the breakup with a light-mass target, where nuclear breakup dominates over the Coulomb breakup. The nuclear breakup accompanied by $1n$ or $2n$ emissions, which we treat here, are categorized into two mechanisms:

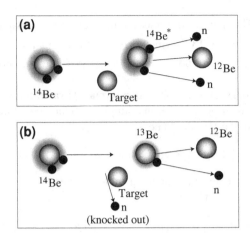

Fig. 2.18 Schematic drawings of the neutron removal processes. *Upper figure* (**a**) shows diffractive dissociation (or inelastic scattering/elastic breakup) for the case of ^{14}Be. For instance, unbound resonances such as the first 2^+ state can be excited. In the forward detectors, the ^{12}Be fragment and the two neutrons are detected. The *bottom figure* (**b**) shows the $1n$ knockout process. In this case, the remaining ^{13}Be is produced as an intermediate unbound state followed by the decay into ^{12}Be and n. In the forward detectors, the ^{12}Be fragment and the neutron are detected

1. Diffractive dissociation (or Elastic breakup/Inelastic scattering): This process is inelastic scattering into the resonant state or non-resonant continuum as shown in (Fig. 2.18a). The term elastic breakup is used since this process corresponds to an elastic scattering of the neutron off the target. In this process, the neutron(s) are basically going in the forward directions which can be covered by the neutron detectors in the standard setup of invariant-mass spectroscopy of neutron-rich nuclei.

2. Knockout reaction (or Stripping/Inelastic breakup): The knockout reaction may be viewed as a quasi-free inelastic scattering of the neutron off the target (Fig. 2.18b). In this process, the neutron is scattered to a large angle due to a relatively large momentum transfer or even be absorbed by the target. As a consequence, the neutron will not appear in the forward direction. This process is sometimes referred to as absorption.

Experiments using the neutron removal reactions of radioactive nuclei with light targets have been extensively studied [147], as a powerful spectroscopic tool as in Coulomb breakup. Inelastic breakup can be used to excite directly unbound resonances. In Sect. 2.3.1.2, the case for ^{11}Be + C was briefly mentioned, where two resonances of ^{11}Be above the $1n$-decay threshold were observed (see Fig. 2.5b). In this section, we show the excitation of ^{14}Be using the inelastic scattering with carbon and proton targets.

In the knockout reaction, the momentum distribution of the core fragment, and the cross section can be used to determine the orbital angular momentum (ℓ) of the removed neutron. At RI-beam facilities using high/intermediate-energy fragmenta-

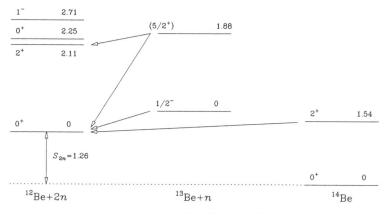

Fig. 2.19 Energy levels of low-lying states of ^{12}Be, ^{13}Be, and ^{14}Be, where the excitation energies are shown in MeV. The energy and the spin-parity of the first excited state of ^{14}Be were obtained from the elastic breakup of ^{14}Be on a carbon target [148], which was later confirmed by the breakup on a proton target. The level scheme of ^{13}Be was obtained from the $1n$ removal reaction of ^{14}Be on the proton target. The S_n value of the ^{13}Be ground state was found $-0.51(1)$ MeV (see text). The arrows from the ^{13}Be to ^{12}Be correspond to the decays observed in this experiment. The direct decay from the 2_1^+ state of ^{14}Be into the ground state of ^{12}Be was also observed, which follows the phase-space decay (see Fig. 2.22)

tion, there have been extensive experimental studies on neutron removal reactions using a variety of radioactive beams on light targets [147]. At MSU, a method of measuring γ rays in coincidence with the core fragment has been developed, which can determine the final bound excited state of the core fragment. Here, we show the case of determining the "unbound" final states of ^{13}Be, following the one-neutron removal of ^{14}Be. This requires extra cares, compared to the bound excited state, as described below.

In what follows, we review experimental studies using the ^{14}Be beam at about 70 MeV/nucleon on carbon and proton targets [51, 148]. The unbound low-lying states of ^{14}Be and ^{13}Be were studied, whose level schemes, including the ones obtained in these experiments are summarized in Fig. 2.19.

2.5.1 Inelastic Scattering of ^{14}Be

2.5.1.1 Issues with ^{14}Be

^{14}Be is the most neutron-rich bound beryllium isotope with the two-neutron separation energy $S_{2n} = 1.26(13)$ MeV [86, 87] (see Figs. 2.1 and 2.19). This nucleus is a Borromean nucleus since ^{13}Be and the two neutrons are unbound, and is known to have two-neutron halo structure due to the weak binding of the two valence neutrons. The structure of ^{14}Be has been investigated theoretically as a ^{12}Be$+n+n$

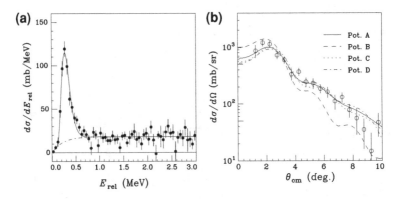

Fig. 2.20 **a** Relative energy spectrum of ^{12}Be $+n+n$ in the inelastic scattering of ^{14}Be on a carbon target at 68 MeV/nucleon. **b** Differential cross sections for the inelastic scattering of ^{14}Be. Curves show the DWBA calculations using optical potential parameter sets. The figures are from Ref. [148]

three-body system [149–153]. This nucleus is also discussed in terms of $\alpha - \alpha$ clustering structure as a common property of the neutron-rich beryllium isotopes [154].

The energies of the first 2^+ states (2_1^+) are important benchmarks to investigate the shell gaps, and their associated nuclear magicity. The 2_1^+ states of most nuclei are bound, and thus can be studied by in-beam γ ray spectroscopy as in the case of ^{12}Be [155]. For ^{12}Be, notably large deformation length was obtained by the (p, p') reaction, despite the fact that ^{12}Be has $N = 8$. Meanwhile, the 2_1^+ state of ^{14}Be has long been controversial, which is lying above the $2n$ decay threshold. A candidate of the 2_1^+ state was reported at around $E_x = 1.6$ MeV [156, 157]. However, subsequent experiments did not show evidence of the state at around 1.6 MeV [19, 122]. In addition, no spin-parity assignment was done in these experiments. This nucleus was thus investigated by the kinematically complete measurement of ^{14}Be breakup on a carbon target with higher statistics at RIKEN [148].

2.5.1.2 Relative Energy Spectrum and Angular Distribution

Figure 2.20a shows the relative energy spectrum of the ^{12}Be $+ n +n$ system measured in the breakup of ^{14}Be on a carbon target at 68 MeV/nucleon [148]. The detector setup used is identical to the one in Fig. 2.15. The two neutrons in coincidence with ^{12}Be at forward detectors were measured, which corresponds to the measurement of the elastic breakup (inelastic scattering/diffractive).

A clear resonance peak has been observed at $E_{rel} = 0.28(1)$ MeV, which was assigned to the first excited state at $E_x = 1.54(13)$ MeV. It should be noted that the uncertainty of E_x is mainly from that of S_{2n} of ^{14}Be. The spin and parity of the state were determined from the analysis of the angular distribution of the inelastically scattered ^{14}Be* (^{12}Be $+ n + n$). This angular distribution was obtained using the same procedure as in the Coulomb breakup experiments of ^{11}Be, ^{15,19}C, and ^{11}Li

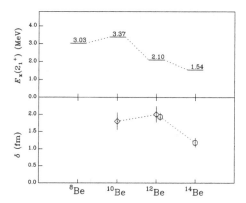

Fig. 2.21 Systematics of the 2_1^+ excitation energies $E_x(2_1^+)$ (*top*) and quadrupole deformation lengths δ for beryllium isotopes (*bottom*) are plotted, including the ^{14}Be results [148]. In the *bottom figure*, the *open diamonds* represent the result from the inelastic scattering with a ^{12}C target, while the *open circles* are those with a proton target. The systematics shows the melting of the $N = 8$ shell gap at ^{12}Be. ^{14}Be has the least excitation energy and the least deformation length among the Be isotopes. The figure is from Ref. [148]

described in the previous sections. Figure 2.20b shows the differential cross sections for populating the 1.54-MeV state as a function of the scattering angle. Based on the angular distribution, which is well fitted with curves of the DWBA calculations for the $\Delta L = 2$ transition, the observed resonance was assigned as a $J^\pi = 2^+$ state. The quadrupole deformation length was extracted from the DWBA analysis to be 1.18(13) fm. It is interesting to note that the observed $E_x(2_1^+)$ value of ^{14}Be is smaller than that of ^{12}Be as compared in Fig. 2.21. Contrary to the naive picture that the lower $E_x(2_1^+)$ indicates a larger collectivity for ^{14}Be, the deformation length was observed to be significantly smaller than that of ^{12}Be. Such small collectivity in ^{14}Be may be related to the quenching of effective charges due to the high isospin value as in neutron-rich B isotopes [158, 159]. It was also found that the shell model calculations with effective interactions more appropriate to neutron-rich nuclei in the p-shell (SFO) [160] reproduced the properties of the first 2_1^+ states better both for ^{12}Be and ^{14}Be, rather than the conventional ones (PSDMK) [161].

2.5.2 Breakup of ^{14}Be with a Proton Target and Spectroscopy of ^{13}Be

^{14}Be was also studied by the reaction ^{1}H (^{14}Be, ^{12}Be + 2n) at 69 MeV/nucleon using a liquid hydrogen target provided by the cryogenic target system CRYPTA [162]. Since the reaction yield per energy loss for liquid hydrogen is larger than any other target with a larger mass number, the use of such a target is advantageous for RI beam experiments. The γ ray detectors (DALI NaI(Tl) array) were also installed, surrounding the target. The experimental setup was essentially similar to the one used for the Coulomb breakup of ^{11}Li (see Fig. 2.15), except for the liquid hydrogen target.

In the Eikonal terminology, the breakup with a proton target at the energy of about 70 MeV is treated as an elastic breakup, irrespective of the amplitude of the neutron momentum transfer, since the proton remains in the ground state through the reaction in any ceases. However, here, we distinguish the case where the two neutrons and the core fragment (^{12}Be) are captured in the range of the forward detectors, from the one where one neutron is deflected with a large momentum transfer and thus only the ^{12}Be and the neutron are in the range of the forward detectors. The former corresponds to the case shown in Fig. 2.18a, while the latter corresponds to the case shown in Fig. 2.18b. In the former case, we find that the major contribution is the excitation to the 2_1^+ state of ^{14}Be, and the latter case corresponds to the production of ^{13}Be in the intermediate state. Both channels can be studied in the same experimental setup. As later described, some events from the $2n+^{12}$Be channel can be mixed with the $1n + ^{12}$Be channel due to finite efficiency of the neutron detection. Hence, a careful subtraction was made for $1n + ^{12}$Be spectrum.

2.5.2.1 Elastic Breakup and Study of the Three-Body Decay

The analysis of the $2n+^{12}$Be channel showed a similar energy spectrum as Fig. 2.20, and confirmed the existence of 2_1^+ state at about $E_x=1.5$ MeV, whose detailed analysis on the spectrum is published elsewhere. It should be noted that there was no coincidence with γ rays associated with the 2_1^+ state of ^{12}Be which supports that the 2_1^+ state decays into the ground state of ^{12}Be.

We also investigated the property of the three-body decay from this state. Figure 2.22 shows the two dimensional plots of the two-body relative energies, so called Dalitz plot, for the decay of the 2_1^+ state. The experimental distribution is well described by the phase-space decay simulation with no signal of any correlation, and hence no evidence for the sequential decay through a ^{13}Be state was provided. This implies that there are no resonance states of ^{13}Be up to about 0.28 MeV $(=E_{rel}(2_1^+))$ above the ^{12}Be $+ n$ decay threshold. In the subsequent subsections, we show that excitation to the 2_1^+ state in ^{14}Be can affect the spectrum of ^{13}Be, which will be properly subtracted.

2.5.2.2 One-Neutron Removal Reaction of ^{13}Be -Introduction

The $1n$ removal reaction of ^{14}Be on a proton target at 69 MeV/nucleon was used to study ^{13}Be [50], whose production process is schematically shown in Fig. 2.18b. Here, we describe how one could extract the spectrum of the intermediate unbound states of ^{13}Be by the nuclear breakup at intermediate energies.

The energy levels of the unbound nucleus ^{13}Be were controversial because several experimental studies [163–171] were not consistent. In particular, there had

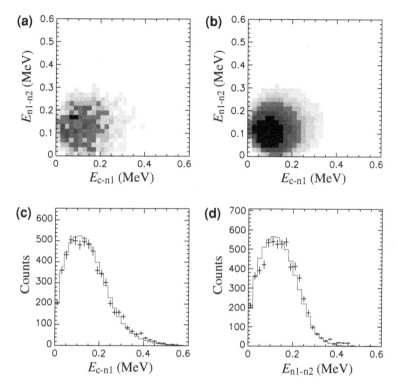

Fig. 2.22 a: Experimental Dalitz plot for the decay of the 2^+ state of ^{14}Be. E_{c-n1} and E_{n1-n2} represent two-body relative energies of ^{13}Be-n and the two neutrons, respectively. **b**: Simulation results of the same Dalitz plot by assuming the decay according to the three-body phase space. **c** and **d**: Projected histograms of **a** on E_{c-n1} and E_{n1-n2}, respectively. Simulations shown by *histograms* reproduce the experimental distributions very well.

been large ambiguities in spin-parity assignments. The resonances of the two-body constituent, ^{13}Be, of the three-body system ^{14}Be is a key to understand its Borromean structure. In addition, theoretically, there is an interesting question as to whether at $N = 9$ the parity inversion as in ^{11}Be occurs [172] or not. The experiment of Ref. [51] was thus performed to disentangle the situation and to clarify the low-lying unbound levels of the ^{13}Be nucleus by determining both the energy and J^{π}.

Similar spectroscopic studies to the case of ^{13}Be were performed for the two-body unbound resonance states in Borromean nuclei, such as ^{5}He [120, 173, 174], ^{7}He [173–175], ^{10}Li [49, 171, 176], by the breakup of nuclei ^{6}He, ^{8}He, and ^{11}Li, respectively. The proton knockout reactions, which may have different selection rules for the production, have been applied for the study of ^{16}B [177] and ^{18}B [178]. Some of the earlier breakup extracted the momentum distribution of ^{5}He [179], ^{10}Li [180] and ^{16}B [177], to provide spectroscopic information such as the amplitude and ℓ of the removed neutron (proton). However, the distribution obtained was only inclusive. Namely, the momentum distributions in those experiments were obtained without selecting the state in the energy (mass) spectrum.

The work on ^{13}Be measured E_{rel} and the momentum distribution, simultaneously. The high statistics of the experiment in Ref. [51] enabled to subtract the effect of the $2n + {}^{12}$Be channel, as described later. The γ-ray coincidence also allowed to distinguish the decay into the daughter fragment ^{12}Be in the excited states. Such a kinematically complete measurement and analysis was realized in this experiment for the first time.

2.5.2.3 Momentum Distribution of the Bound and Unbound State Following the Neutron Removal

The inclusive momentum distribution of the fragment following the neutron removal reaction has been used as a spectroscopic tool to probe the extension of the radial wave function, in particular for halo nuclei since the early work on ^{9}Li fragment from ^{11}Li [2, 3].

At MSU, the method of combining γ-ray detection following the neutron removal has been successfully developed to identify the final state of the core fragment as in Ref. [59] and references in Ref. [147]. For $1n$ removal, the momentum distribution for a specific core state was used to determine the orbital angular momentum (ℓ) of the removed neutron, leading to the spin-parity assignment of the relevant state. In addition, the partial cross sections to individual final states provide the spectroscopic factors by applying reaction theory such as the Eikonal model. The Eikonal models to analyze such experimental observables are reviewed in detail in Ref. [181].

At RIKEN, the one-neutron removal reactions of ^{18}C and ^{19}C on a proton target were studied [182] to probe the single particle states of these nuclei and their daughter nuclei. Figure 2.23 shows the momentum distribution of ^{17}C following the $1n$ removal of ^{18}C at 81 MeV/nucleon on a proton target with identifying the final state of outgoing ^{17}C by detecting de-excitation γ rays [182]. In this work, the experimental momentum distributions were compared with the theoretical ones calculated by the Continuum-Discretized Coupled-Channels (CDCC) method [183], instead of the Eikonal method. The Eikonal calculation always treat the $1n$ removal on the proton target as elastic breakup, and thus it is difficult to distinguish an excitation to a state of the projectile, and the breakup associated with the large momentum transfer to the neutron. The CDCC calculations can treat such processes on equal footing.

In this measurement, the momentum distribution for populating the first excited state of ^{17}C tagged by the 0.21-MeV γ rays (Fig. 2.23a) is well described by the calculation of the neutron removal from the $\nu 2s_{1/2}$ orbital, while the other one (Fig. 2.23b) for the second excited state is well described by the calculation for the neutron removal from the $\nu 1d_{5/2}$ orbital. These results led to the confirmation of the spin-parity assignments of $J^\pi = 1/2^+$ and $J^\pi = 5/2^+$ for the first and second excited states of ^{17}C, as proposed in Ref. [184].

When the final state of the core is an unbound resonance as in ^{13}Be from ^{14}Be, one cannot determine the energy of the state by measuring the γ ray but the decaying

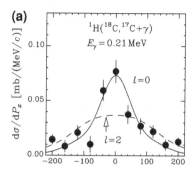 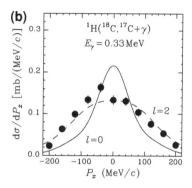

Fig. 2.23 Transverse momentum distributions of ^{17}C in coincidence with de-excitation γ rays following the $1n$ removal from ^{18}C with a proton target at 81 MeV/nucleon [182]. The *solid and dashed curves* correspond to the CDCC calculations for the assumption of removal of s-wave and d-wave neutron, respectively. The figures are from Ref. [182]

particles. Hence, the invariant-mass spectroscopy is suitable for such a case instead of in-beam γ-ray spectroscopy. For the measurement of the unbound nucleus ^{13}Be in Ref. [51], invariant mass spectroscopy was combined with the measurement of the momentum distribution measurement.

2.5.2.4 Extraction of ^{13}Be states: Selection of $1n + {}^{12}$Be Events

In the $1n$ removal process (Fig. 2.18b), one neutron is emitted in the forward direction after the decay of ^{13}Be, while the neutron removed first is deflected in a large angle. As such, the first neutron is out of range of the neutron detectors. Therefore, the $M_n = 1$ events, where M_n represents the neutron multiplicity, were selected for this process.

However, we should note that the $M_n = 1$ events also contain contributions from the inelastic process (Fig. 2.18a), since the efficiency of the neutron detection is not 100% but finite (22%). In the inelastic process, two neutrons are emitted in the forward direction, but could be recorded as an $M_n = 1$ event when only one neutron is detected due to this finite efficiency of the neutron detectors. Here, the amount of mixture of two-neutron events was extracted by analyzing the $M_n = 2$ events since we have such events as in Fig. 2.22.

Figure 2.24 (left) shows the total $1n + {}^{12}$Be spectrum (solid curve), as well as the estimation of the two-neutron events mixed in the $M_n = 1$ events (dashed curve). The latter was obtained by reconstructing the $n - {}^{12}$Be energy for the $2n + {}^{12}$Be events and by normalizing the spectrum properly by using the efficiency values. It was found that the lowest peak corresponds to the decay of the 2_1^+ state of ^{14}Be. The contaminant from the inelastic process in the $M_n = 1$ events, which is mainly from the excitation to the 2_1^+ state of ^{14}Be, is thus subtracted in the analysis.

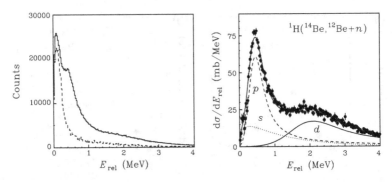

Fig. 2.24 (*Left*) Relative energy spectrum for the $M_n = 1$ events. The *dashed histogram* shows the contribution from the inelastic scattering of ^{14}Be, estimated from the $M_n = 2$ events. The lowest peak thus corresponds primarily to the decay from the 2_1^+ state of ^{14}Be. This figure is adopted from Ref. [51]. (*Right*) Relative energy spectrum for ^{13}Be after subtracting the contribution from $2n + {}^{12}$Be events which are dominated by the excitation of 2_1^+. The peak at 0.5 MeV and the shoulder around 2 MeV are observed

It should be noted that earlier work on ^{13}Be using the inclusive breakup reaction may be affected by the contribution from the 2_1^+ state of ^{14}Be. In this sense, it is important to perform a kinematically complete measurement. Note also that the effect of inelastic excitation to resonances (diffractive dissociation) is expected to be larger for lower incident energies. The eikonal calculations by Hencken et al. [185] predicted that the contributions from diffractive dissociation and stripping (knockout) reaction are comparable below around 80 MeV/nucleon, while above 200 MeV/nucleon the stripping (knockout) reaction is dominant over the diffractive dissociation. This may be suggestive when one uses inclusive nuclear breakup reactions.

2.5.2.5 Relative Energy Spectra and Momentum Distributions

Figure 2.24 (right) shows the relative energy spectrum of ^{12}Be $+n$ after subtracting the inelastic component shown in Fig. 2.24(left). Two peaks at around 0.5 and 2 MeV are observed. The γ-ray coincident events were also analyzed to verify whether the final state of ^{12}Be in the decay is the ground state. Figure 2.25 shows the E_{rel} spectra obtained in coincidence with the 2.1-MeV (filled circles) and 2.7-MeV (open triangles) γ rays, corresponding to the decays to ^{12}Be$(2_1^+)+n$ and ^{12}Be$(1_1^-)+n$ (see also Fig. 2.19). A peak at $E_{rel} \sim 0$ for the 2.1-MeV γ coincidence events is likely to correspond to the resonance at around 2 MeV in Fig. 2.24 (right). The cross sections up to $E_{rel} = 4$ MeV were 16(2) mb and 9(1) mb for the 2.1-MeV and 2.7-MeV γ-ray coincidence events, respectively, which are small compared with 103(7) mb for the inclusive spectrum in Fig. 2.24 (right). Hence, in the discussion hereafter, we neglect the contribution of the γ-ray related events.

The contribution of the decay to the isomeric state ^{12}Be (0_2^+) was not taken into consideration in the analysis, which could not be extracted experimentally due to its

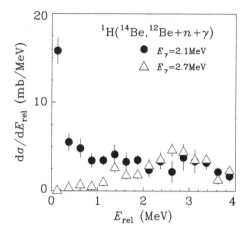

Fig. 2.25 Relative energy spectra of ^{13}Be in coincidence with 2.1-MeV and 2.7-MeV γ ray decay in ^{12}Be in the breakup of ^{14}Be on the proton target [51]. For the coincidence with the 2.1-MeV γ in ^{12}Be, the peak at $E_{rel} \sim 0$ MeV corresponds to the excitation energy of the d-wave peak of ^{13}Be in Fig. 2.24 (*right*). See also Fig. 2.19

long mean life time, 331(12) ns, of this state [184]. However, the 0.5-MeV peak is likely to be relevant to the decay to the ground state but not to the isomeric state, from the consideration of penetration factor [51].

The transverse momentum distributions of ^{13}Be were extracted as shown in Fig. 2.26, depending on the energy region. The difference of widths corresponds to the different orbital angular momentum of the removed neutron. A simultaneous fitting of the relative energy spectrum and the momentum distributions determines the dominant angular momentum of $\ell = 1$ for the 0.5-MeV peak. The extracted resonance energy and width were 0.51(1) MeV and 0.45(3) MeV, respectively. This width is consistent with the single-particle width of 0.55 MeV for the $\ell = 1$ decay. These results led to the spin-parity assignment of $1/2^-$, which is different from the assignment of the earlier experiments. However, we notice that the spectrum of GSI experiment [171] is rather similar to the current result (although their primary assignment was s-wave). Our result is also consistent with the recent theory [172], which predicted the existence of this low-lying p-wave state.

The 2 MeV peak was assigned as the d-wave component in agreement with most of the previous work, although we note that the width is much wider than the single-particle estimate. This could imply that this peak is composed of a few states rather than a single resonance.

The s-wave component is also found to exist in the spectrum, for the analysis of which we assumed this to be a virtual state [187]. The scattering length was thus extracted to be $a_s = -3.4(6)$ fm, which is very small and thus implies a weak correlation between ^{12}Be and the s-wave neutron. This is consistent with the fact that in the three-body decay of the 2_1^+ state of ^{14}Be, no signal of the resonance state (nor strong virtual state) of ^{13}Be below $E_{rel} = 0.28$ MeV was observed. In this

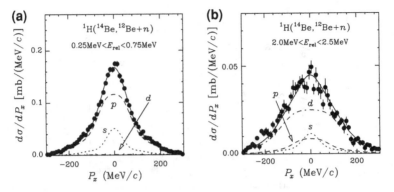

Fig. 2.26 Transverse momentum distributions of ^{13}Be for (**a**) 0.25 MeV $< E_{rel} <$ 0.75 MeV, and (**b**) 2.0 MeV $< E_{rel} <$ 2.5 MeV, in the one-neutron knockout reaction of ^{14}Be. The curves show the result of the simultaneous fitting, assuming that the energy spectrum is composed of s-, p-, and d-wave components. The figures are from Ref. [51]

sense, it is not appropriate to refer the s-wave level as a resonance or virtual state. Instead, it is likely that the weak s-wave continuum is distributed at low-energy range ($E_{rel} \sim 0-2$ MeV), and its level energy is not definite. The level and decay scheme for the structures of ^{13}Be and ^{14}Be, including the three-body decay and the γ-decay branches, are summarized in Fig. 2.19.

2.5.2.6 Disappearance of $N = 8$ Shell Gap

The notable finding in the result of Ref. [51] was the existence of a p-wave resonance as the ground state of ^{13}Be. Taking consideration of the low-lying s-wave strengths, it may be more appropriate to conclude that the p-wave ground state is almost degenerate in energy with the s-wave contribution. We now show the systematics of energy difference ($\Delta\epsilon$) between the $1/2^+$ and $1/2^-$ states for the $N = 7$ and $N = 9$ nuclei as shown in Fig. 2.27. Since the position of the $1/2^+$ state of the unbound nucleus ^{13}Be cannot be defined, only the upper limit is shown for ^{13}Be. The slope of $\Delta\epsilon$ with respect to the atomic number for the $N = 9$ nuclei is nearly identical to the one for $N = 7$. The latter was originally plotted by I. Talmi and I. Una for the difference of the single particle $p_{1/2}$ and $s_{1/2}$ states [28], and was summarized in Ref. [187] with other $N = 9$ nuclei. This results suggest the melting of the $N = 8$ shell gap in ^{13}Be.

Such a shell evolution could be interpreted by modern shell models. One explanation was given by inclusion of the spin-flip p-n monopole interaction [160, 191], which could explain the structure of neutron rich p-shell nuclei. In the stable nuclei ^{13}C and ^{15}C, the $\pi p_{3/2}$ orbital is fully occupied, resulting in the $N = 8$ shell gap due to the strong interaction with the $\nu p_{1/2}$ orbital. Since the interaction becomes weak at ^{11}Be and ^{13}Be by decreasing the occupation number of the $\pi p_{3/2}$, the $\nu p_{1/2}$ goes up and the gap between the neutron sd-shells weakens. The shell model calculation

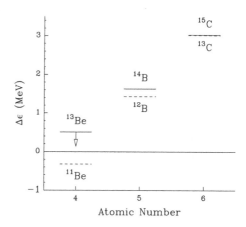

Fig. 2.27 Energy difference of $1/2^+$ and $1/2^-$ states for $N = 7$ (*dashed*) and 9 (*solid*) nuclei as shown in Fig. 8 of Ref. [184], as a function of atomic number. A similar plot for the single particle states was presented for $N = 7$ in Ref. [28]. The upper-limit value is shown for ^{13}Be [51]. The energy values except for ^{13}Be are taken from Refs. [188–190]

including this effect by introducing the enhanced spin-flip *p-n* monopole interaction [160] provides the intruder $1/2^-$ ground state. This agreement may indicate the importance of this effect in the neutron-rich $N = 8$ region. As was mentioned, this shell model calculation reproduced the overall level schemes of the low-lying states of $^{12-14}$Be.

However, we also should note that the shell model may not explain the effect of strong deformation expected in the neutron-rich Be isotopes. In fact, strong deformation in ^{12}Be was observed [155]. For neutron-rich weakly-bound nuclei and unbound nuclei, we also should note that the use of harmonic oscillator wave functions may not be appropriate. Recently, the properties of the neighboring nuclei ^{11,12}Be were investigated in terms of single-particle motion in deformed potential [192], where the two Nilsson levels $[101]\frac{1}{2}^-$ and $[220]\frac{1}{2}^+$ become almost degenerate at large prolate deformation. This can also be true for the 9th neutron of ^{13}Be. As such, the $1/2^-$ state may become the ground state. Note that, in this case, the single particle orbital at the spherical limit has still a significant gap between $\nu 1p_{1/2}$ and $\nu 2s_{1/2}$. It is interesting to note that prolate deformation can be strongly induced in the neutron-rich region even near the magic number due to the nuclear Jahn-Teller effect (or a spontaneous breaking of symmetry). Low-ℓ orbitals as a function of the potential depth has smoother function than the higher-ℓ orbitals, which tend to cause the degeneracy in the unbound region between the different ℓ orbitals, resulting in the Jahn-Teller effect [193]. Hence, the deformation can play major roles in the structure of neutron-rich weakly-bound and unbound nuclei.

It is interesting to investigate how these theoretical approach, such as shell models and deformed potential models, can be understood in a universal way. In this respect, spectroscopic studies of neutron-rich weakly-bound and unbound nuclei are of great importance.

2.6 Concluding Remarks

This lecture note has reviewed breakup reactions of halo nuclei, and their related topics, primarily from the experimental point of view. Halo nuclei are interesting many-body quantum states, where one or two neutrons extend outside of the range of nuclear mean field potential. A neutron halo nucleus breaks up easily with a high cross section in response to Coulomb/nuclear interactions with a target nucleus. Hence, the breakup can be a probe of microscopic structure of halo nuclei.

In Sect. 2.2, we described the method of the Coulomb breakup of halo nuclei at intermediate/high energies. We emphasized that the Coulomb breakup is a suitable tool when we apply this method to radioactive nuclei. This is partly because Coulomb breakup at intermediate/high energies gains yield due to a large photon number, availability of a thick target, and to kinematic focusing. We also note that the relatively good energy resolution can be obtained in the invariant mass method when it is applied to breakup experiments, which is advantageous compared to missing-mass spectroscopy.

In the following sections (Sect. 2.3 and Sect. 2.4), recent Coulomb breakup experiments of several halo nuclei, mainly using the data from our group, were reviewed. We clarified that the main mechanism of the soft $E1$ excitation is attributed to direct breakup, and as such the final state is not a resonance but rather structure-less continuum. The phenomena and spectroscopic significance appear in a different manner between $1n$- and $2n$ halo nuclei. For $1n$ halo nuclei, the final state represents simply a relative motion between the neutron and the core. As such, the spectrum can be described by the matrix element shown in Eq. (2.14). Since the Coulomb breakup is highly sensitive to the tail of the radial wave function. The spectroscopic amplitude and ℓ of the halo configuration, as well as the S_n of the nucleus, can be extracted by a kinematically complete measurement (exclusive measurement). Examples shown were the cases of ^{11}Be and ^{15,19}C. The Coulomb breakup of ^{11}Be provided the basis of the experimental and analysis methods of intermediate-energy Coulomb breakup. The latter experiments were the cases to demonstrate how the Coulomb breakup of $1n$-weakly bound nuclei can be a useful spectroscopic tool. For ^{19}C, we could also extract the S_n value by a novel method using the angular distribution in the Coulomb breakup.

We have also demonstrated the usefulness of inclusive Coulomb breakup, by showing the case of ^{31}Ne, whose experimental work was long scarce due to the small yield of this nucleus. Inclusive measurement is more feasible for very exotic nuclei with the beam yield of the order of particle per second. Here, evidence is shown for the first time for a heavy halo nucleus, ^{31}Ne, much beyond the known halo nuclei.

It should be noted, however, that the exclusive measurement would provide more structure information, as illustrated in the Coulomb breakup experiments of ^{11}Be and ^{15,19}C. In 2012 at RIBF, we plan a commissioning experiment of the SAMURAI facility, which is equipped with the superconducting magnet with the 80 cm gap and the maximum magnetic field of about 3 T, and with the large-area neutron detectors,

NEBULA (**NE**utron-detection system for **B**reakup of **U**nstable Nuclei with **La**rge Acceptance). We plan to perform an exclusive Coulomb breakup experiment of ^{31}Ne in the near future. Such an experiment would clarify the microscopic structure (ℓ, shell configurations, and S_n) of this nucleus.

For Coulomb breakup of $2n$ halo nuclei, the initial and final states are much more complicated. However, we have demonstrated that spatial dineutron correlation may play significant roles in the enhancement of the strong $E1$ strength observed in ^{11}Li. We have also shown the importance of unambiguous $2n$ detection, which has been accomplished by the novel method, using a kinematical condition of velocities associated with the two neutrons. With such high-statistical data combined with the high sensitivity down to $E_{rel} = 0$ MeV, reliable comparisons with theories have become possible for the first time. We showed some of such comparisons. We should note, however, that the theoretical interpretation is still not fully settled, partly due to effects of the core excitation. Inclusion of such effects was attempted theoretically by Myo et al. [31, 32].

Recently, evidence for the two-neutron halo structure in ^{22}C has been provided by the reaction cross section measurement [14]. We have also preliminary results on the Coulomb breakup of this nucleus. Study of such a heavy $2n$ halo nucleus having a different shell configuration is very important to clarify the nature of dineutron correlation. Understanding of dineutron correlations in halo nuclei is essential to investigate such correlations in neutron-skin nuclei in heavier mass regions. In addition, studies of halo nuclei with $4n$ or multiple halo neutrons become more important for the future. Such studies would shed light on the physics of neutron stars, as well.

In the final part of this lecture note, we showed examples of ^{14}Be breakup on the light targets, which undergo the diffractive dissociation to the excitation to the 1st excited state of ^{14}Be, and the $1n$ removal process to produce the ^{13}Be unbound states. We have shown that such a kinematically complete measurement is powerful to determine the spin-parities and the energies of the levels of the $-1n$ unbound system. As such, this method provided the basis of studying the $1n$ removal channel.

Through this lecture note, we showed typical experimental tools for halo nuclei and neighboring states including unbound nuclei. With the advent of new-generation facilities, where RIKEN has already launched such a facility, RIBF, and FAIR, FRIB, KoRIA, and SPIRAL2 will be commissioned in a few to several years, these experimental tools will play more significant roles when we explore closer to the bound limit in heavier mass regions.

Acknowledgments This work has been supported in part by a Grant-in-Aid for Scientific Research (B) (22340053) from the Ministry of Education, Science and Culture (MEXT). We thank the support by the GCOE program "Nanoscience and Quantum Physics". We are grateful to J.A. Tostevin and H. Esbensen for fruitful theoretical discussions.

References

1. Tanihata, I., Hamagaki, H., Hashimoto, O., Shida, Y., Yoshikawa, N., Sugimoto, K., Yamakawa, O., Kobayashi, T., Takahashi, N.: Phys. Rev. Lett. **55**, 2676 (1985)
2. Kobayashi, T., Yamakawa, O., Omata, K., Sugimoto, K., Shimoda, T., Takahashi, N., Tanihata, I.: Phys. Rev. Lett. **60**, 2599 (1988)
3. Orr, N.A et al.: Phys. Rev. Lett. **69**, 2050 (1992)
4. Hansen, P.G, Jonson, B.: Europhys. Lett. **4**, 409 (1987)
5. Kobayashi, T. et al.: Phys. Lett. B **232**, 51 (1989)
6. Yano, Y.: Nucl. Instrum. Methods Phys. Res. B **261**, 1009 (2007)
7. Kubo, T.: Nucl. Instrum. Methods Phys. Res. B **204**, 97 (2003)
8. Ohnishi, T. et al.: J. Phys. Soc. Jpn. **77**, 083201 (2008)
9. Ohnishi, T. et al.: J. Phys. Soc. Jpn. **79**, 073201 (2010)
10. Nakamura, T. et al.: Phys. Rev. Lett. **103**, 262501 (2009)
11. Rosner, G.: Nucl. Phys. B (Proc. Suppl.) **167**, 77 (2007)
12. Gales, S.: Nucl. Phys. A **834**, 717c (2010)
13. Thoennessen, M.: Nucl. Phys. A **834**, 688c (2010)
14. Tanaka, K. et al.: Phys. Rev. Lett. **104**, 062701 (2010)
15. Tanihata, I. et al.: Prog. Part. Nucl. Phys. **35**, 505 (1995)
16. Hansen, P.G, Jensen, A.S., Jonson, B.: Annu. Rev. Nucl. Sci. **45**, 591 (1995)
17. Hansen, P.G., Sherrill, B.M.: Nucl. Phys. A **693**, 133 (2001)
18. Jensen, A.S., Riisager, K., Fedorov, D.V.: Rev. Mod. Phys. **76**, 215 (2004)
19. Jonson, B.: Phys. Rep. **389**, 1 (2004)
20. Riisager, K., Jensen, A.S., Möller, P.: Nucl. Phys. A **548**, 393 (1992)
21. Hamamoto, I.: Lecture note. Tokyo Institute of Technology, Tokyo (2009)
22. Migdal A.B, (1972) Yad. Fiz. 16:427 ; English translation Sov. J. Nucl. Phys., 16:238 (1973).
23. Hagino, K.: Phys. Rev. Lett. **99**, 022506 (2007)
24. Matsuo, M.: Phys. Rev. C **73**, 044309 (2006)
25. Efimov, V.: Phys. Lett. B **33**, 563 (1970)
26. Fedorov, D.V, Jensen, A.S, Riisager, K.: Phys. Rev. Lett. **73**, 2817 (1994)
27. Garrido, E., Fedorov, D.V, Jensen, A.S: Phys. Rev. Lett. **96**, 112501 (2006)
28. Talmi, I., Unna, I.: Phys. Rev. Lett. **4**, 469 (1960)
29. Misu, T., Nazarewicz, W., Åberg, S.: Nucl. Phys. A **614**, 44 (1997)
30. Hamamoto, I.: Phys. Rev. C **69**, 041306(R) (2004)
31. Myo, T., Kato, K., Toki, H., Ikeda, K.: Phys. Rev. C **76**, 024305 (2007)
32. Ikeda, K., Myo, T., Kato, K., Toki, H. Clusters in Nuclei - Vol. 1, In : C. Beck (ed.) Lecture Notes in Physics 818: Springer, berlin (2010)
33. Hamamoto, I., Sagawa, H., Zhang, X.Z.: Phys. Rev. C **55**, 2361 (1997)
34. Suzuki, T., Sagawa, H., Bortignon, P.F.: Nucl. Phys. A **662**, 282 (2000)
35. Mizutori, S. et al.: Phys. Rev. C **61**, 044326 (2000)
36. Bennaceur, K. et al.: Phys. Lett. B **496**, 154 (2000)
37. Meng, J. et al.: Phys. Rev. Lett. **80**, 460 (1998)
38. Nakamura, T. et al.: Phys. Lett. B **331**, 296 (1994)
39. Palit, R. et al.: Phys. Rev. C **68**, 034318 (2003)
40. Fukuda, N. et al.: Phys. Rev. C **70**, 054606 (2004)
41. Nakamura, T. et al.: Phys. Rev. Lett. **83**, 1112 (1999)
42. Horváth, .Á et al.: Astrophys. J. **570**, 926 (2002)
43. Datta Pramanik, U. et al.: Phys. Lett. B **551**, 63 (2003)
44. Nakamura, T. et al.: Phys. Rev. C **79**, 035805 (2009)
45. Ieki, K. et al.: Phys. Rev. Lett. **70**, 730 (1993)
46. Sackett, D. et al.: Phys. Rev. C **48**, 118 (1993)
47. Shimoura, S. et al.: Phys. Rev. C **348**, 29 (1995)
48. Shimoura, S. Private communication.

49. Zinser, M. et al.: Nucl. Phys. A **619**, 151 (1997)
50. Nakamura, T. et al.: Phys. Rev. Lett. **96**, 252502 (2006)
51. Kondo, Y. et al.: Phys. Lett. B **690**, 245 (2010)
52. Jackson, J.D: Classical Electrodynamics, 2nd Edn. Wiley, New York (1975)
53. Bertulani, C., Baur, G.: Phys. Rep. **163**, 299 (1988)
54. Harakeh, M.N, van der Woude, A.: Giant Resonances. Oxford University Press, Oxford (2001)
55. Ikeda K (1988) INS report JHP-7 [in Japanese].
56. Tanihata, I. et al.: Phys. Lett. B **206**, 592 (1988)
57. Fukuda, M. et al.: Phys. Lett. B **268**, 339 (1991)
58. Kelley, J.H et al.: Phys. Rev. Lett. **74**, 30 (1995)
59. Aumann, T. et al.: Phys. Rev. Lett. **84**, 35 (2000)
60. Auton, D.L.: Nucl. Phys. A **157**, 305 (1970)
61. Zwieglinski, B., Benenson, W., Robertson, R.G.H, Coker, W.R: Nucl. Phys. A **315**, 124 (1979)
62. Capel, P., Baye, D., Suzuki, Y.: Phys. Rev. C **78**, 054602 (2008)
63. Baye, D., Capel, P., Clusters in Nuclei Vol. 2, Lecture Notes in Physics vol. 848, Springer, ed. C. Beck (2012)
64. Goldstein, G., Baye, D., Capel, P.: Phys. Rev. C **73**, 024602 (2006)
65. Otsuka, T. et al.: Phys. Rev. C **49**, R2289 (1994)
66. Mengoni, A., Otsuka, T., Ishihara, M.: Phys. Rev. C **52**, R2234 (1995)
67. Nagarajan, M.A., Lenzi, S.M., Vitturi, A.: Eur. Phys. J. A **24**, 63 (2005)
68. Esbensen, H., Bertsch, G.F.: Nucl. Phys. A **542**, 310 (1992)
69. Esbensen, H. Private communication.
70. Bonaccorso, A., Vinh Mau, N.: Nucl. Phys. A **615**, 245 (1997)
71. Alhassid, Y., Gai, M., Bertsch, G.F: Phys. Rev. Lett. **49**, 1482 (1982)
72. Nakamura, T. et al.: Phys. Lett. B **394**, 11 (1997)
73. Vieira, D.J et al.: Phys. Rev. Lett. **57**, 3253 (1986)
74. Wouters, J.M et al.: Z. Phys. A **331**, 229 (1988)
75. Gillibert, A. et al.: Phys. Lett. B **192**, 39 (1987)
76. Orr, N.A et al.: Phys. Lett. B **258**, 29 (1991)
77. Audi, G., Wapstra, A.H: Nucl. Phys. A **565**, 1 (1993)
78. Audi, G., Wapstra, A.H., Dedieu, M. Nucl. Phys. A **565**:193, and references therein (1993)
79. Bazin, D. et al.: Phys. Rev. Lett. **74**, 3569 (1995)
80. Bazin, D. et al.: Phys. Rev. C **57**, 2156 (1998)
81. Baumann, T. et al.: Phys. Lett. B **439**, 256 (1998)
82. Marquès, F.M et al.: Phys. Lett. B **381**, 407 (1996)
83. Warburton, E.K, Brown, B.A: Phys. Rev. C **46**, 923 (1992)
84. Typel, S., Shyam, R.: Phys. Rev. C **64**, 024605 (2001)
85. Maddalena, V. et al.: Phys. Rev. C **63**, 024613 (2001)
86. Wapstra, A.H., Audi, G., Thibault, C.: Nucl. Phys. A **729**, 129 (2003)
87. Audi, G., Wapstra, A.H., Thibault, C.: Nucl. Phys. A **729**, 337 (2003)
88. Baur, G., Bertulani, C.A., Rebel, H.: Nucl. Phys. A **458**, 188 (1986)
89. Baur, G., Rebel, H.: Annu. Rev. Nucl. Part. Sci **46**, 321 (1996)
90. Wiescher, M., Görres, J., Schatz, H.: J. Phys. G Nucl. Part. Phys. **25**, R133 (1999)
91. Sasaqui T., et al. (2005) Astrophys. J. 634:1173
92. Terasawa, M., et al.: Astrophys. J. **562**, 470 (2001)
93. Wiescher, M., Görres, J., Thielemann, F.K.: Astrophys. J. **363**, 340 (1990)
94. Nagai, Y. et al.: Astrophys. J. **372**, 683 (1991)
95. Ohsaki, T. et al.: Astrophys. J. **422**, 912 (1994)
96. Beer, H. et al.: Astrophys. J. **387**, 258 (1992)
97. Descouvemont, P.: Nucl. Phys. A **675**, 559 (2000)
98. Datta Pramanik, U. et al: Phys. Lett. B **551**, 63 (2003)
99. Datta Pramanik, U. et al.: Prog. Theor. Phys. Suppl., **146**, 427 (2002)
100. Reifarth, R. et al.: Phys. Rev. C **77**, 015804 (2008)

101. Timofeyuk, N.K., Baye, D., Descouvemont, P., Kamouni, R., Thompson, I.J.: Phys. Rev. Lett. **96**, 162501 (2006)
102. Summers, N.C., Nunes, F.M.: Phys. Rev. C **78**, 011601(R) (2008)
103. Esbensen, H.: Phys. Rev. C **80**, 024608 (2009)
104. Sakurai, H. et al.: Phys. Rev. C **54**, R2802 (1996)
105. Jurado, B. et al.: Phys. Lett. B **649**, 43 (2007)
106. Warburton, E.K., Becker, J.A., Brown, B.A.: Phys. Rev. C **41**, 1147 (1990)
107. Caurier, E., Nowacki, F., Poves, A., Retamosa, J. et al.: Phys. Rev. C **58**, 2033 (1998)
108. Utsuno, Y., Otsuka, T., Mizusaki, T., Honma, M. et al.: Phys. Rev. C **60**, 054315 (1999)
109. Motobayashi, T. et al.: Phys. Lett. B **364**, 9 (1995)
110. Thibault, C. et al.: Phys. Rev. C **12**, 644 (1975)
111. Yanagisawa, Y. et al.: Phys. Lett. B **566**, 84 (2003)
112. Doornenbal, P. et al.: Phys. Rev. Lett. **103**, 032501 (2009)
113. Serber, R.: Phys. Rev. **72**, 1008 (1947)
114. Hamamoto, I.: Phys. Rev. C **81**, 021304(R) (2010)
115. Urata, Y., Hagino, K., Sagawa, H.: Phys. Rev. C **83**, 041303(R) (2011)
116. Yordanov, D.T. et al.: Phys. Rev. Lett. **99**, 212501(R) (2007)
117. Trippathi, V. et al.: Phys. Rev. Lett. **101**, 142504 (2008)
118. Yordanov, D.T. et al.: Comment, Phys. Rev. Lett **104**, 129201 (2010)
119. Kanungo, R. et al.: Phys. Lett. B **685**, 253 (2010)
120. Aumann, T. et al.: Phys. Rev. C **59**, 1252 (1999)
121. Wang, J. et al.: Phys. Rev. C **65**, 034306 (2002)
122. Labiche, M. et al.: Phys. Rev. Lett. **86**, 600 (2001)
123. Danilin, B.V., Thompson, I.J., Vaagen, J.S., Zhukov, M.V.: Nucl. Phys. A **632**, 383 (1998)
124. Ershov, S.N., Danilin, B.V., Vaagen, J.S.: Phys. Rev. C, **64**, 064609 (2001)
125. Ershov, S.N., Danilin, B.V., Vaagen, J.S.: Phys. Rev. C, **74**, 014603 (2006)
126. Ershov, S.N., Danilin, B.V., Vaagen, J.S.: Phys. Rev. C, **81**, 044308 (2010)
127. Kikuchi, Y. et al.: Phys. Rev. C **81**, 044308 (2010)
128. de Diego, R., Garrido, E., Fedorov, D.V., Jensen, A.S.: Europhys. Lett. **90**, 52001 (2010)
129. Matsumoto, T., Kato, K., Yahiro, M.: Phys.Rev. C **82**, 051602 (2010)
130. Myo, T., Kato, K., Aoyama, S., Ikeda, K.: Phys. Rev. C **63**, 054313 (2001)
131. Hagino, K., Sagawa, H., Nakamura, T., Shimoura, S.: Phys. Rev. C **81**, 031301(R) (2009)
132. Baye, D., Capel, P., Descouvemont, P., Suzuki, Y.: Phys. Rev. C **79**, 024607 (2009)
133. Bachelet, C. et al.: Phys. Rev. Lett. **100**, 182501 (2008)
134. Smith, M. et al.: Phys. Rev. Lett. **101**, 202501 (2008)
135. Esbensen, H., Hagino, K., Mueller, P., Sagawa, H.: Phys. Rev. C **76**, 024302 (2007)
136. Sanchez, R. et al.: Phys. Rev. Lett. **96**, 033002 (2006)
137. Puchalski, M., Moro, A.M., Pachucki, K.: Phys. Rev. Lett. **97**, 133001 (2006)
138. Cobis, A., Fedorov, D.V., Jensen, A.S.: Phys. Lett. B **424**, 1 (1998)
139. Garrido, E., Fedorov, D.V., Jensen, A.S.: Nucl. Phys. A **708**, 277 (2002)
140. Bang, J.M. and RNBT collaboration, Phys. Rep. **264**, 27(1996)
141. Pushkin, A., Jonson, B., Zhukov, M.V.: J. Phys. G **22**, L95 (1996)
142. Thompson, I.J. et al.: J. Phys. G. **24**, 1505 (1998)
143. Forssén, C., Efros, V.D., Zhukov, M.V.: Nucl. Phys. A **706**, 48 (2002)
144. Myo, T., Aoyama, S., Kato, K., Ikeda, K.: Phys. Lett. B **576**, 281 (2003)
145. Lurie, Yu.A., Shirokov, A.M.: Ann. of Phys. **312**, 284 (2004)
146. Hagino, K., Sagawa, H.: Phys. Rev. C **72**, 044321 (2005)
147. Gade, A., Glasmacher, T.: Prog. Part. Nucl. Phys. **60**, 161 (2008)
148. Sugimoto, T. et al.: Phys. Lett. B **654**, 160 (2007)
149. Thompson, I.J. et al.: Phys. Rev. C **53**, 708 (1996)
150. Labiche, M. et al.: Phys. Rev. C **60**, 027303 (1999)
151. Tarutina, T. et al.: Nucl. Phys. A **733**, 53 (2004)
152. Descouvemont, P. et al.: Nucl. Phys. A **765**, 370 (2006)

153. Blanchon, G., Vinh Mau, N., Bonaccorso, A., Dupuis, M., Pillet, N.: Phys. Rev. C **82**, 034313 (2010)
154. Kanada-Enyo, Y. et al.: Phys. Rev. C **52**, 628 (1995)
155. Iwasaki, H. et al.: Phys. Lett. B **481**, 7 (2000)
156. Bohlen, H.G. et al.: Nucl. Phys. A **583**, 775 (1995)
157. Korsheninnikov, A.A. et al.: Nucl. Phys. A **616**, 189c (1997)
158. Kondo, Y. et al.: Phys. Rev. C **71**, 044611 (2005)
159. Ogawa, H. et al.: Phys Rev. C **67**, 064308 (2003)
160. Suzuki, S., Fujimoto, R., Otsuka, T.: Phys. Rev. C **67**, 044302 (2003)
161. Millner, D.J., Kurath, D.: Nucl. Phys. A **255**, 315 (1975)
162. Ryuto, H. et al.: Nucl. Instrum. Methods Res. A **555**, 1 (2005)
163. Aleksandrov, D.V. et al.: Sov. J. Nucl. Phys. **37**, 474 (1983)
164. Ostrowsk, A.N. et al.: Z. Phys. A **343**, 489 (1992)
165. Korsheninnikov, A.A. et al.: Phys. Lett. B **343**, 53 (1995)
166. Marqués, F.M. et al.: Phys. Rev. C **64**, 061301(R) (2001)
167. Belozyorov, A.V. et al.: Nucl. Phys. A **636**, 419 (1998)
168. Lecouey, J.L.: Few Body Syst. **34**, 21 (2004)
169. Thoennessen, M. et al.: Phys. Rev. C **63**, 014308 (2000)
170. Christian, G. et al.: Nucl. Phys. A **791**, 267 (2007)
171. Simon, H. et al.: Nucl. Phys. A **791**, 267 (2007)
172. Blanchon, G., Bonaccorso, A., Brink, D.M., Garcia-Camacho, A., Vinh Mau, N.: . Nucl. Phys. A **784**, 49 (2007)
173. Markenroth, K. et al.: Nucl. Phys. A **679**, 462 (2001)
174. Aksyutina, Yu. et al.: Phys. Lett. B **679**, 191 (2009)
175. Meister, M. et al.: Phys. Rev. Lett. **88**, 102501 (2002)
176. Aksyutina, Yu. et al.: Phys. Lett. B **666**, 430 (2008)
177. Lecouey, J.-L. et al.: Phys. Lett. B **672**, 6 (2009)
178. Spyrou, A. et al.: Phys. Lett. B **683**, 129 (2010)
179. Aumann, T., Chulkov, L.V., Pribora, V.N., Smedberg, M.H.: Nucl. Phys. A **640**, 24 (1998)
180. Simon, H. et al.: Phys. Rev. Lett. **83**, 496 (1999)
181. Hansen, P.G., Tostevin, J.A.: Annu. Rev. Nucl. Part. Sci. **53**, 219 (2003)
182. Kondo, Y. et al.: Phys. Rev. C **79**, 014602 (2009)
183. Kamimura, M. et al.: Prog. Theor. Phys. Suppl. **89**, 1 (1986)
184. Elekes, Z. et al.: Phys. Lett. B **614**, 174 (2005)
185. Hencken, K., Bertsch, G.F., Esbensen, H.: Phys. Rev. C **54**, 3043 (1996)
186. Shimoura, S. et al.: Phys. Lett. B **654**, 87 (2007)
187. Bertsch, G.F, Hencken, K., Esbensen, H.: Phys. Rev. C **57**, 1366 (1998)
188. Aoi, N. et al.: Phys. Rev. C **66**, 014301 (2002)
189. Ajzenberg-Selove, F.: Nucl. Phys. A **506**, 1 (1990)
190. Ajzenberg-Selove, F.: Nucl. Phys. A **523**, 1 (1991)
191. Otsuka, T. et al.: Phys. Rev. Lett. **87**, 082502 (2001)
192. Hamamoto, I., Shimoura, S.: J. Phys. G:Nucl Part. Phys. **34**, 2715 (2007)
193. Hamamoto, I.: Phys. Rev. C **76**, 054319 (2007)

Chapter 3
Breakup Reaction Models for Two- and Three-Cluster Projectiles

D. Baye and P. Capel

Abstract Breakup reactions are one of the main tools for the study of exotic nuclei, and in particular of their continuum. In order to get valuable information from measurements, a precise reaction model coupled to a fair description of the projectile is needed. We assume that the projectile initially possesses a cluster structure, which is revealed by the dissociation process. This structure is described by a few-body Hamiltonian involving effective forces between the clusters. Within this assumption, we review various reaction models. In semiclassical models, the projectile-target relative motion is described by a classical trajectory and the reaction properties are deduced by solving a time-dependent Schrödinger equation. We then describe the principle and variants of the eikonal approximation: the dynamical eikonal approximation, the standard eikonal approximation, and a corrected version avoiding Coulomb divergence. Finally, we present the continuum-discretized coupled-channel method (CDCC), in which the Schrödinger equation is solved with the projectile continuum approximated by square-integrable states. These models are first illustrated by applications to two-cluster projectiles for studies of nuclei far from stability and of reactions useful in astrophysics. Recent extensions to three-cluster projectiles,

D. Baye (✉)
Physique Quantique C.P. 165/82 and
Physique Nucléaire Théorique et Physique Mathématique C.P. 229,
Université Libre de Bruxelles (ULB), 1050 Brussels, Belgium
e-mail: dbaye@ulb.ac.be

P. Capel
National Superconducting Cyclotron Laboratory,
Michigan State University,
East Lansing, MI 48824, USA

P. Capel
Physique Quantique C.P. 165/82,
Université Libre de Bruxelles (ULB), 1050 Brussels, Belgium
e-mail: pierre.capel@ulb.ac.be

C. Beck (ed.), *Clusters in Nuclei, Vol.2*, Lecture Notes in Physics 848,
DOI: 10.1007/978-3-642-24707-1_3, © Springer-Verlag Berlin Heidelberg 2012

like two-neutron halo nuclei, are then presented and discussed. We end this review with some views of the future in breakup-reaction theory.

3.1 Introduction

The advent of radioactive ion beams has opened a new era in nuclear physics by providing the possibility of studying nuclei far from stability. In particular the availability of these beams favoured the discovery of halo nuclei [1]. Due to the very short lifetime of exotic nuclei, this study cannot be performed through usual spectroscopic techniques and one must resort to indirect methods. Breakup is one of these methods. In this reaction, the projectile under analysis dissociates into more elementary components through its interaction with a target. Many such experiments have been performed with the hope to probe exotic nuclear structures far from stability [2–4].

In order to get valuable information from breakup measurements, one must have not only a fair description of the projectile, but also an accurate reaction model. At present, a fully microscopic description of the reaction is computationally unfeasible. Simplifying assumptions are necessary. First, we will discuss only elastic breakup, i.e. a dissociation process leaving the target unchanged in its ground state. Other channels are simulated through the use of optical potentials. Second, we assume a cluster structure for the projectile. The projectile ground state is assumed to be a bound state of the clusters appearing during the breakup reaction. The bound and continuum states of the projectile are thus described by a few-body Hamiltonian involving effective forces between the constituent clusters. Theoretical reaction models are therefore based on this cluster description of the projectile and effective cluster-cluster and cluster-target interactions.

Even within these simplifying model assumptions, a direct resolution of the resulting few-body Schrödinger equation is still not possible in most cases. In this article, we thus review various approximations that have been developed up to now.

We begin with the models based on the semiclassical approximation [5] in which the projectile-target relative motion is described by a classical trajectory. This approximation is valid at high energies. It leads to the resolution of a time-dependent Schrödinger equation. Initially, the time-dependent equation was solved at the first order of the perturbation theory [5]. Then, as computers became more powerful, it could be solved numerically [6–11]. We present both versions indicating their respective advantages and drawbacks.

We then describe the eikonal approximation [12] and its variants. The principle is to calculate the deviations from a plane-wave motion which are assumed to be weak at high energy. By comparison with the semiclassical model, it is possible to derive the dynamical eikonal approximation (DEA) that combines the advantages of both models [13, 14]. The standard eikonal approximation is obtained by making the additional adiabatic or sudden approximation, which neglects the excitation energies of the projectile. With this stronger simplifying assumption, the final state only differs from the initial bound state by a phase factor. This approach is mostly used to model reactions on light targets at intermediate and high energies. Its drawback is that the Coulomb interaction leads to a divergence of breakup cross sections at forward angles. This problem can be solved using a first-order correction of the Coulomb part within the eikonal treatment. A satisfactory approximation of the DEA can then be derived [15, 16]: the Coulomb-corrected eikonal approximation (CCE), which remains valid for breakup on heavy targets. It reproduces most of the results of the DEA, although its computational time is significantly lower [17] which is important for the study of the breakup of three-cluster projectiles.

Finally, we present the continuum-discretized coupled-channel method (CDCC) [18, 19], in which the full projectile-target Schrödinger equation is solved approximately, by representing the continuum of the projectile with square-integrable states. This model leads to the numerical resolution of coupled-channel equations, and is suited for low- as high-energy reactions.

All the aforementioned models have been developed initially for two-body projectiles. However, the physics of three-cluster systems, like two-neutron halo nuclei, is the focus of many experimental studies and must also be investigated with these models. We review here the various efforts that have been made in the past few years to extend breakup models to three-cluster projectiles [20–22].

In Sect. 3.2, we specify the general theoretical framework within which the projectile is described. The semiclassical model and approximate resolutions of the time-dependent Schrödinger equation are described in Sect. 3.3. Section 3.4 presents the eikonal approximation as well as the related DEA and CCE models. Next, in Sect. 3.5, the CDCC method is developed. In Sect. 3.6, we review applications of breakup reactions to two-body projectiles. In particular, we emphasize the use of breakup to study nuclei far from stability and as an indirect way to infer cross sections of reactions of astrophysical interest. Section. 3.7 details the recent efforts made to extend various reaction models to three-body projectiles. We end this review by presenting some views of the future in breakup-reaction theory.

3.2 Projectile and Reaction Models

We consider the reaction of a projectile P of mass m_P and charge $Z_P e$ impinging on a target T of mass m_T and charge $Z_T e$. The projectile is assumed to exhibit a structure made of N clusters with masses m_i and charges $Z_i e$ ($m_P = \sum_i m_i$ and $Z_P = \sum_i Z_i$). Its internal properties are described by a Hamiltonian H_0, depending on a set of $N - 1$ internal coordinates collectively represented by notation ξ. With the aim of preserving the generality of the presentation of the reaction models, we do not specify here the expression of H_0. Details are given in Sects. 3.6 and 3.7, where applications for the breakup of two- and three-body projectiles are presented.

The states of the projectile are thus described by the eigenstates of H_0. For total angular momentum J and projection M, they are defined by

$$H_0 \phi_\tau^{JM}(E, \xi) = E \phi_\tau^{JM}(E, \xi), \qquad (3.2.1)$$

where E is the energy in the projectile centre-of-mass (c.m.) rest frame with respect to the dissociation threshold into N clusters. Index τ symbolically represents the set of all additional quantum numbers that depend on the projectile structure, like spins and relative orbital momenta of the clusters. Its precise definition depends on the number of clusters and on the model selected when defining H_0. We assume these numbers to be discrete, though some may be continuous in some representations when there are more than two clusters. To simplify the notation, the parity π of the eigenstates of H_0 is understood. In the following, any sum over J implicitly includes a sum over parity.

The negative-energy solutions of Eq. (3.2.1) correspond to the bound states of the projectile. They are normed to unity. The positive-energy states describe the broken-up projectile with full account of the interactions between the clusters. They are orthogonal and normed according to $\langle \phi_{\tau'}^{JM}(E', \xi) | \phi_\tau^{JM}(E, \xi) \rangle = \delta(E - E') \delta_{\tau \tau'}$. To describe final states when evaluating breakup cross sections, we also consider the incoming scattering states $\phi_{\hat{k}_\xi}^{(-)}$. They correspond to positive-energy states of H_0 describing the N clusters moving away from each other in the projectile c.m. frame with specific asymptotic momenta and spin projections. These momenta are not independent, since the sum of the asymptotic kinetic energies of the clusters is the positive energy E. However, within that condition, their directions and, if $N > 2$, their norms can vary. By $\hat{k}_\xi$, we symbolically denote these directions and wave numbers, as well as the projections of the spins of the clusters. These incoming scattering states are thus solutions of the Schrödinger equation

$$H_0 \phi_{\hat{k}_\xi}^{(-)}(E, \xi) = E \phi_{\hat{k}_\xi}^{(-)}(E, \xi). \tag{3.2.2}$$

They can be expanded into a linear combination of the eigenstates ϕ_τ^{JM} of Eq. (3.2.1) with the same energy as

$$\phi_{\hat{k}_\xi}^{(-)}(E, \xi) = \sum_{JM\tau} a_\tau^{JM}(\hat{k}_\xi) \phi_\tau^{JM}(E, \xi), \tag{3.2.3}$$

where the coefficients a_τ^{JM} depend on the projectile structure. These scattering states are normed following $\langle \phi_{\hat{k}_\xi'}^{(-)}(E', \xi) | \phi_{\hat{k}_\xi}^{(-)}(E, \xi) \rangle = \delta(E - E') \delta(\hat{k}_\xi - \hat{k}_\xi')$.

The interactions between the projectile constituents and the target are usually simulated by optical potentials chosen in the literature or obtained by a folding procedure. Within this framework the description of the reaction reduces to the resolution of an $(N + 1)$-body Schrödinger equation

$$\left[\frac{P^2}{2\mu} + H_0 + V_{PT}(\xi, \mathbf{R}) \right] \Psi(\xi, \mathbf{R}) = E_T \Psi(\xi, \mathbf{R}), \tag{3.2.4}$$

where $\mathbf{R} = (R, \Omega_R) = (R, \theta_R, \varphi_R)$ is the coordinate of the projectile centre of mass relative to the target, $\mathbf{P}$ is the corresponding momentum, $\mu = m_P m_T / (m_P + m_T)$ is the projectile-target reduced mass, and E_T is the total energy in the projectile-target c.m. frame. The projectile-target interaction V_{PT} is expressed as the sum of the optical potentials (including Coulomb) that simulate the interactions between the projectile constituents and the target,

$$V_{PT}(\xi, \mathbf{R}) = \sum_{i=1}^{N} V_{iT}(R_{iT}),$$ (3.2.5)

where $\mathbf{R}_{iT}$ is the relative coordinate of the projectile cluster i with respect to the target.

The projectile being initially bound in the state $\phi_{\tau_0}^{J_0 M_0}$ of negative energy E_0, we look for solutions of Eq. (3.2.4) with an incoming part behaving asymptotically as

$$\Psi(\xi, \mathbf{R}) \xrightarrow[Z \to -\infty]{} e^{i\{KZ + \eta \ln[K(R-Z)]\}} \phi_{\tau_0}^{J_0 M_0}(E_0, \xi),$$ (3.2.6)

where Z is the component of $\mathbf{R}$ in the incident-beam direction. The wavenumber K of the projectile-target relative motion is related to the total energy E_T by

$$E_T = \frac{\hbar^2 K^2}{2\mu} + E_0.$$ (3.2.7)

The P-T Sommerfeld parameter is defined as

$$\eta = Z_P Z_T e^2 / \hbar v,$$ (3.2.8)

where $v = \hbar K / \mu$ is the initial P-T relative velocity.

A first idea that may come to mind is to solve Eq. (3.2.4) exactly, e.g., within the Faddeev framework or its extensions. However, the infinite range of the Coulomb interaction between the projectile and the target renders the standard equations ill-defined. Only recently significant progress has been made. For example, in Refs. [23, 24], this problem is tackled by using an appropriate screening of the Coulomb force. This technique has been used to successfully describe the elastic scattering and breakup of the deuteron on various targets. However, it has long been limited to light targets (see [25] for a recent extension to a heavier target). To obtain a model that is valid for all types of target, one must still resort to approximations in the resolution of Eq. (3.2.4). These approximations are made in the treatment of the projectile-target relative motion, like in the semiclassical (Sect. 3.3) or eikonal (Sect. 3.4) approximations, or by using a discretized continuum, like in the CDCC method (Sect. 3.5)

3.3 Semiclassical Approximation

3.3.1 Time-Dependent Schrödinger Equation

The semiclassical approximation relies on the hypothesis that the projectile-target relative motion can be efficiently described by a classical trajectory $\mathbf{R}(t)$ [5]. It is thus valid when the de Broglie wavelength is small with respect to the impact parameter b characterizing the trajectory, $Kb \gg 1$, i.e. when the energy is large enough. Along that trajectory, the projectile experiences a time-dependent potential V that simulates the Coulomb and nuclear fields of the target. The internal structure of the projectile, on the contrary, is described quantum-mechanically by the Hamiltonian H_0. This semiclassical approximation leads to the resolution of the time-dependent equation

$$i\hbar \frac{\partial}{\partial t} \Psi(\xi, \mathbf{b}, t) = [H_0 + V(\xi, t)] \Psi(\xi, \mathbf{b}, t). \tag{3.3.1}$$

The time-dependent potential is obtained from the difference between the projectile-target interaction V_{PT} (3.2.5) and the potential V_{traj} that defines the classical trajectory

$$V(\xi, t) = V_{PT}[\xi, \mathbf{R}(t)] - V_{\text{traj}}[R(t)]. \tag{3.3.2}$$

The potential V_{traj} acts as a P-T scattering potential that bends the trajectory, but does not affect the projectile internal structure. Its interest lies in the fact that V decreases faster than V_{PT}. Its effect amounts to changing the phase of the wave function. Usually it is chosen to be the Coulomb potential between the projectile centre of mass and the target, but it may include a nuclear component. At sufficiently high energy, the trajectory is often approximated by a straight line.

For each impact parameter b, Eq. (3.3.1) has to be solved with the initial condition that the projectile is in its ground state,

$$\Psi^{(M_0)}(\xi, \mathbf{b}, t) \xrightarrow[t \to -\infty]{} \phi_{\tau_0}^{J_0 M_0}(E_0, \xi). \tag{3.3.3}$$

For each trajectory, the time-dependent wave function $\Psi^{(M_0)}$ must be calculated for the different possible values of M_0.

3.3.2 Cross Sections

From the output of the resolution of Eq. (3.3.1), the probability of being in a definite state of the projectile can be obtained by projecting the final wave function onto the corresponding eigenstate of H_0. One can for example compute the elastic scattering probability

$$P_{\text{el}}(b) = \frac{1}{2J_0 + 1} \sum_{M_0} \sum_{M_0'} \left| \left\langle \phi_{\tau_0}^{J_0 M_0'}(E_0, \xi) \middle| \Psi^{(M_0)}(\xi, \mathbf{b}, t \to +\infty) \right\rangle \right|^2. \tag{3.3.4}$$

This probability depends only on the norm of the impact parameter b because the time-dependent wave function $\Psi^{(M_0)}$ depends on the orientation of $\mathbf{b}$, i.e. on the azimuthal angle φ_R, only through a phase that cancels out in the calculation of P_{el}.

From this probability, the cross section for the elastic scattering in direction Ω is obtained as

$$\frac{d\sigma_{\text{el}}}{d\Omega} = \frac{d\sigma_{\text{el}}^{\text{traj}}}{d\Omega} P_{\text{el}}[b(\Omega)], \tag{3.3.5}$$

where $b(\Omega)$ is given by the classical relation between the scattering angle and the impact parameter derived from potential V_{traj}. The factor $d\sigma_{\text{el}}^{\text{traj}}/d\Omega$ is the elastic scattering cross section obtained from V_{traj}. In most cases $d\sigma_{\text{el}}^{\text{traj}}/d\Omega$ is generated from the Coulomb interaction and is thus the P-T Rutherford cross section.

Likewise, a general breakup probability density can be computed by projecting the final wave function onto the ingoing scattering states of H_0,

$$\frac{dP_{\text{bu}}}{d\hat{k}_\xi\, dE}(b) = \frac{1}{2J_0 + 1} \sum_{M_0} \left| \left\langle \phi_{k_\xi}^{(-)}(E, \xi) \middle| \Psi^{(M_0)}(\xi, \mathbf{b}, t \to +\infty) \right\rangle \right|^2. \tag{3.3.6}$$

After integration and summation over $\hat{k}_\xi$, the breakup probability per unit energy reads

$$\frac{dP_{\text{bu}}}{dE}(b) = \frac{1}{2J_0 + 1} \sum_{M_0} \sum_{JM\tau} \left| \left\langle \phi_\tau^{JM}(E, \xi) \middle| \Psi^{(M_0)}(\xi, \mathbf{b}, t \to +\infty) \right\rangle \right|^2. \tag{3.3.7}$$

Similarly to Eq. (3.3.5), a differential cross section for the breakup of the projectile is given by

$$\frac{d\sigma_{\text{bu}}}{dE\, d\Omega} = \frac{d\sigma_{\text{el}}^{\text{traj}}}{d\Omega} \frac{dP_{\text{bu}}}{dE}[b(\Omega)]. \tag{3.3.8}$$

The breakup cross section can then be obtained by summing the breakup probability over all impact parameters

$$\frac{d\sigma_{\text{bu}}}{dE} = 2\pi \int_0^\infty \frac{dP_{\text{bu}}}{dE}(b)b\, db. \tag{3.3.9}$$

Because of the trajectory hypothesis of the semiclassical approximation, the impact parameter b is a classical variable. Therefore, no interference between the different trajectories can appear. This is the major disadvantage of that technique since quantal interferences can play a significant role in reactions, in particular in those which are nuclear dominated.

3.3.3 Resolution at the First Order of the Perturbation Theory

In the early years of the semiclassical approximations, Eq. (3.3.1) was solved at the first order of the perturbation theory [5]. This technique, due to Alder and Winther, was applied to analyze the first Coulomb-breakup experiments of halo nuclei [26].

The time-dependent wave function $\Psi^{(M_0)}$ is expanded upon the basis of eigenstates of H_0 in Eq. (3.2.1). At the first order of the perturbation theory, the resulting equation is solved by considering that V is small. With the initial condition (3.3.3), the wave function at first order is given by [5, 27]

$$e^{\frac{i}{\hbar}H_0 t}\Psi^{(M_0)}(\xi, \mathbf{b}, t) = \left[1 + \frac{1}{i\hbar}\int_{-\infty}^{t} e^{\frac{i}{\hbar}H_0 t'} V(\xi, t')e^{-\frac{i}{\hbar}H_0 t'} dt'\right] \phi_{\tau_0}^{J_0 M_0}(E_0, \xi).$$

(3.3.10)

Following Eq. (3.3.6), the general breakup probability density reads

$$\frac{dP_{bu}}{d\hat{k}_\xi dE}(b) = \frac{\hbar^{-2}}{2J_0 + 1}\sum_{M_0}\left|\int_{-\infty}^{+\infty} e^{i\omega t}\left\langle\phi_{\hat{k}_\xi}^{(-)}(E, \xi)\middle| V(\xi, t)\middle|\phi_{\tau_0}^{J_0 M_0}(E_0, \xi)\right\rangle dt\right|^2,$$

(3.3.11)

where $\omega = (E - E_0)/\hbar$. The breakup probability per energy unit reads

$$\frac{dP_{bu}}{dE}(b) = \frac{\hbar^{-2}}{2J_0 + 1}\sum_{M_0}\sum_{JM\tau}\left|\int_{-\infty}^{+\infty} e^{i\omega t}\left\langle\phi_\tau^{JM}(E, \xi)\middle| V(\xi, t)\middle|\phi_{\tau_0}^{J_0 M_0}(E_0, \xi)\right\rangle dt\right|^2.$$

(3.3.12)

With Eq. (3.3.10), exact expressions can be calculated when considering a purely Coulomb P-T interaction for straight-line trajectories in the far-field approximation [28], i.e. by assuming that the charge densities of the projectile and target do not overlap during the collision. One obtains

$$\left\langle\phi_\tau^{JM}(E, \xi)\middle|\Psi^{(M_0)}(\xi, \mathbf{b}, t \to +\infty)\right\rangle =$$

$$Z_T e\frac{e^{-iEt/\hbar}}{i\hbar}\sum_{\lambda\mu}\frac{4\pi}{2\lambda + 1}I_{\lambda\mu}(\omega, b)\left\langle\phi_\tau^{JM}(E, \xi)\middle|\mathscr{M}_\mu^{E\lambda}(\xi)\middle|\phi_{\tau_0}^{J_0 M_0}(E_0, \xi)\right\rangle, \quad (3.3.13)$$

where $\mathscr{M}_\mu^{E\lambda}$ are the electric multipoles operators of rank λ, and $I_{\lambda\mu}$ are time integrals (see, e.g., Eq. (13) of [29]) that can be evaluated analytically as [28]

$$I_{\lambda\mu}(\omega, b) = \sqrt{\frac{2\lambda + 1}{\pi}}\frac{1}{v}\frac{i^{\lambda+\mu}}{\sqrt{(\lambda + \mu)!(\lambda - \mu)!}}\left(-\frac{\omega}{v}\right)^\lambda K_{|\mu|}\left(\frac{\omega b}{v}\right),$$

(3.3.14)

where K_n is a modified Bessel function [30].

If only the dominant dipole term E1 of the interaction is considered, the breakup probability (3.3.12) reads [31]

$$\frac{dP_{bu}^{E1}}{dE}(b) = \frac{16\pi}{9}\left(\frac{Z_T e}{\hbar v}\right)^2$$

$$\times \left(\frac{\omega}{v}\right)^2\left[K_1^2\left(\frac{\omega b}{v}\right) + K_0^2\left(\frac{\omega b}{v}\right)\right]\frac{dB(E1)}{dE}.$$

(3.3.15)

The last factor is the dipole strength function per energy unit [31],

$$
\frac{dB(\text{E1})}{dE} = \frac{1}{2J_0 + 1} \sum_{\mu M_0} \sum \int d\hat{k}_\xi \left| \left\langle \phi_{\hat{k}_\xi}^{(-)}(E, \xi) \left| \mathscr{M}_\mu^{\text{E1}}(\xi) \right| \phi_{\tau_0}^{J_0 M_0}(E_0, \xi) \right\rangle \right|^2
$$

$$
= \frac{1}{2J_0 + 1} \sum_{\mu M_0} \sum_{JM\tau} \left| \left\langle \phi_\tau^{JM}(E, \xi) \left| \mathscr{M}_\mu^{\text{E1}}(\xi) \right| \phi_{\tau_0}^{J_0 M_0}(E_0, \xi) \right\rangle \right|^2.
$$

(3.3.16)

Since modified Bessel functions decrease exponentially, the asymptotic behaviour of $dP_{\text{bu}}^{\text{E1}}/dE$ for $b \to \infty$ is proportional to $\exp(-2\omega b/v)$.

In the case of a purely Coulomb P-T interaction, the first order of the perturbation theory exhibits many appealing aspects. First, it can be solved analytically. Second, the dynamics part ($I_{\lambda\mu}$) and structure part (matrix elements of $\mathscr{M}_\mu^{\text{E}\lambda}$) are separated in the expression of the breakup amplitudes (3.3.13). This first-order approximation has therefore often been used to analyze Coulomb-breakup experiments by assuming pure E1 breakup (see Ref.[4]). However, as will be seen later, higher-order and nuclear-interaction effects are usually not negligible, and a proper analysis of experimental data requires a more sophisticated approximation.

3.3.4 Numerical Resolution

The time-dependent Schrödinger equation can also be solved numerically. Various groups have developed algorithms for that purpose [6–11, 32, 33]. They make use of an approximation of the evolution operator U applied iteratively to the initial bound state wave function following the scheme

$$
\Psi^{(M_0)}(\xi, \mathbf{b}, t + \Delta t) = U(t + \Delta t, t)\Psi^{(M_0)}(\xi, \mathbf{b}, t).
$$

(3.3.17)

Although higher-order algorithms exist (see, e.g., [34]), all practical calculations are performed with second-order approximations of U. Various expressions of this approximation exist, depending mainly on the way of representing the time-dependent projectile wave function. However they are in general similar to [11]

$$
U(t + \Delta t) = e^{-i\frac{\Delta t}{2\hbar} V(\xi, t + \Delta t)} e^{-i\frac{\Delta t}{\hbar} H_0} e^{-i\frac{\Delta t}{2\hbar} V(\xi, t)} + \mathcal{O}(\Delta t^3).
$$

(3.3.18)

With this expression, the time-dependent potential can be treated separately from the time-independent Hamiltonian H_0, which greatly simplifies the calculation of the time evolution when the wave functions are discretized on a mesh [11].

The significant advantage of this technique over the first order of perturbation is that it naturally includes higher-order effects. Moreover, the nuclear interaction between the projectile and the target can be easily added in the numerical scheme [35]. However, the dynamical and structure evolutions being now more deeply entangled, the analysis of the numerical resolution of the Schrödinger equation is less straightforward than in the first-order approximation. The numerical technique is also much

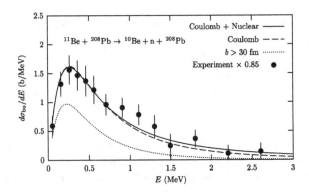

Fig. 3.1 Breakup cross section of ^{11}Be on Pb at 68 MeV/nucleon as a function of the relative energy E between the ^{10}Be core and the neutron. Calculations are performed within the semiclassical approximation with or without nuclear interaction [11]. Experimental data [26] are scaled by 0.85 [36]. Reprinted figure with permission from [11]. Copyright (2003) by the American Physical Society

more time-consuming than the perturbation one. The first order of the perturbation theory therefore remains a useful tool to qualitatively analyze calculations of Coulomb-dominated reactions performed with more elaborate models. Moreover, as will be seen in Sect. 3.4.4, it can be used to correct a divergence in the treatment of the Coulomb interaction within the eikonal description of breakup reactions.

Figure 3.1 illustrates the numerical resolution of the time-dependent Schrödinger equation for the Coulomb breakup of ^{11}Be on lead at 68 MeV/nucleon [11]. It shows the breakup cross section as a function of the relative energy E between the ^{10}Be core and the halo neutron after dissociation. The full line corresponds to the calculation with both Coulomb and nuclear P-T interactions. The dashed line is the result for a purely Coulomb potential, in which the nuclear interaction is simulated by an impact parameter cutoff at $b_{min} = 13$ fm. A calculation performed with an impact parameter cutoff at $b_{min} = 30$ fm simulating a forward-angle cut is plotted as a dotted line. The experimental data from [26] are multiplied by a factor of 0.85 as suggested in [36] after a remeasurement.

This example shows the validity of the semiclassical approximation to describe breakup observables in the projectile c.m. frame for collisions at intermediate energies. It also confirms that for heavy targets the reaction is strongly dominated by the Coulomb interaction. The inclusion of optical potentials to simulate the nuclear P-T interactions indeed only slightly increases the breakup cross section at large energy E. This shows that Coulomb-breakup calculations are not very sensitive to the uncertainty related to the choice of the optical potentials. Nevertheless, since optical potentials can be very easily included in the numerical resolution of the time-dependent Schrödinger equation, they should be used so as to avoid the imprecise impact-parameter cutoff necessary in purely Coulomb calculations.

3.4 Eikonal Approximations

3.4.1 Dynamical Eikonal Approximation

Let us now turn to a purely quantal treatment providing approximate solutions of the Schrödinger equation (3.2.4). At sufficiently high energy, the projectile is only slightly deflected by the target. The dominant dependence of the $(N + 1)$-body wave function Ψ on the projectile-target coordinate $\mathbf{R}$ is therefore in the plane wave contributing to the incident relative motion (3.2.6). The main idea of the eikonal approximation is to factorize that plane wave out of the wave function to define a new function $\widehat{\Psi}$ whose variation with $\mathbf{R}$ is expected to be small [12, 31, 37]

$$\Psi(\xi, \mathbf{R}) = e^{iKZ} \widehat{\Psi}(\xi, \mathbf{R}). \qquad (3.4.1)$$

With factorization (3.4.1) and energy conservation (3.2.7), the Schrödinger equation (3.2.4) becomes

$$\left[\frac{P^2}{2\mu} + vP_Z + H_0 - E_0 + V_{PT}(\xi, \mathbf{R}) \right] \widehat{\Psi}(\xi, \mathbf{R}) = 0, \qquad (3.4.2)$$

where the relative velocity v between projectile and target is assumed to be large.

The first step in the eikonal approximation is to assume the second-order derivative $P^2/2\mu$ negligible with respect to the first-order derivative vP_Z,

$$\frac{P^2}{2\mu} \widehat{\Psi}(\xi, \mathbf{R}) \ll vP_Z \widehat{\Psi}(\xi, \mathbf{R}). \qquad (3.4.3)$$

This first step leads to the second-order equation (but now first-order in Z),

$$i\hbar v \frac{\partial}{\partial Z} \widehat{\Psi}(\xi, \mathbf{b}, Z) = [H_0 - E_0 + V_{PT}(\xi, \mathbf{R})] \widehat{\Psi}(\xi, \mathbf{b}, Z), \qquad (3.4.4)$$

where the dependence of the wave function on the longitudinal Z and transverse $\mathbf{b}$ parts of the projectile-target coordinate $\mathbf{R}$ has been made explicit. This equation is mathematically equivalent to the time-dependent Schrödinger equation (3.3.1) for straight-line trajectories with t replaced by Z/v. It can thus be solved using any of the algorithms cited in Sect. 3.3.4. However, contrary to time-dependent models, it is obtained without the semiclassical approximation. The projectile-target coordinate components $\mathbf{b}$ and Z are thus quantal variables. Interferences between solutions obtained at different b values are thus taken here into account. This first step is known as the dynamical eikonal approximation (DEA) [13, 14].

3.4.2 Cross Sections

The transition matrix element for elastic scattering into direction $\Omega = (\theta, \varphi)$ of the final momentum $\mathbf{K} = (K, \Omega)$ of the projectile in the c.m. frame reads [38]

$$T_{fi} = \left\langle e^{i\mathbf{K}\cdot\mathbf{R}} \phi_{\tau_0}^{J_0 M_0'}(E_0, \xi) \middle| V_{PT}(\xi, \mathbf{R}) \middle| \Psi^{(M_0)}(\xi, \mathbf{R}) \right\rangle, \qquad (3.4.5)$$

where $\Psi^{(M_0)}$ is the exact solution of the Schrödinger equation (3.2.4) with the asymptotic condition (3.2.6). By using Eqs. (3.4.1), (3.2.1), and (3.4.4), one obtains the approximation [13]

$$T_{fi} = \left\langle e^{i\mathbf{K}\cdot\mathbf{R}} \phi_{\tau_0}^{J_0 M_0'}(E_0, \xi) \middle| e^{iKZ} [H_0 - E_0 + V_{PT}(\xi, \mathbf{R})] \middle| \widehat{\Psi}^{(M_0)}(\xi, \mathbf{R}) \right\rangle$$

$$\approx i\hbar v \int d\mathbf{R} e^{-i\mathbf{q}\cdot\mathbf{b}} \frac{\partial}{\partial Z} \left\langle \phi_{\tau_0}^{J_0 M_0'}(E_0, \xi) \middle| \widehat{\Psi}^{(M_0)}(\xi, \mathbf{R}) \right\rangle, \qquad (3.4.6)$$

where the transferred momentum $\mathbf{q} = \mathbf{K} - K\hat{\mathbf{Z}}$ is assumed to be purely transverse, i.e. $\exp[i(\mathbf{K}\cdot\hat{\mathbf{Z}} - K)]$, is neglected. The norm of q is linked to the scattering angle by

$$q = 2K \sin\theta/2. \qquad (3.4.7)$$

Let us define the elastic amplitude

$$S_{el, M_0'}^{(M_0)}(\mathbf{b}) = \left\langle \phi_{\tau_0}^{J_0 M_0'}(E_0, \xi) \middle| \widehat{\Psi}^{(M_0)}(\xi, \mathbf{b}, Z \to +\infty) \right\rangle - \delta_{M_0' M_0}. \qquad (3.4.8)$$

The transition matrix element (3.4.6) reads after integration over Z,

$$T_{fi} = i\hbar v \int d\mathbf{b} e^{-i\mathbf{q}\cdot\mathbf{b}} e^{i(M_0 - M_0')\varphi_R} S_{el, M_0'}^{(M_0)}(b\hat{\mathbf{X}}), \qquad (3.4.9)$$

where φ_R is the azimuthal angle characterizing $\mathbf{b}$. The phase factor $\exp[i(M_0 - M_0')\varphi_R]$ arises from the rotation of the wave functions when the orientation of $\mathbf{b}$ varies [14]. The integral over φ_R can be performed analytically, which leads to the following expression for the elastic differential cross section [14]

$$\frac{d\sigma_{el}}{d\Omega} = K^2 \frac{1}{2J_0 + 1} \sum_{M_0 M_0'} \left| \int_0^\infty b\,db\,J_{|M_0 - M_0'|}(qb) S_{el, M_0'}^{(M_0)}(b\hat{\mathbf{X}}) \right|^2, \qquad (3.4.10)$$

where J_m is a Bessel function [30]. From Eq. (3.4.10), one can see that contrary to the semiclassical approximation (3.3.5), the eikonal elastic cross section is obtained as a coherent sum of elastic amplitudes over all b values. This illustrates that quantum interferences are taken into account in the eikonal framework.

The transition matrix element for dissociation reads

$$T_{fi} = \left\langle e^{i\mathbf{K}'\cdot\mathbf{R}} \phi_{\hat{k}_\xi}^{(-)}(E, \xi) \middle| V_{PT}(\xi, \mathbf{R}) \middle| \Psi^{(M_0)}(\xi, \mathbf{R}) \right\rangle, \qquad (3.4.11)$$

where $\mathbf{K}' = (K', \Omega)$ is the final projectile-target wave vector. One can then proceed as for the elastic scattering. Using Eqs. (3.4.1), (3.2.2), and (3.4.4), taking into account the energy conservation,

$$\frac{\hbar^2 K^2}{2\mu} + E_0 = \frac{\hbar^2 K'^2}{2\mu} + E, \tag{3.4.12}$$

and assuming the transferred momentum $\mathbf{q} = \mathbf{K}' - K\hat{\mathbf{Z}}$ to be purely transverse, the transition matrix element is expressed as

$$T_{fi} \approx i\hbar v \int d\mathbf{b} e^{-i\mathbf{q}\cdot\mathbf{b}} S_{bu}^{(M_0)}(E, \hat{k}_\xi, \mathbf{b}), \tag{3.4.13}$$

with the breakup amplitude

$$S_{bu}^{(M_0)}(E, \hat{k}_\xi, \mathbf{b}) = \left\langle \phi_{\hat{k}_\xi}^{(-)}(E, \xi) \middle| \widehat{\Psi}^{(M_0)}(\xi, \mathbf{b}, Z \to +\infty) \right\rangle. \tag{3.4.14}$$

The differential cross section for breakup is given by

$$\frac{d\sigma}{d\hat{k}_\xi dE d\Omega} \propto \frac{1}{2J_0 + 1} \sum_{M_0} \left| \int d\mathbf{b} e^{-i\mathbf{q}\cdot\mathbf{b}} S_{bu}^{(M_0)}(E, \hat{k}_\xi, \mathbf{b}) \right|^2, \tag{3.4.15}$$

where the proportionality factor depends on the phase space. Like the elastic scattering cross section (3.4.10), it is obtained from a coherent sum of breakup amplitudes (3.4.14), confirming the quantum-mechanical character of the eikonal approximation. Here also, the integral over φ_R can be performed analytically and leads to Bessel functions [14].

By integrating expression (3.4.15) over unmeasured quantities, one can obtain the breakup cross sections with respect to the desired variables, like the internal excitation energy of the projectile. Since these operations depend on the projectile internal structure, we delay the presentation of some detailed expressions to Sects. 3.6 and 3.7 treating of two-body [14] and three-body [20] breakup.

3.4.3 Standard Eikonal Approximation

In most references, the concept of eikonal approximation involves a further simplification to the DEA [31, 39]. This adiabatic, or sudden, approximation consists in neglecting the excitation energy of the projectile compared to the incident kinetic energy. It comes down to assume the low-lying spectrum of the projectile to be degenerate with its ground state, i.e. to consider the internal coordinates of the projectile as frozen during the reaction [31]. This approximation therefore holds only for high-energy collisions that occur during a very brief time. This second assumption leads to neglect the term $H_0 - E_0$ in the DEA equation (3.4.4) which then reads

$$i\hbar v \frac{\partial}{\partial Z} \widehat{\Psi}(\xi, \mathbf{b}, Z) = V_{PT}(\xi, \mathbf{R})\widehat{\Psi}(\xi, \mathbf{b}, Z). \tag{3.4.16}$$

The solution of Eq. (3.4.16) that follows the asymptotic condition (3.2.6) exhibits the well-known eikonal form [12, 37]

$$\widehat{\Psi}^{(M_0)}(\xi, \mathbf{b}, Z) = \exp\left[-\frac{i}{\hbar v} \int_{-\infty}^{Z} V_{PT}(\xi, \mathbf{b}, Z')dZ'\right] \phi_{\tau_0}^{J_0 M_0}(E_0, \xi). \quad (3.4.17)$$

After the collision, the whole information about the change in the projectile wave function is thus contained in the phase shift

$$\chi(\mathbf{s}_\xi, \mathbf{b}) = -\frac{1}{\hbar v} \int_{-\infty}^{+\infty} V_{PT}(\xi, \mathbf{R})dZ. \quad (3.4.18)$$

Due to translation invariance, this eikonal phase χ depends only on the transverse components $\mathbf{b}$ of the projectile-target coordinate $\mathbf{R}$ and $\mathbf{s}_\xi$ of the projectile internal coordinates ξ. Cross sections within this standard eikonal approximation are obtained as explained in Sect. 3.4.2, replacing $\widehat{\Psi}^{(M_0)}$ by $e^{i\chi}\phi_{\tau_0}^{J_0 M_0}$.

Being obtained from the adiabatic approximation, expressions (3.4.17) and (3.4.18) are valid only for short-range potentials. For the Coulomb interaction, the assumption that the reaction takes place in a short time no longer holds, due to its infinite range. The adiabatic approximation thus fails for Coulomb-dominated reactions [31]. Besides imprecise uses of a cutoff at large impact parameters [40], there are two ways to avoid this problem. The first is not to make the adiabatic approximation, i.e. to resort to the more complicated DEA (see Sect. 3.4.1). The second is to correct the eikonal phase for the Coulomb interaction as suggested in [15] (see Sect. 3.4.4). Nevertheless, as shown in Ref. [14], the Coulomb divergence does not affect eikonal calculations performed on light targets at high enough energies. Most of the nuclear-dominated reactions can thus be analyzed within an eikonal model including the adiabatic approximation (see, e.g., Ref. [41]).

Figure 3.2 illustrates the difference between the DEA (full line), the usual eikonal approximation (dashed line) and the semiclassical approximation (dotted line) when Coulomb dominates. It shows the breakup cross section of ^{11}Be on Pb at 69 MeV/nucleon for a ^{10}Be-n relative energy of 0.3 MeV as a function of the P-T scattering angle. As explained above, the usual eikonal approximation diverges for the Coulomb-dominated breakup, i.e. at forward angles. The DEA, which does not include the adiabatic approximation, exhibits a regular behaviour at these angles. Interestingly, the semiclassical approximation follows the general behaviour of the DEA, except for the oscillations due to quantum interferences between different b values. The DEA has therefore the advantage of being valid for describing any breakup observable on both light and heavy targets.

The nuclei studied through breakup reactions being exotic, it may be difficult, if not impossible, to find optical potentials that describe the scattering of the clusters by the target. One way to circumvent that problem is to resort to what is usually known as the Glauber model [31, 37, 39, 43]. This model has been mostly used to calculate total and reaction cross sections. At the optical-limit approximation (OLA) of the Glauber model, correlations in the cluster and target wave functions are neglected. The nuclear component of the eikonal phase shift for cluster i is then expressed as a function of the densities ρ_T of the target and ρ_i of the cluster, and of a profile function $1 - e^{i\chi_{NN}}$ that corresponds to an effective nucleon-nucleon interaction. The nuclear component of the eikonal phase shift is approximated by [31]

Fig. 3.2 Breakup cross section of ^{11}Be on Pb at 69 MeV/nucleon as a function of the P-T scattering angle in the P-T c.m. frame for a ^{10}Be-n energy $E = 0.3$ MeV. Calculations are performed within the DEA, usual eikonal, and semiclassical approximations [42]

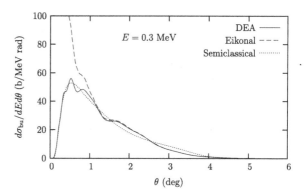

$$\chi_i^N(\mathbf{b}_i) = i \iint \rho_T(\mathbf{r}_T)\rho_i(\mathbf{r}_i)[1 - e^{i\chi_{NN}(|\mathbf{b}_i - \mathbf{s}_T + \mathbf{s}_i|)}]d\mathbf{r}_T d\mathbf{r}_i, \qquad (3.4.19)$$

where $\mathbf{s}_T$ and $\mathbf{s}_i$ are the transverse components of the internal coordinates $\mathbf{r}_T$ of the target and $\mathbf{r}_i$ of cluster i, respectively, and $\mathbf{b}_i$ is the transverse component of the c.m. coordinate of cluster i. The OLA is therefore equivalent to the double-folding of an effective nucleon-nucleon interaction. The density of the target can usually be obtained from experimental data. The cluster density being unknown, it has to be estimated from some structure model, like a mean-field calculation. The profile function is usually parametrized as [31, 44]

$$1 - e^{i\chi_{NN}(b)} = \frac{1 - i\alpha_{NN}}{4\pi\beta_{NN}}\sigma_{NN}^{tot} \exp\left(-\frac{b^2}{2\beta_{NN}}\right), \qquad (3.4.20)$$

where σ_{NN}^{tot} is the total cross section for the N-N collision, α_{NN} is the ratio of the real part to the imaginary part of the N-N scattering amplitude, and β_{NN} is the slope parameter of the N-N elastic differential cross section. These parameters depend on the nucleon type (p or n) and on the incident energy. Their values can be found in the literature (see, e.g., Ref. [44]). The validity of the Glauber approximation is discussed in Ref. [45].

3.4.4 Coulomb-Corrected Eikonal Approximation

The eikonal approximation gives excellent results for nuclear-dominated reactions [14, 31]. However, as mentioned above, it suffers from a divergence problem when the Coulomb interaction becomes significant. To explain this, let us divide the eikonal phase (3.4.18) into its Coulomb and nuclear contributions

$$\chi(\mathbf{s}_\xi, \mathbf{b}) = \chi_{PT}^C(b) + \chi^C(\mathbf{s}_\xi, \mathbf{b}) + \chi^N(\mathbf{s}_\xi, \mathbf{b}). \qquad (3.4.21)$$

In this expression, χ_{PT}^C is the global elastic Coulomb eikonal phase between the projectile and the target. However, Coulomb forces not only act globally on the projectile, they also induce 'tidal' effects due to their different actions on the various clusters. The tidal Coulomb phase χ^C is due to the difference between the cluster-target and projectile-target bare Coulomb interactions. The remaining phase χ^N contains effects of the nuclear forces as well as of differences between Coulomb forces taking the finite size of the clusters into account and the bare Coulomb forces.

At the eikonal approximation, the integral (3.4.18) defining χ^C_{PT} diverges and must be calculated with a cutoff [12, 31]. Up to an additional cutoff-dependent term that plays no role in the cross sections, it can be written as [37]

$$\chi^C_{PT}(b) = 2\eta \ln(Kb), \tag{3.4.22}$$

where appears the projectile-target Sommerfeld parameter η defined in Eq. (3.2.8). The phase (3.4.22) depends only on b.

The tidal Coulomb phase is computed with Eq. (3.4.18) for the difference between the bare Coulomb interactions for the clusters in the projectile and the global P-T Coulomb interaction,

$$\chi^C(s_\xi, \mathbf{b}) = -\frac{\eta}{Z_P} \int_{-\infty}^{+\infty} \left(\sum_{i=1}^{N} \frac{Z_i}{|\mathbf{R}_{iT}|} - \frac{Z_P}{|\mathbf{R}|} \right) dZ. \tag{3.4.23}$$

It can be expressed analytically. Because of the long range of the E1 component of the Coulomb force, this phase behaves as $1/b$ at large distances [14, 17]. In the calculation of the breakup cross sections (3.4.15), the integration over bdb diverges for small q values, i.e. at forward angles, because of the corresponding $1/b$ asymptotic behaviour of the breakup amplitude, as illustrated in Fig. 3.2. This divergence occurs only in the first-order term $i\chi^C$ of the expansion of the eikonal Coulomb amplitude $\exp(i\chi^C)$.

As seen in Sect. 3.3.3, the first order approximation (3.3.15) decreases exponentially at large b and hence does not display such a divergence. A plausible correction is therefore to replace the exponential of the eikonal phase according to [15, 16]

$$e^{i\chi} \rightarrow e^{i\chi^C_{PT}} \left(e^{i\chi^C} - i\chi^C + i\chi^{FO} \right) e^{i\chi^N}, \tag{3.4.24}$$

where χ^{FO} is the result of first-order perturbation theory (3.3.10),

$$\chi^{FO}(\xi, \mathbf{b}) = -\frac{\eta}{Z_P} \int_{-\infty}^{+\infty} e^{i\omega Z/v} \left(\sum_{i=1}^{N} \frac{Z_i}{|\mathbf{R}_{iT}|} - \frac{Z_P}{|\mathbf{R}|} \right) dZ. \tag{3.4.25}$$

Note that because of the phase $e^{i\omega Z/v}$, the integrand in Eq. (3.4.25) does not exhibit a translational invariance. The first-order phase χ^{FO} depends on all internal coordinates of the projectile. When the adiabatic approximation is applied to Eq. (3.4.25), i.e. when ω is set to 0, one recovers exactly the Coulomb eikonal phase (3.4.23). This suggests that without adiabatic approximation the first-order term of $\exp(i\chi^C)$ would

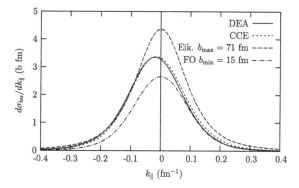

Fig. 3.3 Breakup of ^{11}Be on Pb at 69 MeV/nucleon. The parallel-momentum distribution between the ^{10}Be core and the halo neutron is computed within the DEA, the CCE, the eikonal approximation including the adiabatic approximation, and the first-order of the perturbation theory [17]. Reprinted figure with permission from Ref. [17]. Copyright (2008) by the American Physical Society

be $i\chi^{FO}$ (3.4.25) instead of $i\chi^{C}$ (3.4.23), intuitively validating the correction (3.4.24). Furthermore, since a simple analytic expression is available for each of the Coulomb multipoles (see Sect. 3.3.3), this correction is easy to implement.

With this Coulomb correction, the breakup of loosely-bound projectiles can be described within the eikonal approximation taking on (nearly) the same footing both Coulomb and nuclear interactions at all orders. This approximation has been tested and validated for a two-body projectile in Ref. [17]. Note that in all practical cases [16, 17, 20], only the dipole term of the first-order expansion (3.3.13) is retained to evaluate χ^{FO}.

Figure 3.3 illustrates the accuracy of the CCE for the breakup of ^{11}Be on lead at 69 MeV/nucleon [17]. The figure presents the parallel-momentum distribution between the ^{10}Be core and the halo neutron after dissociation. This observable has been computed within the DEA (full line), which serves as a reference calculation, the CCE (dotted line), the eikonal approximation including the adiabatic approximation (dashed line), and the first-order of the perturbation theory (dash-dotted line). The usual eikonal approximation requires a cutoff at large impact parameter to avoid divergence. The value $b_{max} = 71$ fm is chosen from the value prescribed in Ref. [40]. At the first order or the perturbation theory, the nuclear interaction is simulated by an impact parameter cutoff at $b_{min} = 15$ fm.

We first see that the magnitude of the CCE cross section is close to the DEA one, whereas, the other two approximations give too large (eikonal) or too small (first order) cross sections. Moreover the CCE reproduces nearly perfectly the shape of the DEA distribution. In particular the asymmetry, due to dynamical effects, is well reproduced. This result suggests that in addition to solving the Coulomb divergence problem introduced by the adiabatic approximation, the CCE also restores some dynamical and higher-order effects missing in its ingredients, the usual eikonal approximation and the first order of the perturbation theory.

3.5 Continuum-Discretized Coupled-Channel Method

The CDCC method is a fully quantal approximation which does not imply any restriction on energies. Its main interest lies in low energies where the previous methods are not valid. The principle of the CDCC method is to determine, as accurately as possible, the scattering and dissociation cross sections of a nucleus with a simplified treatment of the final projectile continuum states. To this end, these states describing the relative motions of the unbound fragments are approximately described by square-integrable wave functions at discrete energies. The relative motion between the projectile and target and various cross sections can then be obtained by solving a system of coupled-channel equations. The number of these equations and hence the difficulty of the numerical treatment increase with increasing energy.

The CDCC method was suggested by Rawitscher [46] and first applied to deuteron + nucleus elastic scattering and breakup reactions. It was then extensively developed and used by several groups [18, 47–52]. Its interest has been revived by the availability of radioactive beams of weakly bound nuclei dissociating into two [19, 48–52] or three [21, 22, 53, 54] fragments.

We assume that the breakup process leads to N clusters and that the cluster-target interactions do not depend on the target spin. The projectile wave functions $\phi_\tau^{JM}(E_{\tau B}^J, \xi)$ describing N-body bound states at negative energies $E_{\tau B}^J$ and $\phi_\tau^{JM}(E, \xi)$ describing N-body scattering states at positive energies E are defined with Eq. (3.2.1). Since the total angular momentum of the projectile-target system is a good quantum number, the first step consists in determining partial waves of the $(N + 1)$-body Hamiltonian (3.2.4). The general partial wave function for a total angular momentum J_T can be expanded over the projectile eigenstates as

$$\Psi^{J_T M_T}(\mathbf{R}, \xi) = \sum_{LJ\tau}\sum_{B} [\phi_\tau^J(E_{\tau B}^J, \xi) \otimes \psi_{J_\tau B}^L(\mathbf{R})]^{J_T M_T}$$

$$+ \sum_{LJ\tau} \int_0^\infty [\phi_\tau^J(E, \xi) \otimes \psi_{J_\tau E}^L(\mathbf{R})]^{J_T M_T} dE. \qquad (3.5.1)$$

In this expansion, index B runs over the bound states of the projectile. The total angular momentum J_T results from the coupling of the orbital momentum L of the projectile-target relative motion with the total angular momentum J of the projectile state. The relative-motion partial waves $\psi_{J_\tau B}^L$ and $\psi_{J_\tau E}^L$ are unknown and must be determined. The parity is given by the product of $(-1)^L$ and the parity of ϕ_τ^{JM}. The first term of Eq. (3.5.1) represents the elastic and inelastic channels while the second term is associated with the breakup contribution. However, the presence of the continuum renders this expression intractable.

The basic idea of the CDCC method is to replace wave function (3.5.1) by

$$\Psi^{J_T M_T}(\mathbf{R}, \xi) = \sum_{LJ\tau n} [\phi_{\tau n}^J(\xi) \otimes \psi_{J_\tau E}^L(\mathbf{R})]^{J_T M_T}, \qquad (3.5.2)$$

where the functions $\phi_{\tau n}^{JM}(\xi) \equiv \phi_\tau^{JM}(E_{\tau n}^J, \xi)$ represent either bound states $(E_{\tau B}^J < 0)$ or square-integrable approximations of continuum wave functions $(E_{\tau n}^J > 0)$ at discrete energies

$$E_{\tau n}^J = \left\langle \phi_{\tau n}^{JM}(\xi) \middle| H_0 \middle| \phi_{\tau n}^{JM}(\xi) \right\rangle. \tag{3.5.3}$$

Approximation (3.5.2) resembles usual coupled-channel expansions and can be treated in a similar way.

In practice, two methods are available to perform the continuum discretization. In the "pseudostate" approach, the Schrödinger equation (3.2.1) is solved approximately by diagonalizing the projectile Hamiltonian H_0 either within a finite basis of square-integrable functions or in a finite region of space. In both cases, square-integrable pseudostates $\phi_{\tau n}^{JM}$ are obtained. This approach is simple but there is little control on the obtained energies $E_{\tau n}^J$. Therefore, it is customary to keep only the pseudostates with energies below some limit $E_{\max}$.

The alternative is to separate the integral over E in (3.5.1) into a limited number of small intervals, or "bins", $[E_{n-1}, E_n]$ which may depend on J and to use in each of them some average of the exact scattering states in this range of energies [18, 46–48]. This "bin" method provides the square-integrable basis functions

$$\phi_{\tau n}^{JM}(\xi) = \frac{1}{W_n} \int_{E_{n-1}}^{E_n} \phi_\tau^{JM}(E, \xi) f_n(E) dE, \tag{3.5.4}$$

where the weight functions f_n may also depend on J. Such states are orthogonal because of the orthogonality of the scattering states and they are normed if W_n is the norm of f_n over $[E_{n-1}, E_n]$. Using Eq. (3.5.4), their energy (3.5.3) is given by

$$E_{\tau n}^J = \frac{1}{W_n^2} \int_{E_{n-1}}^{E_n} |f_n(E)|^2 E dE. \tag{3.5.5}$$

Here also, a maximum energy $E_{\max} \equiv E_{n_{\max}}$ is chosen. In practice, these basis states are usually constructed by averaging the scattering states $\tilde{\phi}_\tau^{JM}(k, \xi)$ normalized over the wave number k, often within equal momentum intervals [19].

The total wave function (3.5.2) can be rewritten as

$$\Psi^{J_T M_T}(\mathbf{R}, \xi) = R^{-1} \sum_c \Phi_c^{J_T M_T}(\Omega_R, \xi) u_c^{J_T}(R), \tag{3.5.6}$$

where c represents the channel $LJ\tau n$ and

$$\Phi_c^{J_T M_T}(\Omega_R, \xi) = i^L \left[\phi_{\tau n}^J(\xi) \otimes Y_L(\Omega_R) \right]^{J_T M_T}. \tag{3.5.7}$$

By inserting expansion (3.5.6) in the Schrödinger equation (3.2.4) and using Eq. (3.5.3), the relative wave functions $u_c^{J_T}$ are given by a set of coupled equations

$$
\left[-\frac{\hbar^2}{2\mu} \left(\frac{d^2}{dR^2} - \frac{L(L+1)}{R^2} \right) + E_c - E_T \right] u_c^{J_T}(R) + \sum_{c'} V_{c,c'}^{J_T}(R) u_{c'}^{J_T}(R) = 0,
$$

(3.5.8)

where $E_c \equiv E_{\tau n}^J$. The sum over L is truncated at some value L_{max}. The sum over the pseudo-states or bins is limited by the selected maximum energy E_{max}. The CDCC problem is therefore equivalent to a system of coupled equations where the potentials are given by

$$
V_{c,c'}^{J_T}(R) = \left\langle \Phi_c^{J_T M_T}(\Omega_R, \xi) \middle| V_{PT}(\mathbf{R}, \xi) \middle| \Phi_{c'}^{J_T M_T}(\Omega_R, \xi) \right\rangle.
$$

(3.5.9)

This matrix element involves a multidimensional integral over Ω_R and over the internal coordinates ξ. In general, the potentials are expanded into multipoles corresponding to the total angular momentum operator $\mathbf{J}_T$ of the system. This may allow an analytical treatment of angular integrals.

System (3.5.8) must be solved with the boundary condition for open channels

$$
u_c^{J_T}(R) \xrightarrow[R \to \infty]{} v_c^{-1/2} \left[I_c(K_c R) \delta_{cc_0} - O_c(K_c R) S_{cc_0}^{J_T} \right],
$$

(3.5.10)

where c_0 is the incoming channel. The asymptotic momentum in channel c reads

$$
K_c = \sqrt{2\mu(E_T - E_c)/\hbar^2},
$$

(3.5.11)

and $v_c = \hbar K_c / \mu$ is the corresponding velocity. In Eq. (3.5.10), $I_c = G_c - iF_c$ and $O_c = I_c^*$ are the incoming and outgoing Coulomb functions, respectively [30], and the element $S_{cc_0}^{J_T}$ of the collision matrix is the amplitude for populating channel c from initial channel c_0.

Various methods have been developed to solve system (3.5.8) (see, e.g., Ref. [55]). A convenient approach is the R-matrix formalism [56], which is both simple and accurate. The configuration space is divided into two regions: the internal $(R < a)$ and external $(R > a)$ regions, where a is the channel radius. In the external region, the potential matrix defined by Eq. (3.5.9) can be well approximated by its diagonal Coulomb asymptotic form. Hence the wave function is replaced by combinations of Coulomb functions. In the internal region, the radial wave functions $u_c^{J_T}$ can be expanded over some basis [56]. A significant simplification occurs when using Lagrange functions [52, 57, 58].

A scattering wave function verifying the initial condition (3.2.6) is then constructed with the different partial waves. Inserting this CDCC approximate wave function in Eq. (3.4.11) enables calculating transition matrix elements towards pseudostates or bin states as a function of the collision matrices S^{J_T} (see Eq. (5) of Ref. [19]). Since these transition matrix elements are obtained only at discrete energies $E_{n\tau}^J$, they must be interpolated in order to obtain breakup cross sections at all energies.

The CDCC method has first been applied to two-body projectiles. As an example, Fig. 3.4 shows the convergence of the breakup of ^{8}B on ^{58}Ni at 25.8 MeV. The convergence concerns the set of partial waves l in the ^{7}Be-p continuum of the projectile

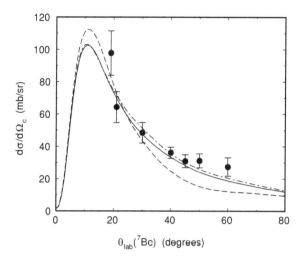

Fig. 3.4 ^{7}Be angular distribution after the breakup of ^{8}B on ^{58}Ni at 25.8 MeV computed within a CDCC model [19]. The convergence of the numerical scheme is illustrated with various maximum values of the ^{7}Be-p relative orbital momentum l in the continuum and various maximum values of multipole order λ of the potential expansion in Eq. (3.5.9): $l \le 3$, $\lambda \le 2$ (*dashed line*), $l \le 3$, $\lambda \le 3$ (*full line*), $l \le 4$, $\lambda \le 4$ (*dash-dotted line*). Experimental data from Ref. [63]. Reprinted figure with permission from Ref. [19]. Copyright (2001) by the American Physical Society

and the number of multipoles in the expansion of the potential appearing in matrix elements (3.5.9). The validity of CDCC has been tested for breakup observables in a comparison with three-body Faddeev calculations [25]. The agreement between both sets of results is good except when the coupling with the transfer channel is important.

Let us also mention extensions beyond the simple two-body model of the projectile allowing the core to be in an excited state [59, 60]. These references present total cross sections for the breakup on a ^{9}Be target of ^{11}Be into ^{10}Be + n and of ^{15}C into ^{14}C + n calculated by including core deformations. This extension of CDCC known as XCDCC leads to very long computational times.

The extension of CDCC to three-body projectiles is more recent [21, 22, 53, 54, 61]. The calculations are still much more time-consuming since the projectile wave functions are much more complicated (see Sect. 3.7). Consequently, the calculation of the potential matrix elements (3.5.9) raises important numerical difficulties. At present, converged calculations are mainly restricted to elastic scattering [53, 54, 61]. Most breakup calculations still involve limited bases and/or simplifying assumptions [21, 22] but these limitations can be overcome [62].

3.6 Breakup Reactions of Two-Body Projectiles

3.6.1 Two-Cluster Model

Most of the reaction models have been applied assuming a two-cluster structure of the projectile. In this section, we specify the expression of the internal Hamiltonian of the projectile and the set of coordinates usually considered in practical applications. We then illustrate the models presented in Sect. 3.2 and the approximations explained in Sects. 3.3, 3.4 and 3.5 with various applications to the study of exotic nuclei and nuclear astrophysics.

We consider here projectiles made up of a single fragment f of mass m_f and charge $Z_f e$, initially bound to a core c of mass m_c and charge $Z_c e$. The core and fragment are assumed to have spins s_c and s_f. The internal structure of these clusters and of the target is usually neglected although some structure effects can be simulated by the effective potentials.

Let us now particularize the general formalism (3.2.1)–(3.2.3) to the present case. The internal coordinates ξ represent the relative coordinate $\mathbf{r} = \mathbf{r}_f - \mathbf{r}_c$. The structure of the projectile is described by the two-body internal Hamiltonian

$$H_0 = \frac{p^2}{2\mu_{cf}} + V_{cf}(\mathbf{r}), \qquad (3.6.1)$$

where $\mu_{cf} = m_c m_f / m_P$ is the reduced mass of the core-fragment pair (with $m_P = m_c + m_f$), $\mathbf{p}$ is the momentum operator of the relative motion and V_{cf} is the potential describing the core-fragment interaction. This potential usually includes a central part and a spin-orbit coupling term in addition to a Coulomb potential. In many cases, the potential is deep enough to contain unphysical bound states below the ground state. These unphysical or forbidden states are useful because they allow the wave function representing the physical ground state to exhibit the number of nodes expected from the Pauli principle, as obtained in microscopic descriptions [64]. Although these forbidden states do not play any role in the core-fragment scattering, they could affect breakup properties. However, as shown in Ref. [65], their presence can be ignored because their effect is negligible.

Let $\mathbf{k}$ be the wave vector describing the asymptotic relative motion between the fragments in the projectile continuum. The corresponding energy is thus $E = \hbar^2 k^2 / 2\mu_{cf}$. Notation τ in Eq. (3.2.1) corresponds here to the coupling mode, i.e. to the total spin S of the projectile and the relative orbital momentum l. The wave functions defined in Eq. (3.2.1) read

$$\phi_{lS}^{JM}(E, \mathbf{r}) = r^{-1} i^l [Y_l(\Omega) \otimes \chi_S]^{JM} u_{lS}^J(k, r), \qquad (3.6.2)$$

where χ_S is a spinor resulting from the coupling of s_c and s_f. The radial functions $u_{lS}^J(k, r)$ are normalized according to $\langle u_{lS}^J(k, r) | u_{lS}^J(k', r) \rangle = \delta(k - k')$ and the scattering wave functions $\phi_{lS}^{JM}(E, \mathbf{r})$ according to $\langle \phi_{lS}^{JM}(E, \mathbf{r}) | \phi_{l'S'}^{J'M'}(E', \mathbf{r}) \rangle = (2E/k)$

Fig. 3.5 Jacobi set of coordinates: $\mathbf{r}$ is the projectile internal coordinate, and $\mathbf{R} = \mathbf{b} + Z\hat{\mathbf{Z}}$ is the target-projectile coordinate

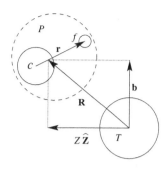

$\delta(E - E')\delta_{JJ'}\delta_{MM'}\delta_{ll'}\delta_{SS'}$. The notation $\hat{k}_\xi$ in Eq. (3.2.2) represents the direction Ω_k of $\mathbf{k}$ and the spin orientations ν_c and ν_f of the core and fragment spins s_c and s_f. Relation (3.2.3) between continuum eigenstates of H_0 becomes

$$\phi_{\Omega_k,\nu_c\nu_f}^{(-)}(E,\mathbf{r}) = \frac{1}{k} \sum_{lSJM} (s_c s_f \nu_c \nu_f | S\nu)(lSM - \nu\nu | JM) Y_l^{M-\nu*}(\Omega_k)\phi_{lS}^{JM}(E,\mathbf{r})$$

(3.6.3)

with the property $\langle \phi_{\Omega_k,\nu_c\nu_f}^{(-)}(E,\mathbf{r})|\phi_{\Omega_k',\nu_c'\nu_f'}^{(-)}(E',\mathbf{r})\rangle = \delta(\mathbf{k} - \mathbf{k}')\delta_{\nu_c\nu_c'}\delta_{\nu_f\nu_f'}$. Notice that notations τ and $\hat{k}_\xi$ are model dependent and would be quite different if a tensor interaction were included in V_{cf}. A detailed description of the simple case $s_c=s_f=0$ can be found in Ref. [27].

Within this framework the description of the reaction reduces to the resolution of a three-body Schrödinger equation (3.2.4) that reads, in the Jacobi set of coordinates illustrated in Fig. 3.5,

$$\left[\frac{P^2}{2\mu} + H_0 + V_{PT}(\mathbf{r},\mathbf{R})\right]\Psi(\mathbf{r},\mathbf{R}) = E_T\Psi(\mathbf{r},\mathbf{R}).$$

(3.6.4)

The projectile-target interaction (3.2.5) then reads

$$V_{PT}(\mathbf{R},\mathbf{r}) = V_{cT}\left(\mathbf{R} - \frac{m_f}{m_P}\mathbf{r}\right) + V_{fT}\left(\mathbf{R} + \frac{m_c}{m_P}\mathbf{r}\right),$$

(3.6.5)

where V_{cT} and V_{fT} are optical potentials that simulate the core-target and fragment-target interactions, respectively.

For a two-body projectile, the DEA breakup cross section (3.4.15) becomes Eq. (46) of Ref. [14]. Integration over Ω_k and summation over ν_c and ν_f lead to the energy and angular distribution of the fragments in the P-T c.m. rest frame. With the normalization of the positive-energy states given above, it reads [14]

$$\frac{d\sigma_{bu}}{dEd\Omega} = \frac{\mu_{cf}}{\hbar^2 k}\frac{KK'}{2J_0+1}\sum_{M_0}\sum_{lJM}\left|\int_0^\infty bdbJ_{|M-M_0|}(qb)S_{klJM}^{(M_0)}(b)\right|^2,$$

(3.6.6)

Fig. 3.6 Breakup of ^{19}C on Pb at 67 MeV/nucleon: semi-classical cross sections for two different binding energies of the projectile: 0.53 MeV (*upper panel*) and 0.65 MeV (*lower panel*) [35]. Experimental data from Ref. [67]. Reprinted figure with permission from Ref. [35]. Copyright (2001) by the American Physical Society

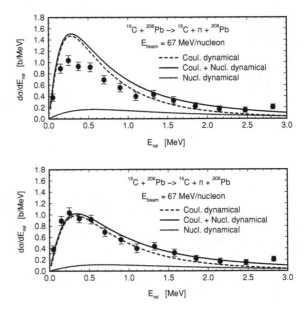

where $S_{klJM}^{(M_0)}$ are coefficients of a partial-wave expansion of the breakup amplitude (3.4.14) (see Eq. (44) of Ref. [14]). Breakup cross sections are mainly expressed as energy distributions $d\sigma_{bu}/dE$ as a function of the energy of the relative motion between the fragments. They are obtained by integrating (3.6.6) over Ω. However, most experimental data concern angular distributions or distributions of the core momentum in the laboratory frame. Note that, in addition, theoretical results should be convoluted with the experimental acceptance and resolution. A change of frame for the theoretical results is thus in general not sufficient to allow a fruitful comparison with experiment.

3.6.2 Two-Body Breakup of Exotic Nuclei

A first information that one can extract from experiment concerns the separation energy of the halo neutrons. Indeed, the shape of the breakup cross section and, in particular, its maximum are sensitive to this energy as can be shown at first order of perturbation theory with rather simple models based on the asymptotic behaviour of the halo wave function [66]. An example is given by the breakup of ^{19}C on lead at 67 MeV/nucleon [4, 67]. In Fig. 3.6, a non-perturbative semi-classical calculation with a ^{18}C$+$n two-body model shows that the shape of the experimental data is much better reproduced if the binding energy of ^{19}C is raised from the recommended value 0.53 MeV to 0.65 MeV [35],

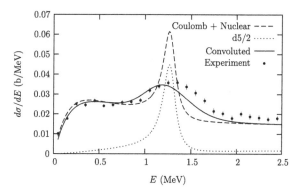

Fig. 3.7 Breakup of ^{11}Be on a C target at 67 MeV/nucleon: calculation performed in a semi-classical model [70]. Experimental data from Ref. [36]

Indirect information can also be obtained on the spin of the ground state of the halo nucleus when few rather different orbital momenta are probable. The magnitude of the cross section is very sensitive to the orbital momentum l of the ground state. A study of the one-neutron removal cross section from ^{31}Ne described in a simple ^{30}Ne + n model allows to rule out the prediction $7/2^-$ of the naive shell model and to confirm the value $3/2^-$ resulting from a shell inversion [4, 68, 69].

Nuclear-induced two-body breakup on light targets is an interesting tool to observe resonances of a halo nucleus and to assess some of their properties. In Fig. 3.7 are displayed experimental data on the ^{11}Be breakup on a C target at 67 MeV/nucleon [4, 36]. These data present a broad bump near the location of a known resonance with an assumed spin-parity $5/2^+$. The bump width is however broader than the known resonance width. A semi-classical calculation (dashed line) based on a ^{10}Be + n model reproduces the shape of the data very well after convolution with the experimental resolution (full line) [70]. Moreover the $d5/2$ component of the theoretical cross section (dotted line) resonates and confirms the $5/2^+$ attribution.

Breakup reactions are also used to infer the spectroscopic factor of the dominant configuration in the core+nucleon structure of halo nuclei [26, 41]. Various theoretical studies have been performed to assess the sensitivity of breakup calculations to the projectile description [71, 72]. These studies have revealed that the breakup cross sections not only depend on the initial bound state of the projectile, but are also sensitive to the description of its continuum [71]. Moreover it has been shown that, for loosely-bound projectiles, only the tail of the wave function is probed in the breakup process and not its whole range [72, 73]. These studies therefore suggest that one should proceed with caution when extracting spectroscopic factors of weakly-bound nuclei from breakup measurements, as other structure properties, like the continuum description, may hinder that extraction.

As mentioned earlier, many Coulomb-breakup experiments have been analyzed within the framework of the first order of the perturbation theory (see Sect. 3.3.3). In order to assess the validity of that approximation, various authors have compared perturbation calculations to numerical resolutions of the time-dependent Schrödinger equation [29, 74–76]. These studies have shown that, in many cases, breakup cannot

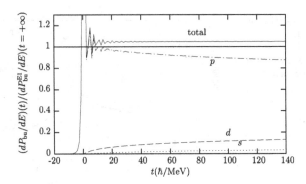

Fig. 3.8 Influence of the couplings inside the continuum [29]. Time evolution of the numerical breakup probability per energy unit (3.3.7) for ^{11}Be impinging on Pb at about 45 MeV/nucleon for a ^{10}Be $-$ n relative energy $E = 1.5$ MeV and an impact parameter $b = 100$ fm. Reprinted figure with permission from Ref. [29]. Copyright (2005) by the American Physical Society

be modelled as a one-step process from the initial bound state towards the continuum and that higher-order effects should be considered for a reliable description of the reaction. In particular, they indicate that significant couplings are at play inside the continuum. To illustrate this, Fig. 3.8 displays the time evolution of the breakup probability per energy unit (3.3.7) for the collision of ^{11}Be on Pb at about 45 MeV/nucleon computed within the time-dependent ^{10}Be+n model of Ref. [11]. The obtained value is divided by its evaluation at the first-order of the perturbation theory (3.3.15) at $t \rightarrow +\infty$. After a sharp increase at the time of closest approach $t = 0$, the breakup probability (full line) oscillates and then stabilizes at a value which differs by about 5% from its first-order estimate. Although the total breakup probability becomes stable, its partial-wave composition still varies: the dominant p wave contribution (dash-dotted line) is depleted towards the s (dotted line) and especially d (dashed line) ones. This signals couplings inside the continuum, which may affect the evaluation of breakup observables [29, 76]. We will see in the next section that it may perturb the analysis of breakup reactions of astrophysical interest [77–80].

3.6.3 Application to Nuclear Astrophysics

Radiative-capture reactions are a crucial ingredient in the determination of the reaction rates in nuclear astrophysics. However the difficulty of their measurement and, in some cases, the scatter of the results has raised interest in indirect methods where the time-reversed reaction is simulated by virtual photons in the Coulomb field of a heavy nucleus [81, 82]. The radiative-capture cross section can be extracted from breakup cross sections if one assumes that the dissociation is due to E1 virtual photons and occurs in a single step. A typical example is the ^{7}Be(p, γ)^{8}B reaction which

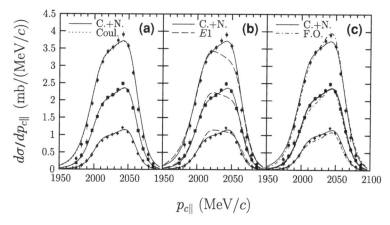

Fig. 3.9 ^{8}B Coulomb breakup on Pb at 44 MeV/nucleon. Parallel-momentum distribution of the ^{7}Be core corresponding to various angular cuts calculated in a DEA model [78]. **a** Influence of nuclear and Coulomb interactions on the calculation. **b** Effects of the various multipoles of the Coulomb interaction. **c** Role of the higher-order effects on the calculation. Experimental data from Ref. [85]

has been studied with the breakup of ^{8}B into ^{7}Be + p on heavy targets at different energies [63, 83–89].

Though appealing, the breakup method also faces a number of difficulties. First, while many reactions are dominated by an E1 transition, an E2 contribution to the breakup cross section may not be negligible [74]. Second, higher-order effects, i.e. transitions from the initial bound state into the continuum through several steps may not be negligible [29, 74–76]. Finally, the nuclear interactions between the projectile and the target may interfere with the Coulomb interaction [11, 35]. Therefore elaborate reaction theories must be used to interpret the experimental data.

The experiments on the breakup of ^{8}B have been analyzed in a number of papers [19, 77, 78, 90–92]. Figure 3.9 shows a comparison between the experimental data of Ref. [85] and DEA calculations [78]. Without adjustable parameters, the calculations (full lines) fairly reproduce the asymmetry exhibited by the data which could not be well explained in earlier works [90, 92]. The three panels of Fig. 3.9 illustrate the influence of various approximations upon the calculation [78]. The left panel illustrates that nuclear P-T interactions can be neglected when data are restricted to forward angles. The central panel confronts a dynamical calculation including only the dipole term of the Coulomb interaction (dashed lines) to the full calculation, indicating that higher multipoles have a significant effect on the breakup process. The right panel compares the dynamical calculation to its first-order approximation (dot-dashed lines), emphasizing the necessity to include higher-order effects in breakup calculations. These results show that some of the assumptions of the breakup method [81, 82] are not valid. It is therefore difficult to infer the accuracy of the S factors extracted from breakup cross sections.

Fig. 3.10 Breakup of ^{15}C on Pb at 68 MeV/nucleon. The experimental energy distribution measured for two scattering-angle cuts [99] is confronted to the time-dependent calculation of Ref. [80]. Reprinted figure with permission from Ref. [80]. Copyright (2009) by the American Physical Society

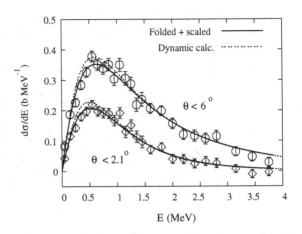

An interesting problem was raised by the ^{14}C(n, γ)^{15}C capture reaction. The measured cross sections for the Coulomb breakup of ^{15}C [93, 94] provided an S factor which disagreed with direct measurements [95, 96]. Moreover, theoretical analyses indicated that the Coulomb-breakup cross sections were inconsistent with information obtained from ^{15}F by charge symmetry and with microscopic models [97]. A new measurement [98, 99] has obtained breakup cross sections that fully agree with properties of the mirror system and with theory [79, 80]. These theoretical analyses show that a fully dynamical calculation, taking proper account of higher-order effects is necessary to correctly analyze the breakup measurements, in agreement with the analysis of the ^{8}B Coulomb breakup of Ref. [78]. They also indicate that including both Coulomb and nuclear interactions as well as their interferences is necessary to correctly reproduce data at large scattering angles. In this way a very good agreement can be obtained between direct and indirect measurements of the S factor. Figure 3.10 displays the breakup cross section of ^{15}C on Pb measured at 68 MeV/nucleon [99] and its comparison to the theoretical calculation of the time-dependent model of Ref. [80]. The dotted lines show the direct results of the calculation, while the full lines correspond to these results folded by the experimental resolution and scaled to the data.

3.7 Breakup Reactions of Three-Body Projectiles

3.7.1 Three-Cluster Model of Projectile

Let us consider a system of three particles, the core with coordinate $\mathbf{r}_c$, mass m_c and charge $Z_c e$ and two fragments with coordinates $\mathbf{r}_1$ and $\mathbf{r}_2$, masses m_1 and m_2, and charges $Z_1 e$ and $Z_2 e$. The projectile mass is $m_p = m_c + m_{12}$ with $m_{12} = m_1 + m_2$. After removal of the c.m. kinetic energy $T_{\text{c.m.}}$, the three-body Hamiltonian of this system in Eq. (3.2.1) can be written as

$$H_0 = \frac{p_c^2}{2m_c} + \frac{p_1^2}{2m_1} + \frac{p_2^2}{2m_2} + V_{c1} + V_{c2} + V_{12} - T_{\text{c.m.}}, \tag{3.7.1}$$

where V_{ij} is an effective potential between particles i and j ($i, j = c, 1, 2$). We assume that these interactions involve central, spin-orbit and Coulomb terms. These potentials may contain unphysical bound states below the two-cluster ground state to simulate effects of the Pauli principle. These forbidden states must be eliminated from the three-body wave functions either with pseudopotentials [100] or with supersymmetric transformations [101, 102].

Various resolution techniques can be considered for obtaining the wave functions of a three-body projectile. A first option is to describe this projectile with an expansion in Gaussian functions depending on Jacobi coordinates [21, 31, 103]. For bound states, the wave functions can be obtained from a variational calculation. Well established techniques allow systematic calculations of the matrix elements [31, 103]. Calculations are then simpler when the interactions are expressed in terms of Gaussians. At negative energies, this type of expansion may however have convergence problems in the description of extended halos. At positive energies, it enables to obtain pseudostates but is not convenient to obtain scattering states.

Let us describe another efficient tool to deal with three-body systems, the formalism of hyperspherical coordinates. It is especially interesting when the two-cluster subsystems are unbound so that only a three-body continuum exists. Notation ξ of Sect. 3.2 represents here five angular variables and one coordinate with the dimension of a length, the hyperradius (see Refs. [104–106] for details). Four angular variables correspond to physical angles and the fifth one is related to a ratio of coordinates defined below in Eq. (3.7.6). The wave functions are expanded in series of hyperspherical harmonics, i.e. a well known complete set of orthonormal functions of the five angular variables. The coefficients are functions of the hyperradius and can be obtained from variational calculations. Scattering states can be obtained from extensions of the R matrix theory [56, 107]. A drawback of this method is that the hyperspherical expansion may converge rather slowly.

With the dimensionless reduced masses $\mu_{c(12)} = m_c m_{12}/m_P m_N$ and $\mu_{12} = m_1 m_2 /m_{12} m_N$ where m_N is the nucleon mass for example, the internal coordinates ξ are scaled Jacobi coordinates defined as

$$\mathbf{x} = \sqrt{\mu_{12}}(\mathbf{r}_2 - \mathbf{r}_1) \tag{3.7.2}$$

and

$$\mathbf{y} = \sqrt{\mu_{c(12)}} \left(\mathbf{r}_c - \frac{m_1 \mathbf{r}_1 + m_2 \mathbf{r}_2}{m_{12}} \right), \tag{3.7.3}$$

i.e., up to a scaling factor, the relative coordinate between the clusters 1 and 2 and the relative coordinate of their centre of mass with respect to the core. With Laplacians Δ_x

and Δ_y with respect to $\mathbf{x}$ and $\mathbf{y}$, the Hamiltonian (3.7.1) of this three-body projectile can be rewritten as

$$H_0 = -\frac{\hbar^2}{2m_N}(\Delta_x + \Delta_y) + V_{c1} + V_{c2} + V_{12}. \tag{3.7.4}$$

To investigate the breakup cross sections for this system, we need wave functions at both positive and negative energies.

In the notation of Refs. [104, 106], the hyperradius ρ and hyperangle α are defined as

$$\rho = \sqrt{x^2 + y^2} \tag{3.7.5}$$

and

$$\alpha = \arctan(y/x). \tag{3.7.6}$$

The hyperangle α and the orientations Ω_x and Ω_y of $\mathbf{x}$ and $\mathbf{y}$ provide a set of five angles collectively denoted as Ω_5. The volume element is $d\mathbf{x}d\mathbf{y} = \rho^5 d\rho d\Omega_5$ with $d\Omega_5 = \sin^2\alpha\cos^2\alpha d\alpha d\Omega_x d\Omega_y$.

The hyperspherical harmonics form an orthonormal basis which verifies a closure relation. The purely spatial hyperspherical harmonics read [104, 106]

$$\mathscr{Y}^{l_x l_y}_{KLM_L}(\Omega_5) = \phi^{l_x l_y}_K(\alpha)\left[Y_{l_x}(\Omega_x) \otimes Y_{l_y}(\Omega_y)\right]^{LM_L}. \tag{3.7.7}$$

where K is the hypermomentum quantum number, l_x and l_y are the orbital quantum numbers associated with $\mathbf{x}$ and $\mathbf{y}$, and L is the quantum number of total orbital momentum. The functions $\phi^{l_x l_y}_K$ depending on the hyperangle α are defined in Eqs. (9) and (10) of Ref. [106]. The hyperspherical harmonics involving spin are defined by

$$\mathscr{Y}^{JM}_{\gamma K}(\Omega_5) = \left[\mathscr{Y}^{l_x l_y}_{KL}(\Omega_5) \otimes \chi_S\right]^{JM}, \tag{3.7.8}$$

where χ_S is a spinor corresponding to a total spin S of the three clusters. Intermediate couplings as, for example, the total spin s_{12} of the fragments are not displayed for simplicity. Index γ stands for $(l_x l_y LS)$.

A partial wave function ϕ^{JM} is a solution of the Schrödinger equation (3.2.1) associated with the three-body Hamiltonian (3.7.4) at energy E. It can be expanded as

$$\phi^{JM}(E, \rho, \Omega_5) = \rho^{-5/2}\sum_{\gamma K}\chi^J_{\gamma K}(E, \rho)\mathscr{Y}^{JM}_{\gamma K}(\Omega_5), \tag{3.7.9}$$

For bound states ($E < 0$), the hyperradial wave functions decrease asymptotically as

$$\chi_{\gamma K}^{J}(E, \rho) \underset{\rho \to \infty}{\sim} \exp(-\sqrt{2m_N|E|/\hbar^2}\rho). \tag{3.7.10}$$

Index τ of Eq. (3.2.1) is irrelevant for bound states within the present assumptions. The normalization of the scattering states $(E > 0)$ is fixed by their asymptotic form. Several choices are possible. The asymptotic form of the hyperradial scattering wave function is for instance given by [20]

$$\chi_{\gamma K(\gamma_\omega K_\omega)}^{J}(E, \rho) \underset{\rho \to \infty}{\to} i^{K_\omega+1}(2\pi/k)^{5/2}$$
$$\times \left[H_{K+2}^{-}(k\rho)\delta_{\gamma\gamma_\omega}\delta_{KK_\omega} - U_{\gamma K,\gamma_\omega K_\omega}^{J} H_{K+2}^{+}(k\rho) \right], \tag{3.7.11}$$

where $k = \sqrt{2m_N E/\hbar^2}$ is the wave number and H_K^- and H_K^+ are incoming and outgoing functions [107–109]. In the neutral case, i.e. when clusters 1 and 2 are neutrons, these functions read $H_K^{\pm}(x) = \pm i(\pi x/2)^{1/2} [J_K(x) \pm iY_K(x)]$ where J_K and Y_K are Bessel functions of first and second kind, respectively. The wave functions $\phi_{(\gamma_\omega K_\omega)}^{JM}(E, \rho, \Omega_5)$ are normalized according to $\langle\phi_{(\gamma_\omega K_\omega)}^{JM}(E, \rho, \Omega_5)|\phi_{(\gamma_\omega' K_\omega')}^{J'M'}(E', \rho, \Omega_5)\rangle = 2E(2\pi/k)^6\delta(E - E')\delta_{JJ'}\delta_{MM'}\delta_{\gamma_\omega\gamma_\omega'}\delta_{K_\omega K_\omega'}$. In the charged case, expression (3.7.11) is only an approximation because the asymptotic form of the Coulomb interaction is not diagonal in hyperspherical coordinates [107, 110]. The indices $\gamma_\omega K_\omega$ where $\gamma_\omega = (l_{x\omega}, l_{y\omega}, L_\omega, S)$ denote the partial entrance channels for this solution. For scattering states, index τ of Eq. (3.2.1) is necessary and rather complicated: it represents the entrance channel $\gamma_\omega K_\omega$. The asymptotic behaviour of a given partial wave depends on the collision matrix. For real interactions, the collision matrix $\mathbf{U}^J$ of each partial wave J is unitary and symmetric. For three-body scattering, it differs from two-body collision matrices in an important aspect: its dimension is infinite since the particles can share the angular momentum in an infinity of ways. In practical calculations, its dimension depends on the number of hypermomenta included in the calculation, limited to a maximum K value, denoted as $K_{\max}$.

The three-body final scattering states are described asymptotically with two relative wave vectors. Let $\mathbf{k}_c$, $\mathbf{k}_1$, $\mathbf{k}_2$ be the wave vectors of the core and fragments in the projectile rest frame. The asymptotic relative motions are defined by the relative wave vector of the neutrons

$$\mathbf{k}_{21} = \sqrt{\mu_{12}}\mathbf{k}_x = \frac{m_1\mathbf{k}_2 - m_2\mathbf{k}_1}{m_{12}} \tag{3.7.12}$$

and the relative wave vector of the core with respect to the centre of mass of the fragments

$$\mathbf{k}_{c(12)} = \sqrt{\mu_{c(12)}}\mathbf{k}_y = \frac{m_{12}\mathbf{k}_c - m_c(\mathbf{k}_1 + \mathbf{k}_2)}{m_P}. \tag{3.7.13}$$

The total internal energy of the projectile with respect to the three-particle threshold is given by

$$E = \frac{\hbar^2}{2m_N}k^2 = \frac{\hbar^2}{2m_N}(k_x^2 + k_y^2). \tag{3.7.14}$$

The orientations Ω_{k_x} of $\mathbf{k}_x$ and Ω_{k_y} of $\mathbf{k}_y$ and the ratio $\alpha_k = \arctan(k_y/k_x)$ form the wave vector hyperangles Ω_{5k}. The hyperangle α_k controls the way the projectile energy E is shared among the fragments. For example, the energy of the relative motion between fragments 1 and 2 is $E\cos^2\alpha_k$. In the scattering states (3.2.2), notation $\hat{k}_\xi$ thus represents Ω_{5k} and the final orientations ν_c, ν_1, ν_2 of the three spins. It is convenient to replace these orientations by the total spin s_{12} of the fragments, the total spin S and its projection ν. Relation (3.2.3) is then given by

$$\phi^{(-)}_{\Omega_{5k}S\nu}(E,\rho,\Omega_5) = (2\pi)^{-3}\sum_{JM}\sum_{l_{x\omega}l_{y\omega}L_\omega K_\omega}(L_\omega S M - \nu\nu|JM)$$

$$\times \mathscr{Y}^{L_\omega M - \nu *}_{l_{x\omega}l_{y\omega}K_\omega}(\Omega_{5k})\mathscr{K}\phi^{J-M}_{(\gamma_\omega K_\omega)}(E,\rho,\Omega_5). \tag{3.7.15}$$

where $\mathscr{K}$ is the time-reversal operator. These functions are normalized with respect to $\delta(\mathbf{k}_x - \mathbf{k}'_x)\delta(\mathbf{k}_y - \mathbf{k}'_y)\delta_{SS'}\delta_{\nu\nu'}$.

The hyperradial wave functions $\chi^J_{\gamma K}$ are to be determined from the Schrödinger equation (3.2.1). The parity $\pi = (-1)^K$ of the three-body relative motion restricts the sum over K to even or odd values. Rigorously, the summation over γK in (3.7.9) should contain an infinite number of terms. In practice, this expansion is limited by the truncation value K_{max}. The l_x and l_y values are limited by $l_x + l_y \le K \le K_{max}$. For weakly-bound and scattering states, it is well known that the convergence is rather slow and that large K_{max} values must be used.

The functions $\chi^J_{\gamma K}$ are derived from a set of coupled differential equations [106, 107]

$$\left[-\frac{\hbar^2}{2m_N}\left(\frac{d^2}{d\rho^2} - \frac{(K+3/2)(K+5/2)}{\rho^2}\right) - E\right]\chi^J_{\gamma K}(E,\rho)$$

$$+ \sum_{\gamma'K'}V^J_{\gamma'K',\gamma K}(\rho)\chi^J_{\gamma'K'}(E,\rho) = 0, \tag{3.7.16}$$

where the potentials matrix elements are defined as

$$V^J_{\gamma'K',\gamma K}(\rho) = \left\langle \mathscr{Y}^{JM}_{\gamma'K'}(\Omega_5)\left|\sum_{i>j=1}^3 V_{ij}(\mathbf{r}_j - \mathbf{r}_i)\right|\mathscr{Y}^{JM}_{\gamma K}(\Omega_5)\right\rangle. \tag{3.7.17}$$

For bound states, approximate solutions can be obtained with an expansion on a finite square-integrable basis. However, using such a basis for scattering states raises problems since they do not vanish at infinity. Their asymptotic form requires a proper treatment. This technical difficulty can be solved within the R-matrix theory [56, 107, 111] which allows matching a variational function over a finite interval with the correct asymptotic solutions of the Schrödinger equation.

In the R-matrix approach, both bound and scattering hyperradial wave functions are approximated over the internal region by an expansion on a set of square-integrable variational functions defined over $[0, a]$. Lagrange-mesh basis functions are quite efficient for describing two-body bound and scattering states [57, 112, 113]. The main advantage of this technique is to strongly simplify the calculation of matrix elements (3.7.17) without loss of accuracy if the Gauss approximation consistent with the mesh is used [106]. This method was extended to three-body bound states in Ref. [106] and to three-body scattering states in Ref. [107]. We refer the reader to those references for details.

3.7.2 Dipole Strength Distribution

The E1 strength distribution for transitions from the ground state to the continuum is a property of the projectile that can be extracted from breakup experiments under some simplifying assumptions for cases where E1 is dominant [4]. In the hyperspherical coordinate system, the multipole operators are given by Eq. (B2) of Ref. [106]. For example, in two-neutron halo nuclei, the E1 strength is given by

$$\mathcal{M}_\mu^{E1}(\rho, \Omega_5) = eZ_c \frac{m_{12}}{m_P} \frac{\rho \sin \alpha}{\sqrt{\mu_{c(12)}}} Y_1^\mu(\Omega_y). \tag{3.7.18}$$

The E1 transition strength (3.3.16) from the ground state at negative energy E_0 with total angular momentum J_0 to the continuum is given by

$$\frac{dB(E1)}{dE} = \frac{4}{2J_0 + 1} \left(\frac{m_N}{\hbar^2}\right)^3 E^2 \sum_{M_0\mu} \sum_{S\nu} \int d\Omega_{5k}$$

$$\left|\left\langle \phi_{\Omega_{5k}S\nu}^{(-)}(E, \rho, \Omega_5) \middle| \mathcal{M}_\mu^{E1}(\rho, \Omega_5) \middle| \phi^{J_0M_0}(E_0, \rho, \Omega_5) \right\rangle\right|^2. \tag{3.7.19}$$

The E1 strength presents the advantage that it can also be calculated in various ways without constructing the complicated three-body scattering states [114]. Most model calculations of the E1 strength for ^{6}He indicate a concentration of strength at low energies E [20, 22, 108, 115–117]. The origin of this low-energy bump remains unclear and can sometimes be attributed to a three-body resonance [20, 108]. The existence of such a bump does not agree with GSI data [118].

This puzzling problem deserves further studies. A first-order description of Coulomb breakup for ^{6}He is probably not very accurate (see Sect. 3.7.3), even at the energies of the GSI experiment [118]. Extracting the E1 strength from breakup measurements is very difficult and not without ambiguities. This is exemplified by the variety of experimental results obtained for the breakup of the ^{11}Li two-neutron halo nucleus. As shown in Fig. 3.11, most early experiments [119–121] did not display a significant strength at low energies in contradiction with data from the more recent RIKEN experiment [4, 122].

Fig. 3.11 Experimental E1
strength for the breakup of
the ^{11}Li two-neutron halo
nucleus: Ref. [122] (*full
circles*), Ref. [119]
(*dash-dotted line*), Ref. [120]
(*histogram*), Ref. [121]
(*dashed line*). Reprinted
figure with permission from
Ref. [122]. Copyright (2006)
by the American Physical
Society

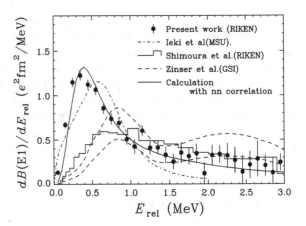

3.7.3 The CCE Approximation for Three-Body Projectiles

We consider a collision between a three-body projectile and a structureless
target with mass m_T and charge $Z_T e$ [20]. The breakup reaction is described by
the four-body Schrödinger equation (3.2.4) where H_0 is given by Eq. (3.7.4). The
effective potential (3.2.5) between projectile and target is defined as

$$V_{PT}(\mathbf{R}, \mathbf{x}, \mathbf{y}) = V_{cT}\left(\mathbf{R} + \frac{m_{12}}{m_P}\frac{\mathbf{y}}{\sqrt{\mu_{c(12)}}}\right) + V_{1T}\left(\mathbf{R} - \frac{m_c}{m_P}\frac{\mathbf{y}}{\sqrt{\mu_{c(12)}}} - \frac{m_2}{m_{12}}\frac{\mathbf{x}}{\sqrt{\mu_{12}}}\right)$$

$$+ V_{2T}\left(\mathbf{R} - \frac{m_c}{m_P}\frac{\mathbf{y}}{\sqrt{\mu_{c(12)}}} + \frac{m_1}{m_{12}}\frac{\mathbf{x}}{\sqrt{\mu_{12}}}\right). \qquad (3.7.20)$$

In this expression, each interaction V_{iT} between a constituent of the projectile and
the target is simulated by a complex optical potential (including a possible Coulomb
interaction taking the cluster extension into account).

In order to obtain breakup cross sections, one must calculate transition matrix
elements for the breakup into three fragments. The transition matrix elements (3.4.11)
read

$$T_{fi} = (\mu_{12}\mu_{c(12)})^{-3/4}$$

$$\times \left\langle e^{i\mathbf{K}'\cdot\mathbf{R}}\phi^{(-)}_{\Omega_{5k}S\nu}(E, \rho, \Omega_5)\middle| V_{PT}(\mathbf{R}, \mathbf{x}, \mathbf{y})\middle|\Psi^{(M_0)}(\mathbf{R}, \rho, \Omega_5)\right\rangle \qquad (3.7.21)$$

for four-body breakup. The factor $(\mu_{12}\mu_{c(12)})^{-3/4}$ appears when the integration
is performed in coordinates ρ and Ω_5 and the bound-state wave function (3.7.9) is
normed in this coordinate system rather than in Jacobi coordinates [20]. At the eikonal
approximation, the exact scattering wave function Ψ in Eq. (3.7.21) is replaced by
its approximation given by Eqs. (3.4.1) and (3.4.17). The transition matrix element
(3.7.21) is then obtained following Eq. (3.4.13) as

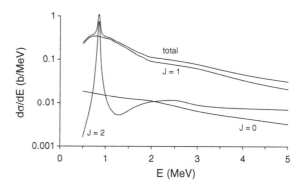

Fig. 3.12 CCE calculation of the total and 0^+, 1^-, 2^+ partial cross sections of ^{6}He breakup on ^{208}Pb at 240 MeV/nucleon [20]. Reprinted figure with permission from Ref. [20]. Copyright (2009) by the American Physical Society

$$T_{fi} = i\hbar v \int d\mathbf{b} e^{-i\mathbf{q}\cdot\mathbf{b}} S_{Sv}^{(M_0)}(E, \Omega_{5k}, \mathbf{b}). \tag{3.7.22}$$

with the eikonal breakup amplitude (3.4.14), that reads here [20]

$$S_{Sv}^{(M_0)}(E, \Omega_{5k}, \mathbf{b}) = (\mu_{12}\mu_{c(12)})^{-3/4}$$
$$\times \left\langle \phi_{\Omega_{5k}Sv}^{(-)}(E, \rho, \Omega_5) \left| e^{i\chi(\mathbf{b},\mathbf{s}_x,\mathbf{s}_y)} \right| \phi^{J_0 M_0}(E_0, \rho, \Omega_5) \right\rangle. \tag{3.7.23}$$

Following Eq. (3.7.20), the eikonal phase shift χ defined in Eq. (3.4.18) is obtained as

$$\chi = \chi_{cT} + \chi_{1T} + \chi_{2T}. \tag{3.7.24}$$

It depends on the transverse part $\mathbf{b}$ of $\mathbf{R}$ as well as on the transverse parts $\mathbf{s}_x$ and $\mathbf{s}_y$ of the scaled Jacobi coordinates $\mathbf{x}$ and $\mathbf{y}$.

From the transition matrix elements (3.7.21), various cross sections can be derived. The differential cross section (3.4.15) with respect to the eight independent variables Ω, $\mathbf{k}_{21}$, $\mathbf{k}_{c(12)}$ reads in the c.m. frame

$$\frac{d\sigma}{d\Omega\, d\mathbf{k}_{21} d\mathbf{k}_{c(12)}} = \frac{1}{2J_0 + 1} \frac{1}{4\pi^2} \left(\frac{\mu}{\hbar^2}\right)^2 \frac{K'}{K} \sum_{SvM_0} |T_{fi}|^2. \tag{3.7.25}$$

The physical wave numbers k_{21} and $k_{c(12)}$ are proportional to k_x and k_y and can thus be expressed from k and α_k [20]. Integrating Eq. (3.7.25) over all angles Ω and Ω_{5k} leads to the energy distribution cross section $d\sigma/dE$.

The CCE approximation has allowed calculating various elastic and breakup cross sections for ^{6}He on ^{208}Pb by treating ^{6}He as an $\alpha + n + n$ three-body system [20]. In Fig. 3.12, the contribution from the different partial waves is displayed at 240 MeV/nucleon. As expected for a transition from a 0^+ ground state, the $J = 1^-$ component is dominant. However the $J = 0^+$ and $J = 2^+$ components are not negligible. The known 2^+ resonance at 0.82 MeV is clearly visible in the total cross

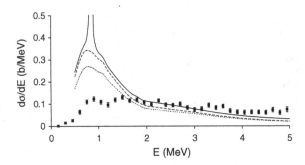

Fig. 3.13 Comparison [20] between the total CCE cross section (*full line*) of ^{6}He breakup on ^{208}Pb at 240 MeV/nucleon with the experimental data of Ref. [118]. The 1^- partial cross sections calculated with two types of elimination of forbidden states (supersymmetry: *dashed line*, projection: *dotted line*) are also displayed. Reprinted figure with permission from [20]. Copyright (2009) by the American Physical Society

section. Extracting an E1 strength from such data is thus not easy, even at this high energy.

A comparison of the CCE cross section (full line) with GSI data [118] is presented in Fig. 3.13. The disagreement already discussed for the E1 strength in Sect. 3.7.2 is clearly visible. The data do not show as large a cross section at low energies as the theory. It is not even clear whether the 2^+ resonance is visible in these data. Nevertheless the agreement is reasonably good above 2 MeV, given that no parameter is fitted to this experiment in the model calculation. The 1^- contribution is calculated with two different ways of eliminating the unphysical bound states in the $\alpha + n$ potentials (dashed and dotted lines). The low-energy peak corresponds to a broad resonance in the lowest 1^- three-body phase shift. Further experimental and theoretical works are needed to explain this discrepancy.

The advantage of the relative simplicity of the CCE is that various types of angular differential cross sections can be calculated. Examples of double differential cross sections showing various energy repartitions between the fragments are presented in Fig. 7 of Ref. [20].

3.7.4 The CDCC Approximation for Three-Body Projectiles

The CDCC method has also been extended to three-body projectiles. In the first works [53, 54, 61], the pseudostate discretization was adopted. Indeed, it avoids the difficult construction of scattering states and allows an accurate treatment using expansions involving Gaussians with various widths. Only recently was the construction of bins attempted [22]. The difficulty of the calculation restricted the first applications to elastic scattering.

The differential cross section for elastic scattering of ^{6}He on ^{12}C at 229.8 MeV is displayed in Fig. 3.14. A single-channel calculation neglecting breakup channels

Fig. 3.14 Ratios of differential cross sections obtained with CDCC to Rutherford cross section for the elastic scattering of ^{6}He on ^{12}C at 229.8 MeV without and with coupling to breakup channels [53]. Experimental data from Ref. [123]. Reprinted figure with permission from Ref. [53]. Copyright (2004) by the American Physical Society

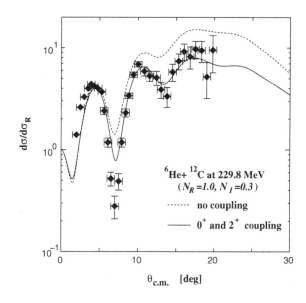

Fig. 3.15 Ratios of differential cross sections to Rutherford cross section for the elastic scattering of ^{6}He on ^{209}Bi at 22.5 MeV: comparison of three- and four-body CDCC without and with coupling to breakup channels [61]. Experimental data from Refs. [124, 125]. Reprinted figure with permission from Ref. [61]. Copyright (2006) by the American Physical Society

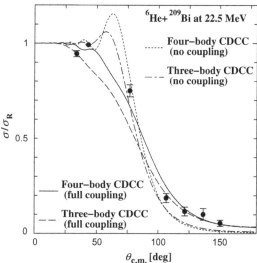

(dotted line) overestimates the experimental data of Ref. [123]. The shape of the data is very well reproduced by introducing 0^+ and 2^+ pseudochannels and taking account of all couplings (full line).

In Fig. 3.15 is displayed a comparison between calculations of ^{6}He elastic scattering on ^{209}Bi at 22.5 MeV involving two-cluster ("Three-body CDCC", dashed line) and three-cluster ("Four-body CDCC", full line) descriptions of ^{6}He. A significant difference appears between calculations neglecting breakup channels ("no coupling") and those including it ("full coupling"). The agreement with experimental

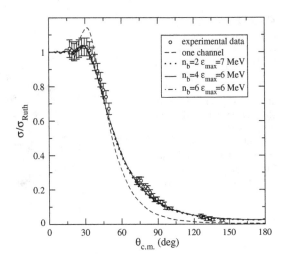

Fig. 3.16 Ratios of differential cross sections to Rutherford cross section for the elastic scattering of ^{6}He on ^{64}Zn at 13.6 MeV: comparison of CDCC calculations with various basis sizes and maximum energies E_{max} with a single-channel calculation [54]. Experimental data from Ref. [126]. Reprinted figure with permission from Ref. [54]. Copyright (2008) by the American Physical Society

data [124, 125] seems better within the four-body treatment including the breakup channels.

Another type of basis functions, based on deformed oscillators, has been used to construct ^{6}He pseudostates in Ref. [54]. This technique also allowed a description of elastic scattering explicitly including breakup channels. In Fig. 3.16, the elastic scattering of ^{6}He on ^{64}Zn at 13.6 MeV is compared with experimental data from Ref. [126]. These results show that including partial waves up to $J = 2$ and taking coupling into account (full line) allow a good agreement with data. Here also, the calculation omitting the coupling to the continuum (dashed line) disagrees with the experimental data. The same basis has recently been extended to the construction of bins [22].

While, for three-body projectiles, the effect of breakup channels has been included for some time in studies of elastic scattering, the determination of breakup cross sections is just starting. Some preliminary calculations have been published recently. Some of them are not fully converged [22] or involve simplifying assumptions [21]. A recent CDCC calculation [62] provides a good agreement with experiment [118] for ^{6}He breakup on ^{12}C. For ^{6}He breakup on ^{208}Pb, it does not agree well with experiment and is about a factor of two lower than the CCE results of Ref. [20] displayed in Fig. 3.13. The reasons of these discrepancies are not yet understood. Nevertheless, the CDCC method should allow a precise treatment of three-body breakup in a near future.

3.8 Perspectives

The theory of breakup reactions offers several accurate non-relativistic approximations covering a broad energy range, that allow an interpretation of various experiments. A good accuracy is reached for some time for the breakup of two-body projectiles and is in view for the breakup of three-body projectiles. Good results can already be obtained with the simplest models of projectile structure, provided that the value of the projectile binding energy is correct. This suggests that only limited spectroscopic information can be extracted from the comparison of theory and experiment. This is partly due to the fact that a comparison of experimental data with results of calculations usually requires complicated convolutions. Nevertheless, breakup has proved to be an efficient alternative probe to measure the separation energies of bound states of exotic nuclei. When performed on light targets, it also provides information about the location and width of resonances of such nuclei. Moreover, some information about the quantum numbers of the ground state of exotic nuclei can be assessed from breakup measurements. The extraction of spectroscopic factors, however, is very sensitive to the accuracy of the absolute normalization of experiments. Moreover, the sensitivity of breakup calculations to the description of the continuum of the projectile indicates that these extractions should be performed with caution. In addition, Coulomb breakup on heavy targets is also used to measure astrophysical S factors. However the accuracy of this indirect technique is uncertain.

Several methods can now be applied to the breakup of three-cluster projectiles (CDCC method, eikonal approximation, ...). They will allow studying coincidence observables that are more difficult to measure but less sensitive to the absolute normalization of cross sections. They should also allow the study of correlations between the emitted fragments. In this respect, efforts should be made at the interface between theory and experiment to facilitate the transformation of the results of model calculations into quantities comparable with the data, taking account of the resolution and acceptances of the detection setup. On the theoretical side, three-cluster bound states can be obtained with good accuracy but the difficult treatment of the three-body continuum still requires progress.

Attempts to improve the model description of the projectile by including excited states of the clusters composing the projectile have started with the extended CDCC. In the future, one can expect a further improvement by using a microscopic description of the projectile within the microscopic cluster model [127–129], involving effective nucleon-nucleon forces and full antisymmetrization. Improvements in the projectile description should first concern bound states. This should reduce the uncertainties appearing in non-microscopic cluster models because of the effective forces between the clusters in the projectile and between the clusters and the target. Using fully antisymmetrized wave functions in breakup calculations seems to be within reach for two-cluster projectiles. This approach should open the way towards ab initio descriptions of the projectile based on fully realistic nucleon-nucleon forces.

All the reaction descriptions presented in this review have been developed within non-relativistic quantum mechanics. However, relativistic effects may be significant and affect the analysis of breakup data, even at intermediate energies of a few tens of MeV/nucleon. Several authors have started analyzing these effects and have proposed ways to take them into account in time-dependent [130] or CDCC [131, 132] frameworks. Since some of the new facilities of radioactive-ion beams will operate at high energies (a few hundreds of MeV/nucleon), these effects will have to be better understood and incorporated in state-of-the-art reaction models.

Acknowledgments We would like to thank P. Descouvemont, G. Goldstein, V. S. Melezhik, F. M. Nunes, and Y. Suzuki for fruitful collaborations concerning the development of some parts of the methods and applications discussed in this text. This text presents research results of BriX (Belgian research initiative on exotic nuclei), the interuniversity attraction pole programme P6/23 initiated by the Belgian-state Federal Services for Scientific, Technical and Cultural Affairs (FSTC). P. C. acknowledges the support of the F.R.S.-FNRS and of the National Science Foundation grant PHY-0800026. The authors also acknowledge travel support of the Fonds de la Recherche Fondamentale Collective (FRFC).

References

1. Tanihata, I., Hamagaki, H., Hashimoto, O., Nagamiya, S., Shida, Y., Yoshikawa, N., Yamakawa, O., Sugimoto, K., Kobayashi, T., Greiner, D.E., Takahashi, N., Nojiri, Y.: Phys. Lett. B **160**, 380 (1985)
2. Tanihata, I.: J. Phys. G **22**, 157 (1996)
3. Jonson, B.: Phys. Rep. **389**, 1 (2004)
4. Nakamura, T., Kondo, Y.: In: Beck C. (ed.) Cluster in Nuclei, Vol. 2. Lecture Notes in Physics, vol. 848, p. 67 (2012)
5. Alder, K., Winther, A.: Electromagnetic Excitation. North-Holland, Amsterdam (1975)
6. Kido, T., Yabana, K., Suzuki, Y.: Phys. Rev. C **50**, R1276 (1994)
7. Esbensen, H., Bertsch, G.F., Bertulani, C.A.: Nucl. Phys. A **581**, 107 (1995)
8. Typel, S., Wolter, H.H.: Z. . Naturforsch. Teil A **54**, 63 (1999)
9. Melezhik, V.S., Baye, D.: Phys. Rev. C **59**, 3232 (1999)
10. Lacroix, D., Scarpaci, J.A., Chomaz, P.: Nucl. Phys. A **658**, 273 (1999)
11. Capel, P., Baye, D., Melezhik, V.S.: Phys. Rev. C **68**, 014612 (2003)
12. Glauber, R.J.: In: Brittin, W.E., Dunham, L.G. (eds.) Lectures in Theoretical Physics, vol. 1, p. 315. Interscience, New York (1959)
13. Baye, D., Capel, P., Goldstein, G.: Phys. Rev. Lett. **95**, 082502 (2005)
14. Goldstein, G., Baye, D., Capel, P.: Phys. Rev. C **73**, 024602 (2006)
15. Margueron, J., Bonaccorso, A., Brink, D.M.: Nucl. Phys. A **720**, 337 (2003)
16. Abu-Ibrahim, B., Suzuki, Y.: Prog. Theor. Phys. **112**, 1013 (2004)
17. Capel, P., Baye, D., Suzuki, Y.: Phys. Rev. C **78**, 054602 (2008)
18. Kamimura, M., Yahiro, M., Iseri, Y., Sakuragi, Y., Kameyama, H., Kawai, M.: Prog. Theor. Phys. Suppl. **89**, 1 (1986)
19. Tostevin, J.A., Nunes, F.M., Thompson, I.J.: Phys. Rev. C **63**, 024617 (2001)
20. Baye, D., Capel, P., Descouvemont, P., Suzuki, Y.: Phys. Rev. C **79**, 024607 (2009)
21. Egami, T., Matsumoto, T., Ogata, K., Yahiro, M.: Prog. Theor. Phys. **121**, 789 (2009)
22. Rodríguez-Gallardo, M., Arias, J.M., Gómez-Camacho, J., Moro, A.M., Thompson, I.J., Tostevin, J.A.: Phys. Rev. C **80**, 051601 (2009)
23. Deltuva, A., Fonseca, A.C., Sauer, P.U.: Phys. Rev. C **71**, 054005 (2005)
24. Deltuva, A., Fonseca, A.C., Sauer, P.U.: Phys. Rev. Lett. **95**, 092301 (2005)
25. Deltuva, A., Moro, A.M., Cravo, E., Nunes, F.M., Fonseca, A.C.: Phys. Rev. C **76**, 064602 (2007)
26. Nakamura, T., Shimoura, S., Kobayashi, T., Teranishi, T., Abe, K., Aoi, N., Doki, Y., Fujimaki, M., Inabe, N., Iwasa, N. et al.: Phys. Lett. B **331**, 296 (1994)
27. Baye, D.: Eur. Phys. J.: Special Topics **156**, 93 (2008)
28. Esbensen, H., Bertulani, C.A.: Phys. Rev. C **65**, 024605 (2002)
29. Capel, P., Baye, D.: Phys. Rev. C **71**, 044609 (2005)

30. Abramowitz, M., Stegun, I.A.: Handbook of Mathematical Functions. Dover, New-York (1970)
31. Suzuki, Y., Lovas, R.G., Yabana, K., Varga, K.: Structure and Reactions of Light Exotic Nuclei. Taylor and Francis, London (2003)
32. Kido, T., Yabana, K., Suzuki, Y.: Phys. Rev. C **53**, 2296 (1996)
33. Fallot, M., Scarpaci, J.A., Lacroix, D., Chomaz, P., Margueron, J.: Nucl. Phys. A **700**, 70 (2002)
34. Baye, D., Goldstein, G., Capel, P.: Phys. Lett. A **317**, 337 (2003)
35. Typel, S., Shyam, R.: Phys. Rev. C **64**, 024605 (2001)
36. Fukuda, N., Nakamura, T., Aoi, N., Imai, N., Ishihara, M., Kobayashi, T., Iwasaki, H., Kubo, T., Mengoni, A., Notani, M. et al.: Phys. Rev. C **70**, 054606 (2004)
37. Bertulani, C.A., Danielewicz, P.: Introduction to Nuclear Reactions. Institute of Physics Publishing, Bristol (2004)
38. Austern, N.: Direct Nuclear Reaction Theories. Wiley, New-York (1970)
39. Hencken, K., Bertsch, G., Esbensen, H.: Phys. Rev. C **54**, 3043 (1996)
40. Abu-Ibrahim, B., Suzuki, Y.: Phys. Rev. C **62**, 034608 (2000)
41. Hansen, P.G., Tostevin, J.A.: Annu. Rev. Nucl. Part. Sci. **53**, 219 (2003)
42. Goldstein, G.: Description de la dissociation de noyaux à halo par l'approximation eikonale dynamique. Ph.D. thesis, Université Libre de Bruxelles, Brussels, Belgium (2007) http://theses.ulb.ac.be/ETD-db/collection/available/ULBetd-09262007-111801/
43. Glauber, R.J., Matthiae, G.: Nucl. Phys. B **21**, 135 (1970)
44. Abu-Ibrahim, B., Horiuchi, W., Kohama, A., Suzuki, Y.: Phys. Rev. C **77**, 034607 (2008)
45. Yahiro, M., Minomo, K., Ogata, K., Kawai, M.: Prog. Theor. Phys. **120**, 767 (2008)
46. Rawitscher, G.H.: Phys. Rev. C **9**, 2210 (1974)
47. Austern, N., Iseri, Y., Kamimura, M., Kawai, M., Rawitscher, G., Yahiro, M.: Phys. Rep. **154**, 125 (1987)
48. Nunes, F.M., Thompson, I.J.: Phys. Rev. C **59**, 2652 (1999)
49. Moro, A.M., Arias, J.M., Gómez-Camacho, J., Martel, I., Pérez-Bernal, F., Crespo, R., Nunes, F.: Phys. Rev. C **65**, 011602 (2001)
50. Rubtsova, O.A., Kukulin, V.I., Moro, A.M.: Phys. Rev. C **78**, 034603 (2008)
51. Moro, A.M., Arias, J.M., Gómez-Camacho, J., Pérez-Bernal, F.: Phys. Rev. C **80**, 054605 (2009)
52. Druet, T., Baye, D., Descouvemont, P., Sparenberg, J.M.: Nucl. Phys. A **845**, 88 (2010)
53. Matsumoto, T., Hiyama, E., Ogata, K., Iseri, Y., Kamimura, M., Chiba, S., Yahiro, M.: Phys. Rev. C **70**, 061601 (2004)
54. Rodríguez-Gallardo, M., Arias, J.M., Gómez-Camacho, J., Johnson, R.C., Moro, A.M., Thompson, I.J., Tostevin, J.A.: Phys. Rev. C **77**, 064609 (2008)
55. Thompson, I.J.: Comp. Phys. Rep. **7**, 167 (1988)
56. Descouvemont, P., Baye, D.: Rep. Prog. Phys. **73**, 036301 (2010)
57. Baye, D., Heenen, P.-H.: J. Phys. A **19**, 2041 (1986)
58. Baye, D.: Phys. Stat. Sol. (b) **243**, 1095 (2006)
59. Summers, N.C., Nunes, F.M., Thompson, I.J.: Phys. Rev. C **73**, 031603 (2006)
60. Summers, N.C., Nunes, F.M., Thompson, I.J.: Phys. Rev. C **74**, 014606 (2006)
61. Matsumoto, T., Egami, T., Ogata, K., Iseri, Y., Kamimura, M., Yahiro, M.: Phys. Rev. C **73**, 051602 (2006)
62. Matsumoto, T., Katō, K., Yahiro, M.: Phys. Rev. C **82**, 051602 (2010)
63. Kolata, J.J., Guimarães, V., Peterson, D., Santi, P., White-Stevens, R.H., Vincent, S.M., Becchetti, F.D., Lee, M.Y., O'Donnell, T.W., Roberts, D.A., Zimmerman, J.A.: Phys. Rev. C **63**, 024616 (2001)
64. Buck, B., Friedrich, H., Wheatley, C.: Nucl. Phys. A **275**, 246 (1977)
65. Capel, P., Baye, D., Melezhik, V.S.: Phys. Lett. B **552**, 145 (2003)
66. Typel, S., Baur, G.: Nucl. Phys. A **759**, 247 (2005)

67. Nakamura, T., Fukuda, N., Kobayashi, T., Aoi, N., Iwasaki, H., Kubo, T., Mengoni, A., Notani, M., Otsu, H., Sakurai, H., Shimoura, S., Teranishi, T., Watanabe, Y.X., Yoneda, K., Ishihara, M.: Phys. Rev. Lett. **83**, 1112 (1999)
68. Nakamura, T., Kobayashi, N., Kondo, Y., Satou, Y., Aoi, N., Baba, H., Deguchi, S., Fukuda, N., Gibelin, J., Inabe, N. et al.: Phys. Rev. Lett. **103**, 262501 (2009)
69. Horiuchi, W., Suzuki, Y., Capel, P., Baye, D.: Phys. Rev. C **81**, 024606 (2010)
70. Capel, P., Goldstein, G., Baye, D.: Phys. Rev. C **70**, 064605 (2004)
71. Capel, P., Nunes, F.M.: Phys. Rev. C **73**, 014615 (2006)
72. Capel, P., Nunes, F.M.: Phys. Rev. C **75**, 054609 (2007)
73. García-Camacho, A., Bonaccorso, A., Brink, D.M.: Nucl. Phys. A **776**, 118 (2006)
74. Esbensen, H., Bertsch, G.F.: Nucl. Phys. A **600**, 37 (1996)
75. Typel, S., Baur, G.: Phys. Rev. C **64**, 024601 (2001)
76. Esbensen, H., Bertsch, G.F., Snover, K.A.: Phys. Rev. Lett. **94**, 042502 (2005)
77. Ogata, K., Hashimoto, S., Iseri, Y., Kamimura, M., Yahiro, M.: Phys. Rev. C **73**, 024605 (2006)
78. Goldstein, G., Capel, P., Baye, D.: Phys. Rev. C **76**, 024608 (2007)
79. Summers, N.C., Nunes, F.M.: Phys. Rev. C **78**, 011601 (2008); Erratum: *ibid.* **78**, 069908 (2008)
80. Esbensen, H.: Phys. Rev. C **80**, 024608 (2009); Erratum: Esbensen, H., Reifarth, R.: Phys. Rev. C **80**, 059904 (2009)
81. Baur, G., Bertulani, C.A., Rebel, H.: Nucl. Phys. A **458**, 188 (1986)
82. Baur, G., Rebel, H.: Annu. Rev. Nucl. Part. Sci. **46**, 321 (1996)
83. Motobayashi, T., Iwasa, N., Ando, Y., Kurokawa, M., Murakami, H., Ruan (Gen), J., Shimoura, S., Shirato, S., Inabe, N., Ishihara, M. et al.: Phys. Rev. Lett. **73**, 2680 (1994)
84. Kikuchi, T., Motobayashi, T., Iwasa, N., Ando, Y., Kurokawa, M., Moriya, S., Murakami, H., Nishio, T., Ruan (Gen), J., Shirato, S. et al.: Phys. Lett. B **391**, 261 (1997)
85. Davids, B., Anthony, D.W., Austin, S.M., Bazin, D., Blank, B., Caggiano, J.A., Chartier, M., Esbensen, H., Hui, P., Powell, C.F. et al.: Phys. Rev. Lett. **81**, 2209 (1998)
86. Guimarães, V., Kolata, J.J., Peterson, D., Santi, P., White-Stevens, R.H, Vincent, S.M., Becchetti, F.D., Lee, M.Y., O'Donnell, T.W., Roberts, D.A., Zimmerman, J.A.: Phys. Rev. Lett. **84**, 1862 (2000)
87. Davids, B., Anthony, D.W., Aumann, T., Austin, S.M., Baumann, T., Bazin, D., Clement, R.R.C., Davids, C.N., Esbensen, H., Lofy, P.A. et al.: Phys. Rev. Lett. **86**, 2750 (2001)
88. Davids, B., Austin, S.M., Bazin, D., Esbensen, H., Sherrill, B.M., Thompson, I.J., Tostevin, J.A.: Phys. Rev. C **63**, 065806 (2001)
89. Schümann, F., Typel, S., Hammache, F., Sümmerer, K., Uhlig, F., Böttcher, I., Cortina, D., Förster, A., Gai, M., Geissel, H. et al.: Phys. Rev. C **73**, 015806 (2006)
90. Mortimer, J., Thompson, I.J., Tostevin, J.A.: Phys. Rev. C **65**, 064619 (2002)
91. Davids, B., Typel, S.: Phys. Rev. C **68**, 045802 (2003)
92. Summers, N.C., Nunes, F.M.: J. Phys. G **31**, 1437 (2005)
93. Horváth, Á., Weiner, J., Galonsky, A., Deák, F., Higurashi, Y., Ieki, K., Iwata, Y., Kiss, Á., Kolata, J.J., Seres, Z. et al.: Astrophys. J. **570**, 926 (2002)
94. Datta Pramanik, U., Aumann, T., Boretzky, K., Carlson, B.V., Cortina, D., Elze, T.W., Emling, H., Geissel, H., Grünschloß, A., Hellström, M. et al.: Phys. Lett. B **551**, 63 (2003)
95. Reifarth, R., Heil, M., Plag, R., Besserer, U., Dababneh, S., Dörr, L., Görres, J., Haight, R.C., Käppeler, F., Mengoni, A. et al.: Nucl. Phys. A **758**, 787 (2005)
96. Reifarth, R., Heil, M., Forssén, C., Besserer, U., Couture, A., Dababneh, S., Dörr L., Görres, J., Haight, R.C., Käppeler, F., et al.: Phys. Rev. C **77**, 015804 (2008)
97. Timofeyuk, N.K., Baye, D., Descouvemont, P., Kamouni, R., Thompson, I.J.: Phys. Rev. Lett. **96**, 162501 (2006)
98. Nakamura, T., Fukuda, N., Aoi, N., Iwasaki, H., Kobayashi, T., Kubo, T., Mengoni, A., Notani, M., Otsu, H., Sakurai, H. et al.: Nucl. Phys. A **722**, C301 (2003)

99. Nakamura, T., Fukuda, N., Aoi, N., Imai, N., Ishihara, M., Iwasaki, H., Kobayashi, T., Kubo, T., Mengoni, A., Motobayashi, T. et al.: Phys. Rev. C **79**, 035805 (2009)
100. Kukulin, V.I., Pomerantsev, V.N.: Ann. Phys. **111**, 330 (1978)
101. Baye, D.: Phys. Rev. Lett. **58**, 2738 (1987)
102. Baye, D.: J. Phys. A **20**, 5529 (1987)
103. Hiyama, E., Kino, Y., Kamimura, M.: Prog. Part. Nucl. Phys. **51**, 223 (2003)
104. Raynal, J., Revai, J.: Nuovo Cim. A **39**, 612 (1970)
105. Zhukov, M.V., Danilin, B.V., Fedorov, D.V., Bang, J.M., Thompson, I.J., Vaagen, J.S.: Phys. Rep. **231**, 151 (1993)
106. Descouvemont, P., Daniel, C., Baye, D.: Phys. Rev. C **67**, 044309 (2003)
107. Descouvemont, P., Tursunov, E., Baye, D.: Nucl. Phys. A **765**, 370 (2006)
108. Danilin, B.V., Thompson, I.J., Vaagen, J.S., Zhukov, M.V.: Nucl. Phys. A **632**, 383 (1998)
109. Thompson, I.J., Danilin, B.V., Efros, V.D., Vaagen, J.S., Bang, J.M., Zhukov, M.V.: Phys. Rev. C **61**, 024318 (2000)
110. Vasilevsky, V., Nesterov, A.V., Arickx, F., Broeckhove, J.: Phys. Rev. C **63**, 034607 (2001)
111. Lane, A.M., Thomas, R.G.: Rev. Mod. Phys. **30**, 257 (1958)
112. Hesse, M., Sparenberg, J.-M., Van Raemdonck, F., Baye, D.: Nucl. Phys. A **640**, 37 (1998)
113. Hesse, M., Roland, J., Baye, D.: Nucl. Phys. A **709**, 184 (2002)
114. Suzuki, Y., Horiuchi, W., Baye, D.: Prog. Theor. Phys. **123**, 547 (2010)
115. Cobis, A., Fedorov, D.V., Jensen, A.S.: Phys. Rev. Lett. **79**, 2411 (1997)
116. Myo, T., Katō, K., Aoyama, S., Ikeda, K.: Phys. Rev. C **63**, 054313 (2001)
117. Hagino, K., Sagawa, H.: Phys. Rev. C **76**, 047302 (2007)
118. Aumann, T., Aleksandrov, D., Axelsson, L., Baumann, T., Borge, M.J.G., Chulkov, L.V., Cub, J., Dostal, W., Eberlein, B., Elze, T.W. et al.: Phys. Rev. C **59**, 1252 (1999)
119. Ieki, K., Sackett, D., Galonsky, A., Bertulani, C.A., Kruse, J.J., Lynch, W.G., Morrissey, D.J., Orr, N.A., Schulz, H., Sherrill, B.M. et al.: Phys. Rev. Lett. **70**, 730 (1993)
120. Shimoura, S., Nakamura, T., Ishihara, M., Inabe, N., Kobayashi, T., Kubo, T., Siemssen, R.H., Tanihata, I., Watanabe, Y.: Phys. Lett. B **348**, 29 (1995)
121. Zinser, M., Humbert, F., Nilsson, T., Schwab, W., Simon, H., Aumann, T., Borge, M.J.G., Chulkov, L.V., Cub, J., Elze, T.W. et al.: Nucl. Phys. A **619**, 151 (1997)
122. Nakamura, T., Vinodkumar, A.M., Sugimoto, T., Aoi, N., Baba, H., Bazin, D., Fukuda, N., Gomi, T., Hasegawa, H., Imai, N. et al.: Phys. Rev. Lett. **96**, 252502 (2006)
123. Lapoux, V., Alamanos, N., Auger, F., Fékou-Youmbi, V., Gillibert, A., Marie, F., Ottini-Hustache, S., Sida, J.L., Khoa, D.T., Blumenfeld, Y. et al.: Phys. Rev. C **66**, 034608 (2002)
124. Aguilera, E.F., Kolata, J.J., Nunes, F.M., Becchetti, F.D., DeYoung, P.A., Goupell, M., Guimarães, V., Hughey, B., Lee, M.Y., Lizcano, D. et al.: Phys. Rev. Lett. **84**, 5058 (2000)
125. Aguilera, E.F., Kolata, J.J., Becchetti, F.D., DeYoung, P.A., Hinnefeld, J.D., Horváth, Á., Lamm, L.O., Lee, H.Y., Lizcano, D., Martinez-Quiroz, E. et al.: Phys. Rev. C **63**, 061603 (2001)
126. Di Pietro, A., Figuera, P., Amorini, F., Angulo, C., Cardella, G., Cherubini, S., Davinson, T., Leanza, D., Lu, J., Mahmud, H. et al.: Phys. Rev. C **69**, 044613 (2004)
127. Wildermuth, K., Tang, Y.C.: A Unified Theory of the Nucleus. Vieweg, Braunschweig (1977)
128. Tang, Y.C.: In: Kuo, T.T.S., Wong, S.S.M. (eds.) Topics in Nuclear Physics II. Lecture Notes in Physics, vol. 145, p. 571. Springer, Berlin (1981)
129. Descouvemont P., Dufour M.: In: Beck C. (ed.) Clusters in Nuclei, Vol. 2. Lecture Notes in Physics, vol. 848, p. 1 (2012)
130. Esbensen, H.: Phys. Rev. C **78**, 024608 (2008)
131. Bertulani, C.A.: Phys. Rev. Lett. **94**, 072701 (2005)
132. Ogata, K., Bertulani, C.A.: Prog. Theor. Phys. **121**, 1399 (2009)

Chapter 4
Clustering Effects Within the Dinuclear Model

Gurgen Adamian, Nikolai Antonenko and Werner Scheid

Abstract The clustering of two nuclei in a nuclear system creates configurations denoted in literature as nuclear molecular structures. A nuclear molecule or a dinuclear system (DNS) as named by Volkov consists of two touching nuclei (clusters) which keep their individuality. Such a system has two main degrees of freedom of collective motions which govern its dynamics: (i) the relative motion between the clusters leading to molecular resonances in the internuclear potential and to the decay of the dinuclear system (separation of the clusters) which is called quasifission since no compound system like in fission is first formed. (ii) the transfer of nucleons or light constituents between the two clusters of the dinuclear system leading to a special dynamics of the mass and charge asymmetries between the clusters in fusion and fission reactions. In this article we discuss the essential aspects of the diabatic internuclear potential used by the di-nuclear system concept and present applications to nuclear structure and reactions. We show applications of the dinuclear model to superdeformed and hyperdeformed bands. An extended discussion is given to the problems of fusion dynamics in the production of superheavy nuclei, to the quasifission process and to multi-nucleon transfer between nuclei. Also the binary and ternary fission processes are discussed within the scission-point model and the dinuclear system concept.

G. Adamian (✉) · N. Antonenko
Bogoliubov Laboratory of Theoretical Physics,
Joint Institute for Nuclear Research,
141980 Dubna (Moscow region), Russia
e-mail: adamian@theor.jinr.ru

N. Antonenko
e-mail: antonenk@theor.jinr.ru

W. Scheid
Institut für Theoretische Physik der Justus-Liebig-Universität Giessen,
Heinrich-Buff-Ring 16, 35392 Giessen, Germany
e-mail: werner.scheid@theo.physik.uni-giessen.de

C. Beck (ed.), *Clusters in Nuclei, Vol.2*, Lecture Notes in Physics 848,
DOI: 10.1007/978-3-642-24707-1_4, © Springer-Verlag Berlin Heidelberg 2012

4.1 Introduction

A nuclear molecule [1] or a dinuclear system (DNS) is a cluster configuration consisting of two nuclei which touch each other and keep their individuality, i.e. $^8Be \rightarrow \alpha + \alpha$. The dinuclear system concept was first introduced by Volkov [2–5]. First observable evidences for nuclear molecules were detected by Bromley et al. [6] in the scattering of $^{12}C + ^{12}C$ and $^{16}O + ^{16}O$ which has some importance for the element synthesis in astrophysics.

The dinuclear system model has far reaching applications in nuclear structure physics, namely for the explanation of normal-, super- and hyper-deformation of heavy nuclei, in fusion and incomplete fusion reactions for the production of super-heavy nuclei, in mass and charge transfer reactions between clusters, in quasifission, when no compound nucleus is formed, and in binary and ternary fission. All these subjects will be discussed in this article.

The main degrees of freedom of the dinuclear model are the relative motion of the clusters and the transfer of nucleons between them. The relative motion describes the formation of the dinuclear system in heavy ion collisions, the properties of the nuclear molecular resonances and the decay of the dinuclear system leading to fission, quasifission and emission of clusters. The transfer of nucleons between the clusters changes the mass and charge asymmetries and can be phenomenologically treated by the mass and charge asymmetry coordinates, defined as $\eta = (A_1 - A_2)/(A_1 + A_2)$ and $\eta_Z = (Z_1 - Z_2)/(Z_1 + Z_2)$ where A_1, A_2 and Z_1, Z_2 are the mass and charge numbers of the clusters, respectively. The case $\eta = 0$ means two equal clusters $(A_1 = A_2)$ and at $\eta = \pm 1$ only one cluster exists because of $A_1 = 0$ or $A_2 = 0$. So the coordinate η runs in the interval $(1, -1)$ and one is able to describe also a completely fused system with $|\eta| = 1$. Further degrees of freedom needed for a more realistic representation of the nucleus-nucleus system are the deformations (vibrations) of the clusters, their orientations (rotation–oscillations), the neck degree of freedom between them, and the single-particle motion in the individual clusters.

4.2 Adiabatic or Diabatic Potentials Between Nuclei

The description of the dynamical way of heavy ion fusion strongly depends on the potential taken between the nuclei. We discriminate between adiabatic and diabatic potentials. Adiabatic potentials represent the minimum of energy of the system for a given set of collective coordinates and a given internuclear distance. The potential energy can be calculated with the Strutinsky method

$$U = U_{LD} + \delta U_{shell}, \tag{4.1}$$

where U_{LD} is the energy of the system obtained with a liquid drop model for the shape coordinates and δU_{shell} includes the effects of the shells. The shell effects

Fig. 4.1 The used
parameters of the two-center
shell model. Here, $2R_0\lambda$
measures the length of the
system. The deformation
parameters are given by
$\beta_i = a_i/b_i$ with $i = 1, 2$.
The neck parameter is
$\varepsilon = E_0/E'$

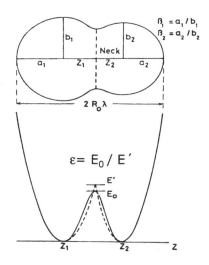

in nucleus-nucleus collisions are obtained with a two-center shell model which is
shortly discussed in the following subsection.

4.2.1 Two-Center Shell Model

In most calculations of potentials one uses the two-center shell model of Maruhn and
Greiner [7]. It is based on the two-center oscillator. The parameters of the Maruhn–
Greiner model (see Fig. 4.1) are the length ℓ of the system, expressed by the ratio
$\lambda = \ell/(2R_0)$ where R_0 is the radius of the spherical compound nucleus, the mass
asymmetry η defined by the masses left and right to the plane through the neck,
$\eta = (A_1 - A_2)/(A_1 + A_2)$, the neck parameter $\epsilon = E_0/E'$ (ratio of barriers, see
Fig. 4.1), and the deformations $\beta_i = a_i/b_i$ with $i = 1$ and 2 (ratio of semiaxes).

Recently, Diaz Torres [8] proposed a two-center shell model with a greater vari-
ability. The potentials are superpositions of two shifted and rotated Woods–Saxon
potentials

$$V = \exp(-\mathbf{R}_1\delta)\hat{U}(\Omega_1)V_1\hat{U}^{-1}(\Omega_1)\exp(\mathbf{R}_1\nabla)$$
$$+ \exp(-\mathbf{R}_2\nabla)\hat{U}(\Omega_2)V_2\hat{U}^{-1}(\Omega_2)\exp(\mathbf{R}_2\nabla) \tag{4.2}$$

with the Woods–Saxon potentials

$$V_{i=1,2} = -V_{0i}f^i(r) + \frac{1}{2}\lambda_i\left(\frac{\hbar}{mc}\right)^2 V_{0i}\frac{1}{r}\frac{df_{so}^i}{dr}\boldsymbol{\ell}\mathbf{s} \tag{4.3}$$

with $f^i, f_{so}^i = \left(1 + \exp([r - R_{0(so)}^i]/a_{0(so)}^i)\right)^{-1}. \tag{4.4}$

The centers are positioned at $\mathbf{R}_1$ and $\mathbf{R}_2$, the relative coordinate is $\mathbf{r} = \mathbf{R}_1 - \mathbf{R}_2$, and $\hat{U}(\Omega_1)$ and $\hat{U}(\Omega_2)$ are operators for rotation by the Euler angles Ω_1 and Ω_2, respectively. To get the single-particle levels of $T + V$, Diaz Torres used two non-orthogonal sets of oscillator functions around the centers of the two parts of (4.2) as basis for diagonalizing the Hamiltonian. He considered the two-center potentials of the system consisting of the spherical nuclei ^{16}O and ^{40}Ca for different relative distances and two oblately deformed ^{12}C nuclei with different relative orientations. This two-center shell model is realistic with respect to bound and continuum levels, but difficult to evaluate for heavier systems.

4.2.2 Calculation of Adiabatic and Diabatic Potentials

In order to calculate adiabatic and diabatic potentials one has to consider both types of single-particle energies for the shell effects. Diabatic TCSM states can be calculated by the maximum overlap method according to Lukasiak et al. [9, 10] or by the method of maximum symmetry where one diagonalizes a two-center Hamiltonian with maximum symmetry excluding the neck potential and terms proportional to ℓ_x, ℓ_y and s_x, s_y in the spin-orbit potential. The maximum overlap method and the method of maximum symmetry which is numerically simpler yield nearly the same results for the single-particle energies. The nucleus-nucleus potentials are obtained by the Strutinsky method [11]:

$$V_{adiab} = V_{LD} + \delta U_{shell},$$

$$V_{diab} = V_{adiab} + \sum_\alpha \left(\varepsilon_\alpha^{diab}(R) n_\alpha^{diab}(R) - \varepsilon_\alpha^{adiab}(R) n_\alpha^{adiab}(R) \right). \qquad (4.5)$$

Diabatic potentials are strongly repulsive and forbid fusion via the internuclear coordinate. They are similar to potentials calculated with double folding methods by using frozen densities. The latter type of potentials is also denoted as sudden potentials. Fig. 4.2 shows diabatic potentials for the $^{100}Mo + ^{100}Mo$ and $^{110}Pd + ^{110}Pd$ systems. One finds that the double folding potential in the $^{110}Pd + ^{110}Pd$ case (dotted–dashed curve in Fig. 4.2) leads to a very similar potential (solid curve in Fig. 4.2) obtained with diabatic single-particle energies of the TCSM and the Strutinsky formalism.

The time-dependence of the transition from a diabatic potential to an adiabatic can be related to the characteristic relaxation time for the shape degrees of freedom of the system. The potential is given by

$$V(\lambda, t) = V_{adiab}(\lambda) + \Delta V_{diab}(\lambda, t) \qquad \text{with}$$

$$\Delta V_{diab} \approx \sum_\alpha \left(\varepsilon_\alpha^{diab}(\lambda, t) n_\alpha^{diab}(\lambda, t) - \varepsilon_\alpha^{adiab}(\lambda) n_\alpha^{adiab}(\lambda) \right). \qquad (4.6)$$

Here, λ is the dimensionless internuclear distance parameter of the TCSM (see Fig. 4.1). $n_\alpha^{adiab}(\lambda)$ varies with λ according to a Fermi distribution with temperature $T(\lambda) = \sqrt{E^*(\lambda)/a}$, where $E^*(\lambda)$ is the excitation energy of the system.

Fig. 4.2 Diabatic potentials for the systems $^{100}\text{Mo} + ^{100}\text{Mo}$ (*dotted line, curve 1*) and $^{110}\text{Pd} + ^{110}\text{Pd}$ (*solid line, curve 2*). The phenomenological double folding potential for the system $^{110}\text{Pd} + ^{110}\text{Pd}$ is shown by the *dashed–dotted line (curve 3)*. The discrepancy between this potential and the diabatic one becomes smaller if, by starting at the minimum of the pocket, the neck parameter ε is diminished by decreasing λ (*dashed line, curve 4*) for $^{110}\text{Pd} + ^{110}\text{Pd}$

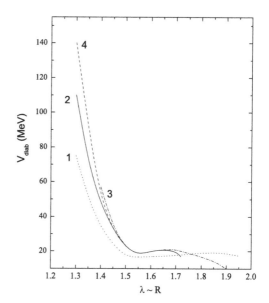

The diabatic occupation numbers n_α^{diab} follow relaxation equations

$$\frac{dn_\alpha^{diab}(\lambda, t)}{dt} = -\frac{1}{\tau(\lambda, t)}\left(n_\alpha^{diab}(\lambda, t) - n_\alpha^{adiab}(\lambda)\right) \tag{4.7}$$

with $\quad \tau(\lambda, t) = \frac{2\hbar}{\langle\Gamma\rangle} \approx 5 \times 10^{-21}\text{s} \quad$ and

$$\langle\Gamma(\lambda, t)\rangle = \sum_\alpha \bar{n}_\alpha^{diab}(\lambda, t)\Gamma_\alpha(\lambda) / \sum_\alpha \bar{n}_\alpha^{diab}(\lambda, t), \tag{4.8}$$

$$\bar{n}_\alpha^{diab} = n_\alpha^{diab} \quad \text{for } \varepsilon_\alpha^{diab} > \varepsilon_F,$$

$$\bar{n}_\alpha^{diab} = 1 - n_\alpha^{diab} \quad \text{for } \varepsilon_\alpha^{diab} \leq \varepsilon_F.$$

Fig. 4.3 shows a calculation of the potential for the $^{110}\text{Pd} + ^{110}\text{Pd}$ system as a function of λ. The dashed potential lying between the diabatic (solid line) and adiabatic (points) potentials results from the above equations for a time $t_0 = 8 \times 10^{-21}$ s which is roughly the time for forming the compound nucleus. The astonishing outcome is that a quite high barrier of about 60 MeV remains towards smaller internuclear distances and hinders the direct fusion to ^{220} U along the internuclear coordinate. From these calculations we conclude that in heavier collision systems one has to consider diabatic or modified diabatic potentials with high barriers to the inside hindering a direct fusion process.

With a microscopic approach based on the formalism of irreducible representations of the SU(3) group one finds an influence of the structural forbiddenness on the

Fig. 4.3 The diabatic (*solid curve*), the diabatic time-dependent (*dashed curve*) and the adiabatic (*dotted curve*) potentials for ^{110}Pd $+^{110}$Pd as a function of λ

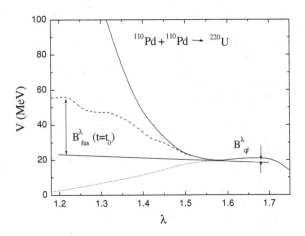

fusion of heavy nuclei and can estimate the energy thresholds for complete fusion in relative distance and mass asymmetry degrees of freedom. This effects are similar as the shown diabaticity of the internuclear potential [12].

4.2.3 The Motion of the Neck

Here we consider the dynamics of the neck degree of freedom between the touching nuclei [13, 14]. The neck dynamics is described by the neck parameter $\varepsilon = E_0/E'$ defined by the ratio of the actual barrier height E_0 to the barrier height E' of the two-center oscillator (see Fig. 4.1). The neck grows with decreasing ε. In order to learn about the neck dynamics we calculated the potential energy surface as a function of λ and ε for the case of the ^{110}Pd $+^{110}$Pd system in the adiabatic approach (see Fig. 4.4) and carried out dynamical, time-dependent calculations which have an adiabatic character because of the adiabatic potential energy surface. The kinetic energy is written

$$T = \frac{1}{2}\sum_{i,j} B_{i,j}\dot{q}_i\dot{q}_j, \quad i = 1, 2, \quad q_1 = \lambda, \quad q_2 = \epsilon, \qquad (4.9)$$

and dissipative forces are included by use of a Raleigh dissipation function

$$\Phi = \frac{1}{2}\sum_{i,j} \gamma_{i,j}\dot{q}_i\dot{q}_j, \qquad (4.10)$$

where the friction coefficients are calculated with

$$\gamma_{i,j} = 2\Gamma B_{i,j}/\hbar \qquad (4.11)$$

Fig. 4.4 Potential energy surface (units MeV), calculated in the (λ, ε)-plane for the reaction ^{110}Pd $+^{110}$Pd with shell corrections and $\beta_i = 1$ (*lowest part*), without shell corrections and $\beta_i = 1$ (*middle part*), and with shell corrections and $\beta_i = 1.2$ (*upper part*). The dynamical trajectories in the lowest part starting from the touching configurations and with initial kinetic energies 0, 40 and 60 MeV are presented by *solid, dashed* and *dotted lines*, respectively

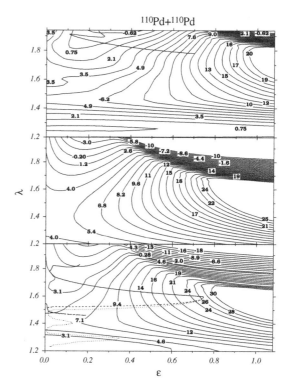

according to the linear response theory and Γ is the average width of single particle states. The classical equations are derived from the Lagrangian $L = T - V$ and Raleigh dissipation function Φ. Starting with $\lambda = 1.59$ and $\epsilon = 0.75$ for ^{110}Pd$+^{110}$Pd and with mass parameters obtained within the Werner–Wheeler approximation under the assumption of an incompressible and irrotational flow of the nuclear matter we reached the fission-type valley already after a very short time of $3 - 4 \times 10^{-22}$ s at $\lambda \sim 1.68$ and then found oscillations in this valley in case of a small kinetic energy. The characteristic time of all processes results $\sim 5 \times 10^{-21}$ s. This has as consequence that fusion may occur easier in reactions with heavier isotopes in contradiction to the experimental data. For the system ^{110}Pd $+^{110}$ Pd we draw the dependence of λ and ϵ as functions of time and the fusion probability with different starting values at and above the Bass barrier in Fig. 4.5, respectively. The experimental value is about $P_{CN} \approx 10^{-4}$.

With the described method one calculates a wrong dependence of the fusion probability on the isotope composition and of the mass asymmetry of target and projectile [14]. There must exist a hindrance for a fast growth of the neck and the motion to smaller values of λ. We found as an essential hindrance large microscopically calculated mass parameters for the neck motion. We obtained the microscopical mass parameters with the cranking formula [15], where the main contributions to the mass

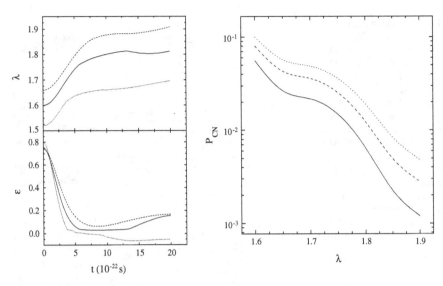

Fig. 4.5 *Left* Time dependence of λ and ε for different initial values of λ and ε at which the ^{110}Pd $+^{110}$Pd system has the same potential energy 26.5 MeV. The initial kinetic energy is zero. The calculated results with $\lambda(0) = 1.6$ and $\varepsilon(0) = 0.75$, $\lambda(0) = 1.66$ and $\varepsilon(0) = 0.74$, and $\lambda(0) = 1.52$ and $\varepsilon(0) = 0.8$ are presented by *solid, dashed* and *dotted lines,* respectively. *Right* Dependence of the fusion probability on λ_v (λ-value in the fusion valley of the potential energy surface) for the ^{110}Pd $+^{110}$Pd reaction at different excitation energies. A fast dissipation of kinetic energy in the entrance channel is assumed. The results of calculations for incident energies corresponding to the Bass barrier, 10 MeV above the Bass barrier and 20 MeV above the Bass barrier are presented by *solid, dashed* and *dotted lines,* respectively

resulted from (cr=cranking)

$$B_{i,j}^{cr} \approx \hbar^2 \sum_{\alpha} \frac{-\dfrac{dn_\alpha}{dE_\alpha}}{\Gamma_\alpha^2} \frac{\partial E_\alpha}{\partial q_i} \frac{\partial E_\alpha}{\partial q_j}, \qquad (4.12)$$

where E_α and n_α are TCSM single-particle eigenvalues and occupation numbers, respectively, and Γ_α is the width of the decaying single-particle states. This formula yields (*WW*= Werner–Wheeler)

$$B_{\lambda\lambda}^{cr} = B_{\lambda\lambda}^{WW}, \quad B_{\epsilon\epsilon}^{cr} \approx 30 \times B_{\epsilon\epsilon}^{WW}, \quad B_{\lambda\epsilon}^{cr} \approx 0.35 \times B_{\lambda\epsilon}^{WW}. \qquad (4.13)$$

The much larger neck mass parameter $B_{\epsilon\epsilon}^{cr}$ has as consequence that the system stays nearly fixed at the entrance configuration, which is the typical dinuclear system configuration, for a sufficient long time. Fig. 4.6 shows the motion of ϵ as functions of the time and the internuclear distance λ obtained with the Werner–Wheeler masses (pointed curves) and with microscopical masses (solid curves). The latter masses lead to a slow growth of the neck and justify the assumption of a fixed neck as we presume in the DNS model [15]. Beside large neck mass parameter we found also

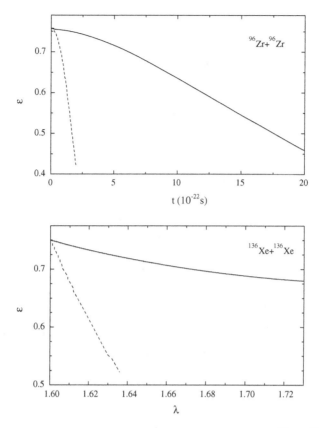

Fig. 4.6 *Upper part* Time-dependence of the neck parameter ε in the system ^{96}Zr + ^{96}Zr calculated with microscopical (*solid curve*) and Werner–Wheeler (*dashed curve*) mass parameters. *Lower part* Trajectories in the (λ, ε)-plane calculated for the system ^{136}Xe + ^{136}Xe with microscopical (*solid curve*) and Werner–Wheeler (*dashed curve*) mass parameters. The end points of the *solid* and *dashed curves* are at time $t = 2 \times 10^{-21}$ s and $t = 2 \times 10^{-22}$ s, respectively

other dynamical restrictions for a fast growth of the neck which are caused by the potential energy surface intermediate between the adiabatic and diabatic limits.

4.2.4 Repulsive Potentials by the Quantization of Kinetic Energy

Also a consequent quantization of the kinetic energy of a collectively described nucleus–nucleus system can lead to a repulsive potential as shown by the work of Fink et al. [16]. Assume the classical Hamiltonian of the nucleus-nucleus system as

$$H = T(x^i, \dot{x}^i) + V(x^i) \quad \text{with} \quad x^i = x^1, x^2, x^3, \dots \quad \text{and} \quad (4.14)$$

$$T = \frac{1}{2} g_{ik} \dot{x}^i \dot{x}^k. \tag{4.15}$$

The quantization yields for the operator of kinetic energy:

$$\hat{T} = -\frac{\hbar^2}{2} g^{-1/2} \frac{\partial}{\partial x^i} g^{ik} g^{1/2} \frac{\partial}{\partial x^k} \tag{4.16}$$

with $g = \det(g_{ik})$, $g^{ik} = (g^{-1})_{ik}$, $g^{ik} g_{kl} = \delta_{il}$.

Choosing the coordinate $x^1 = R$ as the internuclear distance and $x^{\mu=2,3,4,\cdots}$ as the other coordinates (Greek letters) we obtain after some transformations for the Hamiltonian:

$$\hat{H} = -\frac{\hbar^2}{2} g^{11} \frac{\partial^2}{\partial R^2} - \frac{\hbar^2}{2} g^{-1/2} \frac{\partial}{\partial x^\mu} \tilde{g}^{\mu\nu} g^{1/2} \frac{\partial}{\partial x^\nu}$$
$$+ V\left(R, x^\mu + \int_\infty^R \frac{g^{1\mu}}{g^{11}} dR'\right) + V_{add}(R) \tag{4.17}$$

The additional potential V_{add} will be given below. As example was chosen the scattering of two ^{12}C nuclei: ^{12}C + ^{12}C. As coordinates we took the coordinate of the relative motion $\mathbf{r} = \mathbf{R}$ and the quadrupole deformations of both nuclei by using symmetrical and asymmetrical coordinates as follows

$$\alpha_{2\mu}^{(s)} = \frac{1}{\sqrt{2}} (\alpha_{2\mu}^{(1)} + \alpha_{2\mu}^{(2)}) \quad \alpha_{2\mu}^{(a)} = \frac{1}{\sqrt{2}} (\alpha_{2\mu}^{(1)} - \alpha_{2\mu}^{(2)}). \tag{4.18}$$

The transformation of $\hat{H}$ to a constant reduced mass μ_0 for the relative motion yields

$$\hat{H}' = -\frac{\hbar^2}{2\mu_0} \frac{\partial^2}{\partial r^2} + \cdots \frac{1}{g^{11}\mu_0} V_{add} + \frac{g^{11}\mu_0 - 1}{g^{11}\mu_0} E \tag{4.19}$$

with $V_{add} = \frac{\hbar^2}{4} g^{11} \left(\frac{\partial^2}{\partial r^2} \ln(g^{11} g^{1/2}) + \frac{1}{2} \left(\frac{\partial}{\partial r} \ln(g^{11} g^{1/2}) \right)^2 \right)$ (4.20)

The additional potential results from the r-dependence of the mass tensor. Since this potential depends not only on g^{11}, but also on the determinant g of the mass tensor, all mass elements contribute equivalently. Fig. 4.7 shows the result for $\tilde{V}_{add} = V_{add}/(g^{11}\mu_0)$ which has the main contributions from the $\alpha_{2\mu}$ masses. In conclusion, we learn that the correct inclusion of more degrees of freedom leads to repulsive potentials which screen the inside from the outside like the diabatic potentials perform it. If in calculations with adiabatic potentials the number of explicitly treated degrees of freedom is increased, then these calculations give a relative motion which proceeds essentially in a diabatic potential since the treated degrees of freedom consume the content of the kinetic energy of the relative motion and bring it perhaps to a stop.

Fig. 4.7 The additional potential $\mathscr{V}_{add} = \tilde{V}_{add}(r) = (1/\mu_0 g^{11}) V_{add}(r)$ is plotted for ^{12}C + ^{12}C. In the *broken curve*, only the mass of the r degree of freedom is taken into consideration; in the *full curve*, the masses of the $\alpha_{2\mu}$ motion are also considered. The comparison of the two curves shows that the main contributors to $\tilde{V}_{add}$ are the $\alpha_{2\mu}$ masses

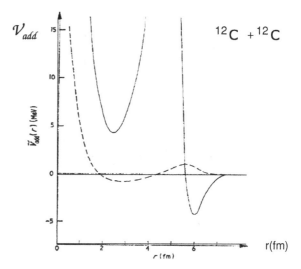

The nucleus–nucleus potentials have a potential pocket resulting by the attractive nuclear and repulsive Coulomb interactions. With increasing Coulomb repulsion in the DNS the depth of the pocket decreases. Due to a quite shallow potential pocket in the diabatic regime, about 0.5 MeV, in the ^{58}Ni + ^{208}Pb reaction, the capture cross section in this reaction would have the maximum at low bombarding energies about (15–25) MeV above the Coulomb barrier [17]. The experimental verification of this effect would allow us to discriminate between the adiabatic and diabatic regimes of nucleus-nucleus interaction and determine the depth of the potential pocket.

4.3 Nuclear Molecules, Hyperdeformed Nuclear Structures

The evidence of low-spin hyperdeformed (HD) states in actinides has been experimentally established only in induced fission reactions [18–22]. Usually these states are explained by a third minimum in the potential energy surface corresponding to a nuclear ellipsoid shape with a ratio of axes of 1 : 3. An interesting observation in shell model calculations was made that the third minimum of the potential energy surface of actinide nuclei belongs to a molecular configuration of two touching nuclei (clusters). Therefore, interpreting a HD configuration as a dinuclear system we showed that the dinuclear systems have quadrupole moments and moments of inertia as those measured for superdeformed states and estimated for HD states. The cluster states of light nuclei and the possible existence of the necked-in shaped nuclei were considered [23–30].

The indications of the population of high-spin HD states in fusion-evaporation reactions with heavy-ions have been discussed by Herskind et al. (2006). In the publication of Adamian et al. [31] it was suggested to directly populate these states in heavy ion collisions without going through the stage of compound nucleus formation. Such states in the nucleus-nucleus potential are discussed in connection with nuclear molecular states first observed in the ^{12}C $+^{12}$C collision by Bromley et al. (1960) and then seen up to the system Ni + Ni by Cindro et al. (1996).

4.3.1 Hyperdeformed States Directly Formed in the Scattering of ^{48}Ca $+^{140}$Ce and ^{90}Zr $+^{90}$Zr

In the following we discuss the dinuclear systems ^{48}Ca $+^{140}$Ce and ^{90}Zr $+^{90}$Zr as possible candidates for explaining the hyperdeformed states [31]. First, we calculate the potentials $V(R, \eta, L)$ as functions of the relative distance for various angular momenta [32]:

$$V(R, \eta, \eta_Z, L) = V_C(R, \eta_Z) + V_N(R, \eta) + V_{rot}(R, \eta, L) \qquad (4.21)$$

This potential consists of the Coulomb potential, the nuclear part and the centrifugal potential $V_{rot} = \hbar^2 L(L + 1)/(2\Im)$. The nuclear part is calculated by a double folding procedure with a Skyrme-type effective density dependent nucleon-nucleon interaction taken from the theory of finite Fermi systems [33]. The potentials for the two above mentioned systems are depicted in Fig. 4.8. They have a minimum around a distance $R_m \approx R_1 + R_2 + 0.5$fm ≈ 11 fm where R_1 and R_2 are the radii of the nuclei. The position of the Coulomb (outer) barrier corresponds to $R = R_b \approx R_m + 1$ fm. The depth of this molecular minimum decreases with growing angular momentum and vanishes for $L > 100$ in the considered systems.

In the potential around the minimum there are situated virtual and quasibound states above and below the barrier, respectively. Replacing the potential near the minimum by a harmonic oscillator potential we find the positions of one to three quasibound states with an energy spacing of $\hbar\omega \approx 2.2$ MeV for $L > 40$. For example, the ^{90}Zr $+^{90}$Zr system has the lowest quasibound state for $L = 50$ lying 1.1 MeV above the potential minimum.

The charge quadrupole moments of $(40 - 50) \times 10^2 e$ fm^2 and the moments of inertia of $(160–190) \times \hbar^2 /$ MeV of the quasibound dinuclear configurations ^{48}Ca + ^{140}Ce and ^{90}Zr + ^{90}Zr are near to those estimated for hyperdeformed states. Therefore, we assume the quasibound states as HD states which can in principle produced in the heavy ion reactions ^{48}Ca on ^{140}Ce and ^{90}Zr on ^{90}Zr. However, the following conditions should be fulfilled: (1) The quasibound states should be directly excited and the DNS should have no extra excitation energy. The excitation of the quasibound states can proceed directly via a tunneling through the outer barrier including the centrifugal potential. (2) The DNS should not change the mass and charge asymmetries and stay fixed in the potential minimum for some time. Spherical and stiff nuclei (magic and double magic nuclei) have this property.

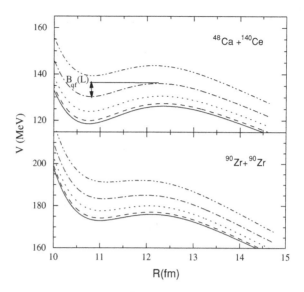

Fig. 4.8 The potential $V(R, L)$ for the systems ^{48}Ca + ^{140}Ce (*upper part*) and ^{90}Zr + ^{90}Zr (*lower part*) as a function of R for $L = 0, 20, 40, 60, 80$ presented by *solid, dashed, dotted, dashed–dotted* and *dashed–dotted–dotted curves*, respectively

The cross section for penetrating the barrier and populating quasibound states is given as

$$\sigma(E_{c.m.}) = \frac{\pi \hbar^2}{2\mu E_{c.m.}} \sum_{L=L_{min}}^{L_{max}} (2L + 1) T_L(E_{c.m.}). \tag{4.22}$$

Here, $E_{c.m.}$ is the bombarding energy in the center of mass system and $T_L(E_{c.m.})$ the transmission probability through the entrance barrier at $R = R_b$ approximated by a parabola with frequency ω'

$$T_L(E_{c.m.}) = 1/\left(1 + \exp[2\pi(V(R_b, \eta, \eta_Z, L) - E_{c.m.})/(\hbar\omega')]\right). \tag{4.23}$$

The angular quantum numbers L_{min} and L_{max} fix the interval of angular momenta which contribute to the excitation of the HD states. Only an interval of angular momenta leads to a stronger excitation of quasibound states. This effect is known in the theory of nuclear molecules in the case of light heavy ions and there it is called the molecular window. The lower limit L_{min} is determined by the absorption of the partial wave to fusion, transfer and excitation channels and the higher limit L_{max} is fixed by the upper border of quasibound states.

In the reaction ^{48}Ca on ^{140}Ce, cold and long living DNS can be formed at an incident energy $E_{c.m.} = 147$ MeV and $90 \leq L \leq 100$, and in the reaction ^{90}Zr on ^{90}Zr at $E_{c.m.} = 180$ MeV and $40 \leq L \leq 50$. For both reactions we expect a cross section of smaller than 1 μ b. Also the reaction ^{58}Ni + ^{58}Ni ($E_{c.m.} = 117$MeV, $70 \leq L \leq 80$, and the moment of inertia is 96 $\hbar^2$/MeV) is a possible candidate for populating the cluster-type HD states [34–36].

The spectroscopic investigation of HD structures is difficult. One can observe the γ-radiation caused by transitions between the collective rotational HD states or one can measure their return-decay into the entrance channel. For high values of $L \geq 70$ ($L \geq 90$), electromagnetic $E2-$ transitions in the $^{58}\text{Ni}+^{58}\text{Ni}$ ($^{48}\text{Ca}+^{140}\text{Ce}$) system have an energy of about 1.4–2 MeV (0.7–1.10 MeV). The formation cross sections are small, and a high background is produced by fusion-fission, quasifission and other processes. However, the latter processes have characteristic times much shorter than the life-time of the HD states. The HD states should show up as sharp resonance lines as a function of the incident energy.

4.3.2 Hyperdeformed States Formed by Neutron Emission from the Dinuclear System

Another possibility to populate the HD states is the excitation after neutron emission [37]. Now the incident energy is slightly larger than the Coulomb (outer) barrier. With neutron-rich isotopes as projectiles, one lowers the neutron separation energy and increases the probability of neutron emission from the DNS. Thus one produces HD states with larger cross sections.

In our model we treat the formation and decay of the HD states as a three-step process. First, an excited initial DNS is formed in the entrance channel. Here the excitation energy of the DNS means the intrinsic excitation energy. Second, the DNS loses its excitation energy by the emission of a neutron and forms a HD state. Third, this rotating DNS emits γ-quanta and/or decays into two fragments.

The cross section $\sigma_{HD}(E_{c.m.})$ for the formation of the HD state depends on the capture cross section or the capture probability P_{cap} and on the probability P_{HD} of the transformation of the DNS under neutron emission into a HD state which is a cold DNS:

$$\sigma_{HD}(E_{c.m.}) = \sum_{L=L_{min}}^{L_{max}} \frac{\pi \hbar^2}{2\mu E_{c.m.}}(2L+1)P_{cap}(E_{c.m.}, L)P_{HD}(E_{c.m.}, L). \quad (4.24)$$

In order to calculate the value P_{cap}, we use the formalism of the reduced density matrix and, therefore, take into account the influence of dissipation and fluctuations in the relative coordinate. The details of the calculation of P_{cap} are presented in [17].

The decrease of the excitation energy of the hot DNS is provided by the emission of a neutron which carries away the energy $\varepsilon_{n_k} + B_{n_k}$ where ε_{n_k} is the kinetic energy of the neutron and B_{n_k} the neutron separation energy. The index $k = 1$ or 2 corresponds to the DNS fragment from where the neutron can be evaporated. We apply the statistical approach and describe the evolution of the excited DNS by the competition between the neutron emission from the system and the DNS transition over the quasifission barrier B_R^{qf} in R or over barriers $B_{\eta_Z}^{sym}$ and $B_{\eta_Z}^{asym}$ in η_Z in the direction to more symmetric and more asymmetric configurations, respectively.

Taking into account the competition between different deexcitation channels by the probability P_{n_k}, one can write the probability $P_{HD}(E_{c.m.}, L)$ of the formation of the HD state under neutron emission from the excited initial DNS state as:

$$P_{HD} = \sum_{k=1}^{2} P_{n_k}(E^*, B_{n_k}, L) w_{n_k}(E^*, B_{n_k}), \qquad (4.25)$$

where w_{n_k} is the probability to emit the neutron with such kinetic energy to cool the excited DNS to the deepest HD state in the potential well.

The experimental method to identify the HD states is to measure rotational γ-quanta between the HD states in coincidence with the decay into the fragments which built up the HD configuration in the entrance channel. Then the formed cold system has to fulfil the following conditions

$$T_\gamma \lesssim T_R \lesssim T_{\eta_Z}. \qquad (4.26)$$

Here, T_γ is the time of the collective E2 transition, and T_R and T_{η_Z} are the tunneling times through the barriers in the R and η_Z coordinates, respectively. Measuring the consecutive E2 γ transitions, one can determine the moments of inertia and electric quadrupole moments of the HD states.

The isotopic dependence of the cross section for the formation of the HD state in the reactions ^{48}Ca $+^{A_2}$ Sn at $L = 20 - 40$ is shown in Fig. 4.9. The calculated cross sections result between 10 and 100 nb and can be understood by considering the closed neutron shell $N = 82$ of the tin isotopes. The above condition on the various times restricts the interval of angular momenta at which it is possible to identify HD states. In Fig. 4.9 we present the values of T_γ, T_R and T_{η_Z} as functions of the angular momentum L for the reactions ^{60}Ni $+^{60}$Ni and ^{48}Ca $+^{140}$Ba. The condition T_γ smaller or about T_R is only satisfied in some intervals of L. At very large L the value of T_R becomes very small because the quasifission barrier vanishes. In addition, the values of $B_{\eta_Z}^{sym}$ and $B_{\eta_Z}^{asym}$ should be quite large to provide the condition T_R smaller or about T_{η_Z}. The reactions ^{48}Ca $+^{124,128,130,132,134}$Sn, ^{48}Ca $+^{136,138}$ Xe, ^{48}Ca $+^{137,138,140}$Ba, ^{48}Ca $+^{83,84,86}$Kr, ^{40,48}Ca $+^{40,48}$Ca, ^{58,60}Ni $+^{58,60}$Ni, and ^{40}Ca$+^{58}$Ni are good candidates for the production and experimental identification of HD states. The estimated identification cross sections $\sigma_{x\gamma R}$ of at least x ($x \geq 1 - 3$) γ quanta from the HD state before its decay in R are of the order of 1 nb to 2.5 µb for optimal incident energies.

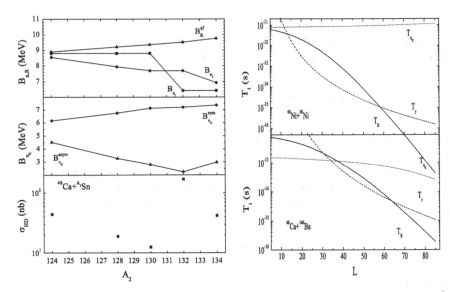

Fig. 4.9 *Left* Isotopic dependence of neutron separation energies and quasifission barriers B_R^{qf} (*upper part*), barriers $B_{\eta_Z}^{sym}$ and $B_{\eta_Z}^{asym}$ (*middle part*), and cross sections for the formation of the HD state in the reactions ^{48}Ca + ^{A_2}Sn at $L = 20 - 40$ (*lower part*). *Right* Time of collective $E2$-transition and tunneling times through the barriers in R and η_Z for the hyperdeformed states formed in the entrance channel of the indicated reactions as a function of the angular momentum L

4.4 Normal Deformed and Superdeformed Nuclei

4.4.1 Internuclear Potential, Moments of Inertia, Quadrupole and Octupole Moments of the Dinuclear Shape

First we want to consider the calculation of the internuclear potential. The potential of the DNS is strongly repulsive for smaller distances and hinders the nuclei to amalgamate together in the relative coordinate as already stated above. Under a small overlap of the nuclei, one usually calculates the potential energy semi-phenomenologically

$$U(R, \eta, \eta_Z, I) = B_1 + B_2 + V(R, \eta, \eta_Z, I) - B_{12}. \qquad (4.27)$$

Here, B_i ($i = 1, 2$) are the asymptotic negative binding energies of the nuclei, B_{12} the negative binding energy of the compound nucleus and $V(R, \eta, \eta_Z, I)$ is the interaction between the nuclei, described already earlier,

$$V(R, \eta, \eta_Z, I) = V_C(R, \eta_Z) + V_N(R, \eta) + V_{rot}(R, \eta, I). \qquad (4.28)$$

Also deformations are taken into consideration by assuming the clusters in a pole-to-pole or other orientations with minimal potential energy.

The moment of inertia of the DNS can be taken in the sticking limit

$$\Im = \Im_1 + \Im_2 + \mu R^2, \tag{4.29}$$

where μ is the reduced mass of relative motion. The moments of inertia, $\Im_i$ ($i = 1, 2$) of the nuclei may be calculated in the rigid-body approximation ($\Im_i = \Im_i^r$).

The mass and charge multipole moments of the DNS shape can be obtained with the expression

$$Q_{\lambda\mu} = \sqrt{\frac{16\pi}{2\lambda + 1}} \int \rho(\mathbf{r}) r^\lambda Y_{\lambda\mu}(\Omega) d\tau. \tag{4.30}$$

·Here, $\rho(\mathbf{r})$ is the sum of the mass or charge densities of the clusters forming the DNS: $\rho(\mathbf{r}) = \rho_1(\mathbf{r}) + \rho_2(\mathbf{r} + \mathbf{R})$ which is the frozen density approximation. The shape of the dinuclear system is described by the internuclear distance $\mathbf{R}$ and the mass and/or the charge asymmetry coordinates (also deformation coordinates of the individual clusters can additionally be used). This shape of the system can be compared [38] with a multipole expansion of the nuclear surface with deformation parameters β_λ: $\bar{R} = R_0(1 + \beta_0 Y_{00} + \beta_1 Y_{10} + \beta_2 Y_{20} + ...)$, where R_0 is the spherical equivalent radius. If we equate the mass multipole moments calculated with deformation parameters and with the parameters of the DNS,

$$Q_{\lambda 0}(\beta_\lambda) = Q_{\lambda 0}(R, \eta), \tag{4.31}$$

we find a relation between these parameters. In Fig. 4.10 we depict the mass multipole moments $Q_\lambda = Q_{\lambda 0}$ for two touching clusters forming the ^{152}Dy system (at $R = R_m$) and the deformation parameters β_λ as a function of the mass asymmetry η. Thus, the DNS with $0.7 \leq \eta \leq 0.8$ ($0.5 \leq \beta_2 \leq 0.75$) and $\eta < 0.7$ ($\beta_2 > 0.75$) can be related to the superdeformed and HD nuclear states, respectively.

4.4.2 Parity Splitting in Heavy Nuclei

A low-lying band with negative parity states is found near the positive parity ground-state band in even-even actinide nuclei such as Ra, Th, U, and Pu which is caused by reflection-asymmetric shapes of these nuclei. The negative parity states are shifted upwards with respect to the positive parity states, denoted as parity splitting. The band with negative parity and the parity splitting are explained by the dynamics of the octupole shape degree of freedom or by assuming vibrations in the mass asymmetry degree of freedom describable within the dinuclear system model [39–42]. The latter type of approach will be explained in the following.

The mass asymmetry coordinate η is used as the relevant collective variable. The ground state wave function in η is thought as a superposition of different cluster-type configurations including the mono-nucleus configuration at $|\eta| = 1$. If we calculate

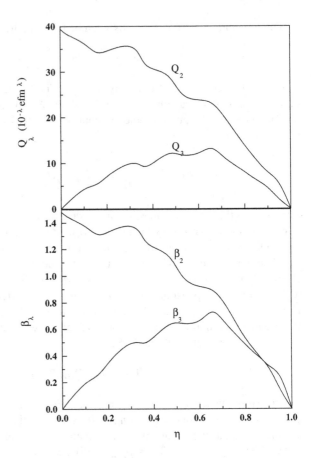

Fig. 4.10 Dependence of the quadrupole (Q_2) and octupole (Q_3) moments (*upper part*) and of the deformation parameters β_2 and β_3 (*lower part*) on the mass asymmetry η for the compound nucleus ^{152}Dy

the potential energy of the dinuclear system, we find that the α-cluster configuration with $\eta_\alpha = \pm(1 - 8/A)$ can have the lowest potential energy, lower than the energy of the mono-nucleus. The schematic picture (Fig. 4.11) shows the potential in the mass asymmetry and the reflection-asymmetric shapes of the α-cluster configurations.

For the description we substitute the coordinate η by $x = \eta - 1$ if $\eta > 0$ and $x = \eta + 1$ if $\eta < 0$. Then the Schrödinger equation is written

$$\left(-\frac{\hbar^2}{2} \frac{d}{dx} \frac{1}{B_x} \frac{d}{dx} + U(x, I) \right) \psi_n(x, I) = E_n(I)\psi_n(x, I). \qquad (4.32)$$

The mass $B_x = B_\eta$ is the effective mass parameter in the η coordinate. A method of calculating of B_η is described in Ref. [43]. The value of B_η can be estimated by relating the mass-asymmetry coordinate η to the octupole deformation coordinate β_3. It can be derived:

$$\beta_3 = \sqrt{\frac{7}{4\pi}} \frac{\pi}{3} \eta(1 - \eta^2) \frac{R^3}{R_0^3} \qquad (4.33)$$

Fig. 4.11 Schematic picture
of the potential in the mass
asymmetry and of the two
states with different parities
(*parallel lines* lower state
with positive parity, higher
state with negative parity)

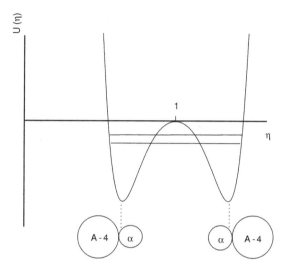

with R_0 as the spherical equivalent radius of the corresponding compound nucleus.
With a value of $B_{\beta_3} = 200\,\hbar^2/\,\text{MeV}$ known from the literature, we calculate $B_\eta \simeq (d\beta_3/d\eta)^2 B_{\beta_3} = 9.3 \times 10^4\,M\text{fm}^2$ (M is the nucleon mass), a value compatible with
the one used in the calculations.

The potential $U(x, I)$ is obtained by setting $U(x, I) = U(R = R_m, \eta, I)$ with
the touching distance R_m of the clusters. The moment of inertia $\Im$ in the centrifugal
potential is expressed for cluster configurations with α and Li as light clusters as

$$\Im(\eta) = c_1 \left(\Im_1^r + \Im_2^r + M \frac{A_1 A_2}{A} R_m^2 \right). \tag{4.34}$$

The rigid-body moments $\Im_i^r$ and the constant $c_1 = 0.85$ were used for all considered
nuclei. For $|\eta| = 1$ we assumed $\Im = c_2 \Im^r(|\eta| = 1)$ where $\Im^r$ was calculated for a
deformed mono-nucleus and $c_2 = 0.1 - 0.3$ was fixed by the energy of the first 2^+
state. For example, for the isotopes ^{220,222,224,226}Ra we have $\Im(|\eta| = 1) = 12, 17,$
22, 32 $\hbar^2/\,\text{MeV}$, respectively.

We calculated the parity splitting for even-even isotopes of the actinides Ra, Th, U
and Pu and of the nuclei Ba, Ce and Nd for different values of the nuclear spin I [39,
40]. In Fig. 4.12 we show the rotational spectra for ^{232,234,236,238}U. The points are
the experimental energies of the states [44], the lines connect the calculated points.

The calculated alternative parity states of ground-state rotational bands in
$N = 150$ isotones are listed in Table 4.1 [42]. They agree well with the available
experimental data [44] for the nuclei ^{244}Pu (positive and negative parity states) and
^{246}Cm, ^{248}Cf, ^{252}No (positive parity states). One can see in the rotational band of
^{244}Pu, that there is an appreciable shift of the negative parity states with respect
to the positive parity states. The alternative parity states are predicted in nuclei
^{246}Cm, ^{248}Cf, ^{250}Fm, ^{252}No, and ^{254}Rf. In these nuclei there is also an appreciable shift of the negative parity states with respect to the positive parity states.

Table 4.1 Calculated (E) and experimental (E_{exp}) energies (in keV) of the levels of ground-state rotational band ($K^p = 0^+$) in $N = 150$ isotones

I^p	^{244}Pu		^{246}Cm		^{248}Cf		^{250}Fm		^{252}No		^{254}Rf	
	E	E_{exp}	E	E_{exp}	E	E_{exp}	E	E_{exp}	E	E_{exp}	E	E_{exp}
0^+	0	0	0	0	0	0	0		0	0	0	
1^-	804		797		797		782		692		596	
2^+	44	44	43	43	42	42	42		46	46	45	
3^-	870	957	862		861		845		760		660	
4^+	145	155	143	142	141	138	139		154	154	150	
5^-	989	1068	978		976		957		880		775	
6^+	303	318	298	295	294	285	290		321	321	313	
7^-	1157	1206	1144		1139		1117		1050		937	
8^+	515	535	507	500	500		492		545	545	530	
9^-	1373	1395	1356		1348		1322		1268		1143	
10^+	779	802	765		755		743		822	822	799	
11^-	1634	1628	1612		1600		1569		1589		1390	
12^+	1090	1116	1071		1057		1041		1148	1150	1115	

Experimental data are taken from Ref. [44]

The maximal uncertainty of calculated energies is estimated to be about 100 keV and mainly related to uncertainty of the calculation of the potential energy of the α-cluster configuration. Therefore, within the model [42] the unknown collective states can be predicted with high accuracy.

One should note that the alternative parity states in the yrast rotational band of heaviest nuclei ($Z \geq 100$) are not yet found in the experiments. However, there is known the 3^- state of non-yrast structure at the excitation energy of 987 keV in ^{254}No [44] which is close to our predicted 3^- yrast state. Perhaps, the lack of other negative parity states in the present experimental yrast rotational bands is explained by the difficulties to detect these states due to the small production cross sections, large background, strong competition between the channels of γ-decay and emission of conversion electrons, and appreciable shift of the states with different parities. Further experimental and theoretical investigations of the predicted negative and positive parity partners are desirable.

Also we considered spectra of odd–even nuclei, e. g. Ra, Th, U, Pu, Fm, No and Rf isotopes, with an odd number of neutrons which leads to a doubling of states of fixed nuclear spin I [41, 42]. A good test for the quality of the calculations are the reduced matrix elements of the electric multipole moments $Q(1)$, $Q(2)$ and $Q(3)$. The electric multipole operators can be calculated for the DNS and it results the expression, for example, for the charge dipole moment with respect to the center of mass

$$Q_{10} = e\frac{A}{2}(1 - \eta^2)R_m \left(\frac{Z_1}{A_1} - \frac{Z_2}{A_2}\right) \qquad (4.35)$$

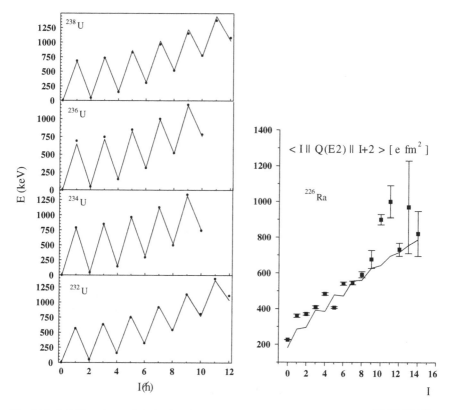

Fig. 4.12 *Left* Experimental (*points*) and theoretical (*lines*) rotational spectra of ^{232,234,236,238}U. *Right* Reduced matrix elements of the electric quadrupole operator (*solid curve*) for ^{226}Ra in comparison with experimental data (*squares*)

The calculated transition multipole matrix elements with effective charges are in good agreement with the experimental data (see Fig. 4.12, [39–42]).

4.4.3 Cluster Effects in the Ground State and Superdeformed Bands of ^{60}Zn

More than 200 superdeformed (SD) bands have been investigated in various mass regions ($A = 60$, 80, 130, 150 and 190). In the following we consider the SD band of the ^{60}Zn nucleus with the dinuclear model [45]. At low energies the structure of ^{60}Zn is built up by a double magic ^{56}Ni and an α-particle. We can assume that the ground state band of ^{60}Zn contains an α component since the threshold for α-decay is only 2.7 MeV. The observed SD band decays into the states of the ground state band in the spin region $I = 8 - 12$. The moment of inertia in the SD band

Fig. 4.13 Potential energy of
^{60}Zn as a function of x is
presented in a stepwise
manner. Absolute squares of
the wave functions of the
ground state (*solid curve*)
and of the lowest state of the
SD band (*dashed curve*) with
$I = 0$ as a function of x

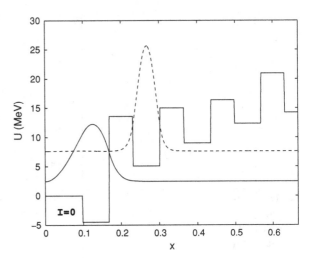

varies between 692 and 795 M fm^2 depending on spin. These values are close to
the sticking moment of inertia of the ^{52}Fe $+^8$Be cluster configuration which is 750
Mfm^2. The threshold energies for the decay of ^{60}Zn into ^{52}Fe $+^8$Be and ^{48}Cr $+^{12}$C
are 10.8 and 11.2 MeV, respectively. This values are close to the estimated value of
the SD band head at about 7.5 MeV. Therefore, we assume that two different cluster
configurations mainly occur in the states of ^{60}Zn, namely the states of the ground
state band have an α-cluster configuration and the states of the SD band contain a
Be cluster configuration as an important component.

Let us describe the clusterization of ^{60}Zn with the dinuclear system model. Since
the mass fragmentations are near $|\eta| = 1$, we replace the coordinate η by $x = \eta - 1$
if $\eta > 0$ and $x = 1 + \eta$ if $\eta < 0$. Fig. 4.13 presents the potential energy $U(x, I = 0)$
of ^{60}Zn in a stepwise manner calculated for the fragmentations with an α, ^{6}Li,
^{8}Be, ^{10}B, ^{12}C (and so on) clusters. We find the α-minimum by 4.5 MeV deeper than
the energy of the mono-nucleus at $x = 0$. The further two minima belonging to the
^{8}Be and ^{12}C cluster configurations have the values 5.1 and 9.0 MeV above the zero
value at $x = 0$. The widths of all intermediate barriers and the minima are separately
equal to each other in order to minimize the number of free parameters. These two
widths were determined by the experimental energy of the 3$^-$ state at 3.504 MeV.

Fig. 4.13 shows also the absolute squares of the wave functions of the ground state
and of the lowest state of the SD band for $I = 0$. The calculated and experimental
spectra of the ground state and SD bands are given in Fig. 4.14. The moments of
inertia $\Im(x)$ are set as the rigid body moments of inertia for cluster configurations
with $x \neq 0$. But the known energies of the states of the ground state band are fitted in
order to reach the correct γ-transition energies in the ground state band. We describe
this band as a soft rotor with the moment of inertia depending linearly on the spin
for $I \leq 8$. Between the states with $I = 8$ and 10 a larger energy gap arises since
the nucleons occupy higher lying single particle states, namely first of all the $1g_{9/2}$

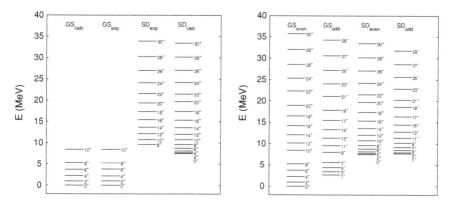

Fig. 4.14 *Left hand side* Experimental and calculated energies of the states of the ground state (GS) and superdeformed (SD) bands of ^{60}Zn. *Right hand side* Calculated energies of positive and negative parity bands of ^{60}Zn

state. This effect can be described by an increase of the potential $U(x = 0, I = 10)$ for reproducing the experimental energy of the 10^+ state. For higher values of I we take the same angular momentum dependence of the moment of inertia of the ground state band as for $I \leq 8$. The details of the estimation of the mass parameter $B_x = B_\eta$ used in the Schrödinger equation can be read in [45].

The energies of the SD band have a variation of the moment of inertia with spin reproduced in our calculations. One finds a crossing of the ground state band and the SD band around $I = 20$. Then the SD band is the yrast band above $I = 20$. Also negative parity states are predicted with the dinuclear system model (see Fig. 4.14). The parity splitting practically disappears in the SD band. There is no experimental information about low lying collective states with negative parity in ^{60}Zn with exception of the 3^- state at 3.504 MeV.

We treated electromagnetic transitions inside the SD band and from this band to the ground state band. The branching ratio $I(12^+_{sd} \rightarrow 10^+_{gs})/I(12^+_{sd} \rightarrow 10^+_{sd})$ of the E2 transitions, where the lowest experimental 10^+ state is taken as the 10^+_{gs}, is found 0.54 in experiment [46] and obtained as 0.42 in our calculations. For the ratio $I(10^+_{sd} \rightarrow 8^+_{gs})/I(10^+_{sd} \rightarrow 8^+_{sd})$, the experimental value is 0.60 [46] and the calculated one 0.63.

4.4.4 Decay Out Phenomenon of Superdeformed Bands in the Mass Region $A \approx 190$

While the rotational transitions between the SD states are easy to detect with modern Ge arrays, it is hard to localize the SD bands in excitation energy, spin and parity and to link them to the normal deformed (ND) bands [47–52]. This is because of the

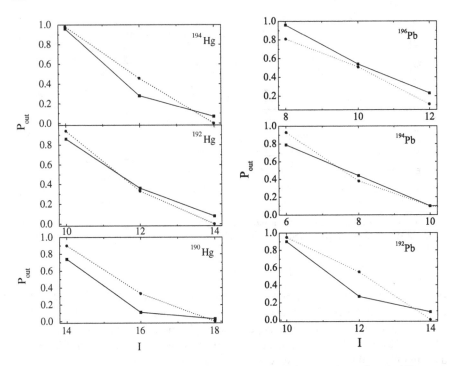

Fig. 4.15 The calculated (*solid circles*) probability P_{out} that the SD state decays into the ND states as a function of spin I for different isotopes of mercury and lead. Experimental data (*solid squares*) are taken from Ref. [44]. The solid and dashed lines are to guide the eye

decay out phenomenon of the SD band: the whole population of the SD band goes practically to zero within two transitions at some critical spin. The decay out of SD bands in the mass region $A \approx 190$ was explained in Ref. [53] in the framework of the DNS approach. One can see in Fig. 4.15 that the calculated total probabilities P_{out} of decay out are in good agreement with the experimental ones. Our analysis indicates that the sudden decay out of the SD band (^{8}Be-cluster configuration) into the ND band (related to the α-particle clusterization) is because of the crossing of the SD band with the nearest neighboring excited ND band. The main reasons for the decay out at the crossover point are (i) the perceptible square of the amplitude of the SD wave function component in the ND well and (ii) the reduction of the in-band SD collective $E2$ decay rate and the increase in the ND statistical E1 transition rate due to the large excitation energy of the SD state with respect to the ND yrast line. Near the band-crossing point, the statistical E1 decay to the ND configurations competes successfully against the collective $E2$ decay along the SD band.

The decay width of a light cluster is one of the most important physical quantities to identify a cluster-like structure. During γ-emission, the SD cluster states can decay into two fragment clusters. Therefore, one can try to identify the SD states by measuring the rotational γ-quanta in coincidence with the decay fragments of the DNS.

The sub-barrier fission resonances in Th, U, and Pu nuclei can be interpreted as arising from the reflection-asymmetric SD and HD cluster states. The low-spin SD and HD isomers are described mainly as Mg, Al, and Ca cluster configurations, respectively. Then, we can propose a new experimental method to identify the SD isomers by measuring the decay fragments of the isomer DNS, where the light nucleus is Mg or Al, at the energies of transmission resonances in sub-barrier induced fission.

4.5 Complete Fusion and Quasifission in the Dinuclear Model

4.5.1 Reaction Models for Fusion With Adiabatic and Diabatic Potentials

Reaction models which use adiabatic potentials describe the nuclear fusion as a melting of the clusters into a compound nucleus. Since the adiabatic potential barrier to the inside is usually smallest for two equal nuclei, such models have the property that the two clusters exchange nucleons in a touching configuration up to the point they are nearly equal (same mass, $\eta \approx 0$) and then they fuse to the compound nucleus along the internuclear distance R. This process yields large cross sections for fusion with similar target and projectile nuclei ($\eta \approx 0$) [14, 54], which contradicts the experimental data in the production of superheavy nuclei [55].

In contrast to this picture the dinuclear system concept, which is based on the ideas of Volkov and also von Oertzen, makes use of diabatic potentials which are strongly repulsive behind the touching point of the clusters. Therefore, the nuclei can not melt together along the internuclear coordinate [56]. In heavier systems they remain for some time in a touching configuration and form the dinuclear system. Then they start to exchange nucleons up to the point when the smaller nucleus is eaten up by the larger one and the compound nucleus is formed [57–63]. This process prefers the formation of the compound nucleus between an asymmetric system of two clusters which is in agreement with the experience in the production of heavy and superheavy nuclei.

As one can note, the dynamics of fusion is very different if described by adiabatic or diabatic potentials. The adiabatic potentials prefer the dynamics of fusion in the internuclear coordinate R, whereas the diabatic potentials describe the fusion by the dynamics in the mass asymmetry coordinate η. The question arises which of these very different reaction mechanisms for describing the production of superheavy nuclei is realized in nature. A possible answer would be given for example by a detailed measurement of the quasifission process accompanying the fusion. Quasifission means the direct decay of the dinuclear system without forming the compound nucleus and proceeds always in competition with the exchange of nucleons between the clusters.

It is clear that an adiabatic description with many explicitly treated collective coordinates finally leads to a diabatic description, since the kinetic energy of the relative cluster motion, first available in the system, gets then transferred into other degrees of freedom and the nuclear system will stop its internuclear motion around the touching point. However, this is the starting point of the description with the dinuclear system concept.

In the following we will describe the fusion to superheavy nuclei with the dinuclear system concept. There are different methods like statistical procedures and master equations to calculate production cross sections for superheavy nuclei. Many interesting new treatments using the dinuclear system concept have recently carried out in China and are described elsewhere [64].

4.5.2 Problems of Adiabatic Treatment of Fusion

In order to calculate the fusion probability within the adiabatic treatment, we started from a value of $\lambda = \lambda_v$ in the fission-type valley obtained from the dynamical calculation of the descent into this valley [14]. The phenomenological Werner–Wheeler mass tensor was used. The necessary condition for the compound nucleus formation is that the fusing system passes inside the fission saddle point at $\lambda = \lambda_{sd}$. The fusion probability is, thus, determined by the leakage $\Lambda_{fus}^{\lambda}(t)$ through the barrier in λ which separates the strongly deformed configuration with an equilibrated large neck and the compound nucleus:

$$P_{CN} = \int_0^{t_0} \Lambda_{fus}^{\lambda}(t)dt, \qquad (4.36)$$

where t_0 is the lifetime of the system and $\Lambda_{fus}^{\lambda} \sim \exp(-B_{fus}^{\lambda}/(k\Theta))$ (Kramers expression). Here, $B_{fus}^{\lambda} = V_{adiab}(\lambda_{sd}) - V_{adiab}(\lambda_v)$ is the barrier for fusion in λ and $k\Theta = \sqrt{12E^*/A}$ the local thermodynamic temperature with the excitation energy E^* of the system. For nearly symmetric reactions $^{100}Mo+^{100}Mo$, $^{100}Mo+^{110}Pd$, $^{110}Pd+^{110}Pd$ and others, the fusion probabilities in λ are much larger [14] than the values found from the experimental data. While the experimental fusion probability decreases with mass asymmetry in the entrance channel [65], the calculated data have the opposite tendency. Experimental evidence for a hindrance of fusion has been raised mainly by the impossibility to produce fermium evaporation residues with nearly symmetric projectile-target combinations. The adiabatic treatment of fusion in λ mostly gives the wrong dependence of the fusion probability on the isotope composition of the colliding nuclei [14]. The qualitative and quantitative contradictions obtained with the adiabatic scenario of the fusion point to the existence of an additional hindrance for the fast growth of the neck and the motion to smaller λ.

Since the microscopical mass parameter of the neck degree of freedom is rather large (see Sect. 4.2.3), the neck parameter can be taken approximately fixed ($\varepsilon = 0.75$) during the fusion. As in the previous case, the adiabatic treatment of fusion in λ (at the fixed neck parameter) yields fusion probabilities P_{CN} which are still considerably overestimated in comparison to the experimental data. The reason of this overestimate is clearly seen in Fig. 4.16. In contrast to experiment the hindrance of fusion in these potentials is almost absent ($P_{CN} \approx 1$) because there is no internal fusion barrier for the motion to smaller elongations. So, the fusion probabilities will be considerably overestimated in any model of fusion in λ where only the neck parameter is taken fixed.

Since the adiabatic potential (with and without fixed neck parameter) is not adequate for the description of fusion, we have to answer how fast is the transition between the initial diabatic and the asymptotically adiabatic regimes during the fusion process. The main question is the use of an adiabatic potential from the beginning of the fusion process. The dynamical (time-dependent) diabatic potential (see Eq. 4.6) at the lifetime t_0 of the initial DNS (Fig. 4.16) has a very large fusion barrier in λ and, correspondingly, the fusion probability in λ is negligible for combinations leading to ^{246}Fm. It should be noted that these dynamical potentials were calculated by using the smallest possible relaxation time for the transition between diabatic and adiabatic potentials [15]. The calculated energy thresholds for the complete fusion in the λ-and η-channels lead to the conclusion that the DNS evolution to the compound nucleus proceeds in the mass asymmetry degree of freedom. For example, the average fusion barriers B_η^{fus} in mass asymmetry are about 10, 12 and 15 MeV for the reactions ^{76}Ge $+^{170}$Er($\eta = 0.4$), ^{86}Kr $+^{160}$Gd($\eta = 0.3$) and ^{110}Pd $+^{136}$Xe($\eta = 0.1$), respectively [59, 60]. The fusion barrier B_λ^{fus} in λ is about 3–4 times larger than the fusion barrier B_η^{fus} in η. As shown in Fig. 4.17, the fusion probability P_{CN} in η strongly increases with mass asymmetry in the entrance channel. The same behaviour was experimentally established [65]. For the reactions ^{40}Ar $+^{206}$Pb and ^{76}Ge $+^{170}$Er, the values of P_{CN} in η are in good agreement with experimental data from evaporation residue cross sections.

In compound systems heavier than ^{246}Fm the difference between the fusion barriers and fusion probabilities in both λ-and η-channels is even larger [56]. Our analysis with the diabatic dynamics demonstrates that a structural forbiddenness exists for a direct motion of the nuclei to smaller internuclear distances during the fusion process. Fusion of heavy nuclei along the internuclear distance in the coordinates R or λ is practically impossible. These facts strongly support our standpoint that the correct model of fusion of heavy nuclei is the dinuclear system model where fusion is described by the transfer of nucleons, i.e., by a motion in η.

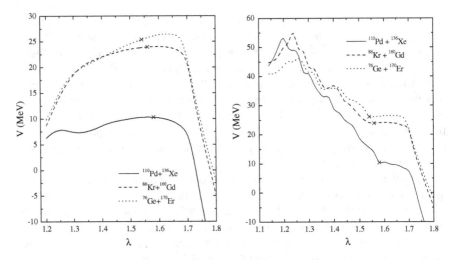

Fig. 4.16 *Left* Adiabatic potential in different reactions leading to ^{246}Fm as a function of λ for a fixed $\varepsilon = 0.75$. The crosses denote the touching configurations. *Right* Dynamical diabatic potential

Fig. 4.17 Fusion probability P_{CN} in the reactions leading to ^{246}Fm with excitation energy 30 MeV as a function of the mass asymmetry in the entrance channel. The result of the adiabatic treatment of the fusion in λ is presented by the *dotted line*. The upper limit of the fusion probability in λ in the dynamical diabatic treatment is presented by the *dashed line*. The fusion probability in the η channel with a closed fusion channel in λ is presented by the *solid line*

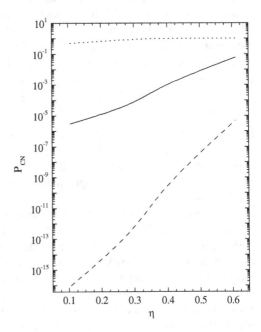

4.5.3 Fusion to Superheavy Nuclei

The evaporation residue cross section can be written as a sum over partial contributions [66]

$$\sigma_{ER}(E_{c.m.}) = \sum_{J=0}^{J_{max}} \sigma_{cap}(E_{c.m.}, J) P_{CN}(E_{c.m.}, J) W_{sur}(E_{c.m.}, J). \qquad (4.37)$$

The factors are the partial capture cross sections, the fusion and survival probabilities. The contributing angular momenta in σ_{ER} are limited by the survival probability W_{sur} with $J_{max} \approx 10 - 20$ when highly fissile superheavy nuclei are produced with energies above the Coulomb barrier. We approximate the evaporation residue cross section by

$$\sigma_{ER}(E_{c.m.}) = \sigma_{cap}^{eff}(E_{c.m.}) P_{CN}(E_{c.m.}, J = 0) W_{sur}(E_{c.m.}, J = 0) \qquad (4.38)$$

with an effective capture cross section $\sigma_{cap}^{eff} = (\lambda^2/(4\pi))(J_{max} + 1)^2 T(E_{c.m.}, J = 0)$ (see also [17]). For reactions leading to optimal cross sections for superheavy nuclei, the bombarding energy $E_{c.m.}$ is above the outer Coulomb barrier, and we set $T(E_{c.m.}, J = 0) = 0.5$ near this barrier. The effective capture cross section results in the order of a few mb. Whereas the capture cross section and the survival probability are largely similarly formulated in all the models, the fusion probability is treated along very different trajectories through a different potential energy surface there. Here we want to present our approach proposed and applied for the fusion of clusters to superheavy nuclei within the dinuclear system concept.

4.5.3.1 Fusion Probability Within the DNS Concept

After the system is captured in a DNS configuration, the total relative kinetic energy is transferred into potential and excitation energies. Then the dinuclear system statistically evolves in time by diffusion in the mass asymmetry and relative coordinates. The fusion probability P_{CN} is the probability that the dinuclear system crosses the inner fusion barrier B_η^{fus} in η and an excited compound nucleus is formed (see Fig. 4.18). This barrier is measured with respect to the potential $U(R_m, \eta_i)$ of the initial dinuclear configuration with the mass fragmentation η_i at the touching radius R_m. There are different methods to calculate the fusion probability: Diffusion equations can be solved with Fokker–Planck equations [66] or with the Kramers approximation [59–61]. Also master equations in the coordinates η and η_Z were used. In the diffusion equations the mean value $\bar{\eta}(t)$ mostly tends to the symmetric fragmentation $\eta = 0$ with an increasing probability for quasifission, determined by the quasifission barrier $B_{qf}^R(\eta)$ measured with respect to the minimum of the potential $U(R, \eta)$ at $R = R_m$.

The minima in the potential $U(R_m, \eta)$ play an important role for selecting optimum target and projectile combinations for producing superheavy elements. Sandulescu et al. (1976) argued that the nuclei fuse with higher probabilities along the valleys in an adiabatic potential in the R coordinate and pointed to the experimentally successful choice of target-projectile combinations with a Pb nucleus as target as proof for their hypothesis. This idea can simply transferred to the DNS concept. A certain initial system in a minimum of the potential is hindered by the barrier B_η^{sym}

Fig. 4.18 Potential energy of
the dinuclear system in the
reaction
^{54}Cr $+^{208}$Pb $\rightarrow$ ^{262}Sg
($|\eta_i| = 0.59$) as a function
of η for $J = 0$. Both curves
are obtained with
experimental binding
energies. The *dotted curve* is
calculated for spherical
shapes of the nuclei, the *solid
curve* for deformed shapes in
pole-to-pole orientations

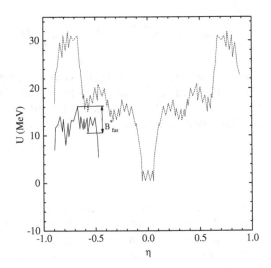

of the potential in η to move to more symmetric systems which would lead to a fast
decay by quasifission. Therefore, an asymmetric DNS in a potential minimum lives
a longer time with respect to its decay by quasifission than outside of the minimum
and has a larger chance to fuse by diffusion via nucleon transfer.

Let us assume that the initial configuration is in a minimum of the driving potential
$U(R_m, \eta)$. Then the probability for complete fusion depends on the quasi-stationary
rate λ_η^{fus} for fusion, on λ_η^{sym} for going to a symmetric DNS which is easily decaying
into two fragments and on λ_R^{qf} for the decay by quasifission of the initial DNS
[59–61].

$$P_{CN} = \lambda_\eta^{fus} / \left(\lambda_\eta^{fus} + \lambda_\eta^{sym} + \lambda_R^{qf} \right). \tag{4.39}$$

The rates can be calculated with two-dimensional Kramers-type formulas falling off
exponentially with the fusion barrier B_η^{fus} in η, with the barrier B_η^{sym} in η in the
direction to more symmetric configurations and with the quasifission barrier B_{qf}^R,
respectively. The probability P_{CN} to overcome B_η^{fus} can be approximately written as

$$P_{CN} \sim \exp\left(-(B_\eta^{fus} - \min[B_\eta^{sym}, B_{qf}^R])/T\right). \tag{4.40}$$

The barriers, following from the potential $U(R_m, \eta)$, have heights strongly influenced by shell and deformation effects. The temperature T is the local temperature
of the initial DNS and obtained from the excitation energy E^*: $T = \sqrt{E^*/a}$ with
$a = (A_1 + A_2)/12 \, \text{MeV}^{-1}$.

The main hindrance for complete fusion is the evolution of the initial DNS to
more symmetric configurations and the subsequent quasifission. In cold fusion the

Fig. 4.19 Calculated fusion probabilities P_{CN} for cold fusion reactions (^{A}X $+^{208}$Pb) and hot fusion reactions (^{48}Ca $+^A$Y) as a function of the charge number Z_{CN} of the superheavy compound nucleus

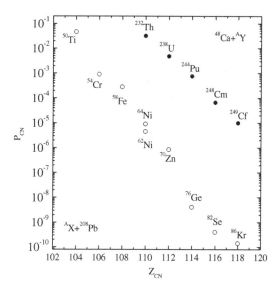

quasifission mainly arises from the initial DNS because of $B_{qf}^R < B_\eta^{sym}$, whereas in hot fusion reactions which have $B_{qf}^R > B_\eta^{sym}$, the DNS prefers to go to symmetric systems and to decay. Fig. 4.19 shows calculated probabilities P_{CN} for cold (^{A}X $+ ^{208}$Pb) and hot (^{48}Ca $+ ^A$Y) fusion reactions.

4.5.3.2 The Survival Probability

The survival probability under the evaporation of x neutrons is calculated [67, 68] with the expression

$$W_{sur} = P_{xn}(E_{CN}^*) \prod_{i=1}^{x} \frac{\Gamma_n((E_{CN}^*)_i)}{\Gamma_n((E_{CN}^*)_i) + \Gamma_f((E_{CN}^*)_i)}. \tag{4.41}$$

The factor $P_{xn}(E_{CN}^*)$ is the probability for the realization of the xn channel at the excitation energy $E_{CN}^* = E_{c.m.} + Q_{fus}$ of the compound nucleus. The index i denotes the evaporation step. $(E_{CN}^*)_i$ is the mean excitation energy of the compound nucleus at the beginning of step i with the initial condition $(E_{CN}^*)_1 = E_{CN}^*$. There is an analytical expression for the ratio of the partial widths Γ_n and Γ_f for neutron emission and fission, respectively, depending on level densities. An approximate expression for P_{1n} is given by

$$P_{1n} = \exp\left(-(E_{CN}^* - B_n - 2T)^2/(2\sigma^2)\right), \tag{4.42}$$

where the parameter σ is set to $\sigma = 2.5$ MeV.

Fig. 4.20 Measured and
calculated excitation
functions for xn evaporation
channels in the ^{48}Ca $+^{208}$Pb
reaction. The experimental
data from Refs. [93,
133–135] are presented by
circles, squares, triangles
and *diamonds*, respectively.
The closed symbols
correspond to the 1n and 2n
channels. The open symbols
correspond to the 3n and 4n
channels. The *solid* and
dashed curves show the
results obtained with the
Fermi-gas model and with
the model accounting a
collective enhancement of
the level density,
respectively. The predictions
of nuclear properties from
Refs. [100–103] are used

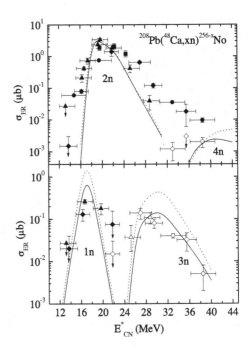

Also it is possible to calculate W_{sur} with statistical codes, e.g. GROGIF [69]. As
an example we present a comparison between calculated and experimental excitation
functions of the reaction ^{208}Pb(^{48}Ca, $xn)^{256-x}$No in Fig. 4.20 [70]. Comparing the
results with measurements we conclude that the systematic uncertainty of our calculations of σ_{ER} is up to a factor 3. Taking into account the experimental uncertainties
and the differences between various measurements, the calculated values of σ_{ER} are
in good agreement with the experimental data for the most of evaporation channels,
especially near the maxima of the excitation functions (see also [71]).

4.5.3.3 Results for Fusion With Lead-Based Reactions

Let us first consider lead- and bismuth-based complete fusion reactions:

$$^AZ +^{208}Pb(^{209}Bi) \rightarrow \text{superheavy nucleus} + 1n.$$

To get the largest cross section for producing a compound nucleus by fusion, one has
to choose the incident energy sufficient that the dinuclear system can cross the inner
fusion barrier B_η^{fus}. Since the potential energy $U(R, \eta)$ counts the energy above the
ground state energy of the compound nucleus, the optimal, i.e. the smallest excitation
energy of the compound nucleus, is given by

$$E_{CN}^* = U(\eta_{initial}, R_t) + B_\eta^{fus}. \tag{4.43}$$

Table 4.2 Cold fusion reactions with ^{208}Pb target

Reaction	E^*_{CN} (MeV)	P_{CN}	σ_{cap} (mb)	W_{sur}	σ^{th}_{ER}	σ^{exp}_{ER}
^{50}Ti $+^{208}$Pb $\rightarrow ^{257}$Rf $+$ n	16.1	3×10^{-2}	5.3	9×10^{-5}	14.3 nb	15.6 ± 1.9nb
^{70}Zn $+^{208}$Pb $\rightarrow ^{277}$Cn $+$ n	9.8	1×10^{-6}	3.0	6×10^{-4}	1.8 pb	$0.5^{+1.1}_{-0.4}$pb
^{86}Kr $+^{208}$Pb $\rightarrow ^{293}$118 $+$ n	13.3	1.5×10^{-10}	1.7	2×10^{-2}	5.1 fb	< 0.5pb

Excitation energy E^*_{CN}, fusion probability P_{CN}, capture cross section σ_{cap}, survival probability W_{sur} and theoretical and experimental [72–75] evaporation residue cross section $\sigma^{th,exp}_{ER}$.

The so calculated optimal excitation energies vary in Pb- and Bi-based reactions with the emission of one neutron between 18 and 10 MeV in accordance with the experimentally used incident energies. The calculated values depend sensitively on deformation effects.

In the Table 4.2 we listed effective capture cross sections, fusion probabilities, survival probabilities and evaporation residue cross sections for three typical reactions [54]. Whereas the capture cross section is nearly constant, the fusion probability falls exponentially down with increasing projectile mass The reason for this strong decrease is that the mass asymmetry of the initial dinuclear system gets smaller with growing projectile mass with the consequence that the inner fusion barrier B^{fus}_{η} increases and the fusion probability P_{CN} decreases exponentially towards symmetric projectile and target combinations in lead-based reactions. Since the variation of the survival probability is moderate, the evaporation cross section drops from nb over pb to fb. The calculated evaporation residue cross sections agree with the experimental data [72–77] as shown in Fig. 4.21.

4.5.3.4 Hot Fusion With ^{48}Ca Projectiles

The actinide-based complete fusion reactions ^{48}Ca $+ ^{232}$Th, ^{238}U, ^{237}Np, ^{242,244}Pu, ^{243}Am, ^{248}Cm and ^{249}Cf were used at JINR in Dubna to synthesize the elements 110, 112–116 and 118 [78]. The evaporation residue cross sections with the emission of 3 and 4 neutrons are on the level of 1 pb. Calculated and experimental evaporation residue cross sections are shown in Fig. 4.21 at the maxima of the excitation functions of the compound nuclei. Since the initial DNS is more asymmetric than in the lead-based reactions, the fusion probability is larger. However, the survival probability is diminished because the compound nucleus has an excitation energy of about 30–40 MeV and, therefore, 3–4 neutrons have to be emitted to reach the ground state.

The main factor which prohibits the complete fusion of heavy nuclei is the evolution of the initial DNS to more symmetric configurations ($B^{sym}_{\eta} \approx 0.5 - 1.5$ MeV and 4–5 MeV for hot and cold fusion, respectively) and the decay of the DNS during this process or the decay of the initial DNS. In hot fusion reactions, the decay of

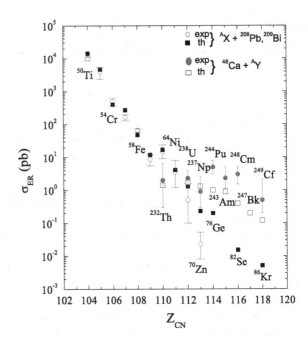

Fig. 4.21 Calculated and experimental [72–78] maximal evaporation residue cross sections for cold fusion with ^{208}Pb and ^{209}Bi targets and for hot fusion with ^{48}Ca projectiles. For the meaning of the symbols see the notations in the figure. The predictions of nuclear properties from Ref. [79] are used

the DNS takes place mainly outside of the initial conditional minimum because of $B_{qf}^{R} > B_{\eta_{sym}}$ in contrast to the cold fusion reactions.

In Fig. 4.22 we show cross sections and probabilities for the reaction ^{48}Ca + ^{248}Cm → 116 calculated before the experiment was carried out [80]. This picture clearly illustrates the different factors yielding the excitation functions: the capture cross section, the fusion probability, which is the most complex, partly not yet definitely determined quantity, and the ratio Γ_n/Γ_f which is lastly responsible for the survival probability.

4.5.3.5 Isotopic Trends

Here we discuss some results on the isotopic variation of complete fusion cross sections [81]. Fig. 4.23 shows that the production of the superheavy nucleus with $Z = 112$ does not profit from the higher isospin. There result quite large cross sections in the ^{208}Pb-based reactions with ^{67}Zn and ^{68}Zn projectiles. As found for the reactions treated, the value of P_{CN} increases with decreasing mass number of the projectile. The odd nucleus ^{275}Cn has a larger P_{1n}, a larger fission barrier and a smaller neutron separation energy than the neighboring even-even nuclei which lead to a larger survival probability W_{sur}. The fission barrier of the isotopes of the element Cn ($Z = 112$) slightly increases with decreasing mass number from $A = 278–274$ due to a large level spacing at $N = 162$ for deformed nuclei. Going from ^{66}Zn to ^{68}Zn

Fig. 4.22 Calculated capture, fusion and *xn*-evaporation residue cross sections, the fusion probability P_{CN} and Γ_n/Γ_f for the reaction $^{48}Ca+^{248}Cm$ leading to element $Z = 116$ as a function of the excitation energy of the compound nucleus $^{296}116$

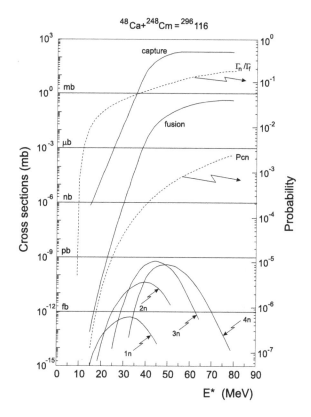

projectiles we see a small decrease of P_{CN}, but a stronger increase of the survival probability. The available experimental data [72–75] are well described (Fig. 4.23).

Calculated evaporation residue cross sections at the maxima of their excitation functions, the corresponding excitation energies of the compound nuclei in the 3n and 4n evaporation channels calculated with the predictions [79, 82, 83] by the macroscopic–microscopic models are shown in Figs. 4.24 and 4.25 for the actinide-based fusion reactions $^{48}Ca +^A U,^A Np,^A Pu,^A Am,^A Cm$ and $^A Bk$ [84, 85]. In this interval of the mass number the value of P_{CN} becomes larger with decreasing A in most cases. The larger P_{3n} and fission barriers and smaller neutron separation energies give a larger W_{sur} for odd nuclei such as ^{241}Pu and $^{243,245,247}Cm$ in comparison to the neighboring even-even nuclei. One can conclude quite small uncertainties in the calculated isotopic trends of σ_{ER} due to the choice of predicted properties of superheavies by the macroscopic–microscopic models. In the most cases the absolute values of σ_{ER} with the predictions [82, 83] differ within the factor of 1–3 from the values of σ_{ER} calculated with the predictions [79]. This difference does not exceed the accuracy of the present experimental measurements and is within the estimated inaccuracy of our calculations. For the $^{48}Ca+^{249}Cf$ reaction, the calculated cross section ($\sigma_{ER} = 0.12\,pb$, Fig. 4.21) with the predictions [82] is about six times larger

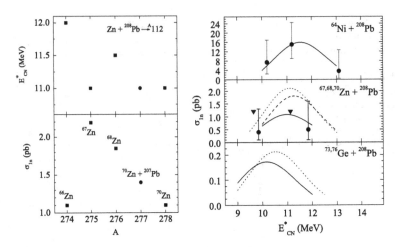

Fig. 4.23 *Left* The calculated maximal evaporation residue cross sections in the 1n channel (*lower part*) at the corresponding excitation energies of the compound nuclei (*upper part*) for the fusion reactions $Zn +^{208}Pb \rightarrow^{(A-1)} Cn + n$ and $^{70}Zn +^{207}Pb \rightarrow^{276}Cn + n$. The predictions of nuclear properties from Ref. [79] are used. *Right* The calculated excitation functions for the 1n channel of the fusion reactions ^{64}Ni (*solid line*), $^{67,68,70}Zn$ (*dotted, dashed* and *solid lines*, respectively), $^{73,76}Ge$ (*dotted* and *solid lines*, respectively) $+^{208}Pb$. The experimental data [72–75] of the reactions $^{64}Ni, ^{70}Zn +^{208}Pb$ and upper limits for the reaction $^{68}Zn +^{208}Pb$ are shown by closed circles with error bars and triangles, respectively. The predictions of nuclear properties from Ref. [79] are used

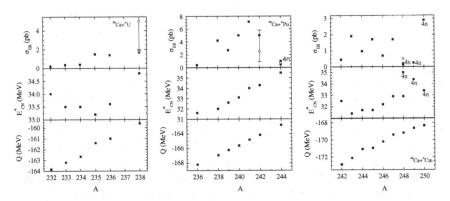

Fig. 4.24 The calculated maximal and experimental (*open circles* with *error bars*) [78] evaporation residue cross sections in the 3n and 4n channels (*upper part*) at the corresponding excitation energies of the compound nuclei (*middle part*) and Q values (*lower part*) for the indicated fusion reactions as functions of the mass number A of the target. The predictions of nuclear properties from Ref. [79] are used

than the cross section calculated with the predictions [79]. The main reason for this is the significant difference between the values of neutron separation energies which leads to larger and more realistic W_{sur} with the predictions [82].

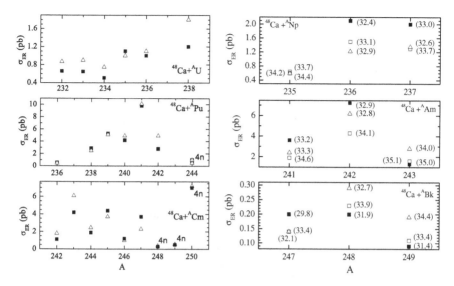

Fig. 4.25 *Left* The maximal evaporation residue cross sections in the 3n and 4n (indicated) channels for the indicated fusion reactions calculated with the predictions of nuclear properties from Ref. [82] (*closed symbols*) and Ref. [83] (*open symbols*). *Right* The calculated maximal evaporation residue cross sections at the corresponding optimal excitation energies of the compound nuclei (in parentheses) for the indicated hot fusion reactions as a function of A. The results obtained with the predictions of Refs. [79, 82, 83], are shown by *closed squares*, *open squares*, and *open triangles*, respectively

Calculations for many other reactions with heavy ions leading to complete fusion with the emission of neutrons from the excited compound nucleus were carried out in the framework of the DNS model [54, 69, 85–92]. The calculated $\sigma_{4n} = 1$ pb for element 114 was reported before the experiment [59, 60]). Note that the calculations for all reactions were performed with the same parameters.

4.5.3.6 Feature of Production of New Superheavy Nuclei in Actinide-Based Complete Fusion Reactions

The cold Pb- and Bi-based and hot actinide-based complete fusion reactions [72–78, 93–99] were carried out in order to approach to "the island of stability" of the heaviest nuclei predicted at charge number $Z = 114$ and neutron number $N = 184$ by the macroscopic-microscopic models [79, 82, 83, 100–106]. The experimental systematic of cross sections and half-lives of the superheavies produced in Dubna with ^{48}Ca-induced reactions reveals the increasing stability of nuclei approaching the spherical closed shell $N = 184$. No discontinuity is observed yet when the proton number 114 is crossed at the neutron numbers 172–176 [76].

As known, the shell at $Z = 114$ disappears in the relativistic and nonrelativistic mean field models [107]. The island of stability close to the element $Z = 120$, or

122, or 124, or 126 and $N = 184$ was predicted within these models. If these predictions are correct, there is some hope to synthesize new superheavy nuclei with $Z \geq 120$ by using the present experimental set ups and the actinide-based reactions with neutron-rich stable projectiles heavier than ^{48}Ca. The survival probability of compound nucleus with $Z \geq 120$ may be much higher than the one of compound nucleus with $Z = 114$ if the shell closure at $Z = 120$, or 122, or 124, or 126 has a stronger influence on the stability of the superheavy nuclei than the subshell closure at $Z = 114$.

The predictions of macroscopic–microscopic models and phenomenological model [108, 109] presently provide us all values which are necessary to calculate σ_{ER}. In the macroscopic–microscopic model [82] the height of the fission barrier B_f of the nucleus with the fixed neutron number and $Z > 116$ decreases with increasing deviation of Z from 114. Instead of the $Z = 114$ magic number, the model [108, 109] relies on the $Z = 126$ closed shell.

Since the lower fission barriers and, correspondingly, the smaller values of $B_f - B_n$ are predicted in the macroscopic-microscopic model [82] for $Z \geq 118$, the expected evaporation residue cross sections of the nuclei with $Z = 118$–126 should be smaller than those of the isotopes of nuclei with $Z = 114$–116. However, the model [108, 109] with the closed proton shell at $Z = 126$ predicts the growth of the values of $B_f - B_n$ for $Z = 118$–126 nuclei which might result in a larger production cross sections for the xn-evaporation channels.

The evaporation residue cross sections at the maxima of $(2 - 4)n$ excitation functions and corresponding optimal excitation energies E^*_{CN} calculated with the mass table [82] are presented in Fig. 4.26 for the reactions ^{50}Ti, ^{54}Cr, ^{58}Fe, ^{64}Ni + ^{238}U, ^{244}Pu, ^{248}Cm, ^{249}Cf [87]. The small excitation energy in the actinide-based reaction with ^{48}Ca is due to the gain in the Q-value. With projectiles heavier than ^{48}Ca the values of E^*_{CN} becomes larger. The values of σ_{ER} decreases by about two orders of magnitude with increasing the charge number of the target from 92 to 98. The main reason of fall-off of σ_{ER} with Z of compound nucleus is the strong decrease of fusion probability P_{CN}, i.e. the increasing role of quasifission with $Z_1 \times Z_2$. Only for the projectiles ^{50}Ti and ^{54}Cr the production cross section of $Z = 114$, 116, and 118 results on the level of the present experimental possibilities.

With the mass table [108, 109] the calculated cross sections (Fig. 4.26) for producing the evaporation residues with $Z \geq 114$ are larger than the cross sections calculated with the mass table [82]. With any mass table the value of σ_{ER} decreases with increasing Z in the interval $Z = 114$–120. However, the slopes of the decrease are different because the survival probability with the mass table [108, 109] is larger than the one with the mass table [82]. Using the mass table [108, 109], the calculated values of σ_{ER} for $Z = 114$, 116 and 118 in the reactions with ^{50}Ti and ^{54}Cr can be even larger than those in the reactions with ^{48}Ca because of the dependence of $B_f - B_n$ on A at fixed Z. The dependencies of σ_{ER} on Z in Fig. 4.26 demonstrate that $Z = 114$ is not a proper magic number in [108, 109]. The calculated production cross sections of element $Z = 120$ with the mass table [108, 109] are about two orders of magnitude larger than those calculated with the mass table [82].

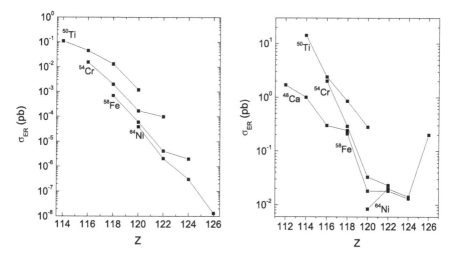

Fig. 4.26 Evaporation residue cross sections calculated with the mass tables of Ref. [82] (*left-hand side*) and of Refs. [108, 109] (*right-hand side*) at the maxima of $(2-4)n$ excitation functions of the reactions ^{50}Ti, ^{54}Cr, ^{58}Fe, ^{64}Ni $+^{238}$U, ^{244}Pu, ^{248}Cm, ^{249}Cf

A series of experiments are desirable to answer the question where the next spherical proton shell after ^{208}Pb occurs [87–89]. The answer can be obtained from the trend and values of evaporation residue cross sections. If the experimental cross sections in reactions ^{50}Ti $+^{238}$U, ^{244}Pu, ^{248}Cm $\rightarrow$ 114, 116, 118 and ^{54}Cr $+^{238}$U, ^{244}Pu $\rightarrow$ 116, 118 are larger than 0.1 pb, one can conclude that $Z = 114$ is not a proper magic number and the next magic nucleus beyond ^{208}Pb is a nucleus with $Z \geq 120$. Based on the experimental results one could define the best model describing the structure and properties of the superheavy nuclei.

In order to produce a new nucleus with $Z = 120$, in the optimal reaction ^{50}Ti$+^{249}$Cf one needs to reach the level of the cross section of about $(0.8–0.09)$ pb with magic number $Z = 126$ and of about $(0.5–1.2)$ fb with magic number $Z = 114$. However, the cross section might be larger in the case of magic number $Z = 120$. Recently, the upper limits of the evaporation residue cross section of about 0.1 and 0.4 pb were reached in the complete fusion reactions ^{64}Ni $+^{238}$U $\rightarrow$ 120 [76] and ^{58}Fe $+^{244}$Pu $\rightarrow$ 120 [110], respectively. These results are along our results based on different mass tables predicting the proton magic number $Z = 114$ or 126.

4.5.3.7 Production of Superheavy Nuclei With Radioactive Beams

In Pb-based reactions with neutron-rich nuclei ^{70,74,78}Ni, ^{80}Zn, ^{86}Ge and ^{92}Se a decrease of P_{CN} can be compensated by W_{sur} increasing with the number of neutrons (Table 4.3). For example, in the ^{62}Ni $+^{208}$Pb reaction the yield of the $Z = 110$ element is comparable with the yields in the ^{70,74}Ni$+^{208}$Pb reactions. The calculated

Table 4.3 Excitation energy, probabilities and cross sections for selected reactions

Reaction	E^*_{CN} (MeV)	σ_{cap} (mb)	P_{CN}	W_{sur}	σ^{th}_{1n}	σ^{exp}_{1n}
$^{66}Zn + ^{174}Yb \rightarrow ^{238}Fm + 2n$	26.0	9.6	4×10^{-2}	8×10^{-7}	0.3 nb	
$^{76}Zn + ^{174}Yb \rightarrow ^{248}Fm + 2n$	23.0	8.8	2×10^{-3}	6×10^{-4}	10.6 nb	
$^{76}Ge + ^{170}Er \rightarrow ^{244}Fm + 2n$	24.6	8.4	5×10^{-4}	3×10^{-4}	1.3 nb	$1.6^{+1.3}_{-1.6}$nb
$^{62}Ni + ^{208}Pb \rightarrow ^{269}110 + 1n$	12.3	3.5	4.5×10^{-6}	5×10^{-4}	7 pb	$3.5^{+2.7}_{-1.8}$pb
$^{64}Ni + ^{208}Pb \rightarrow ^{271}110 + 1n$	10.7	3.4	1×10^{-5}	5×10^{-4}	17 pb	15^{+9}_{-6}pb
$^{70}Ni + ^{208}Pb \rightarrow ^{277}110 + 1n$	13.5	3.1	7×10^{-8}	5×10^{-3}	1.1 pb	
$^{74}Ni + ^{208}Pb \rightarrow ^{281}110 + 1n$	15.0	3.0	6×10^{-8}	2×10^{-2}	3.6 pb	
$^{78}Ni + ^{208}Pb \rightarrow ^{284}110 + 2n$	17.5	3.0	2×10^{-7}	6×10^{-2}	36 pb	
$^{64}Ni + ^{209}Bi \rightarrow ^{272}111 + 1n$	10.5	3.4	2×10^{-6}	6×10^{-4}	4.1 pb	$3.5^{+4.6}_{-2.3}$pb

values of P_{CN} in the cold fusion reactions are maximal when the neutron number of the projectile is a magic number [54]. As follows from our model [54], intensive beams of neutron-rich nuclei will be useful for producing heavy actinides, for example Fm as listed in Table 4.3. In the Pb-based reactions the use of neutron-rich projectiles leads to values of σ_{ER} comparable with evaporation residue cross sections for reactions with stable projectiles (Table 4.3).

4.5.4 Production of Neutron-Deficient Isotopes of Pu

The complete fusion reactions usually result in the neutron-deficient isotopes of compound nuclei. One can use these reactions to produce some unknown neutron-deficient isotopes with high efficiency. The search of the shell effects related to the neutron number $N = 126$ motivates the study of neutron-deficient actinides as well. For example, producing the unknown nuclei $^{220-227}Pu$, one can investigate the role of the neutron magic number $N = 126$ in the region of the neutron-deficient nuclei [86].

The nuclei ^APu with $A \leq 227$ are unknown. These nuclei can be produced in the complete fusion reactions $^{24}Mg + ^{204,206,208}Pb$, $^{26}Mg + ^{204,206}Pb$, $^{32}S + ^{192}Pt$, and $^{40,44,48}Ca + ^{184}W$. The asymmetric reactions with Mg have the preference because of the large ($P_{CN} \approx 1$) fusion probability. The excitation functions in the $(3-5)n$ evaporation channels are shown in Fig. 4.27 for the reactions $^{24}Mg + ^{204,206}Pb$. The products of the 3n evaporation channels have the largest yields. As seen, the unknown neutron-deficient nuclei $^{223-227}Pu$ can be produced with rather large cross sections (0.1–20) nb. The reactions $^{24}Mg + ^{206}Pb$ and $^{24}Mg + ^{204}Pb$ seem to be the best for producing the isotopes $^{226,227}Pu$ and $^{223-225}Pu$, respectively.

In the reactions $^{40,44,48}Ca + ^{184}W$ one can obtain the nuclei $^{219,220,225,227}Pu$ with the cross sections larger than 0.2 nb. In the reactions with Ca the compound

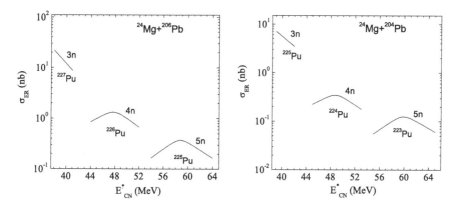

Fig. 4.27 Calculated excitation functions for the indicated xn evaporation channels in the reactions ^{24}Mg $+ ^{204,206}$Pb at bombarding energies above the Coulomb barriers. The mass table [79] is used

nuclei would have smaller number of neutrons than in the ^{24}Mg $+ ^{204}$Pb reactions and one can reach the evaporation residues with more neutron deficit like in ^{218}Pu (cross section $\sim$2 pb). In the reaction with Ca the fusion probabilities are smaller than those in the reactions with Mg mentioned above. However, for such evaporation residues like $^{218-220}$Pu the smaller P_{CN} is overcompensated by larger W_{sur} because a smaller number of neutrons is needed to be evaporated.

As found, the nuclei $^{218-220}$Pu can be also produced in the ^{32}S $+ ^{192}$Pt reaction with the cross sections of 1.5 times larger than those in the ^{40}Ca $+ ^{184}$W reaction. In order to conclude whether the reactions with ^{40}Ca or ^{32}S are preferable, one should take into account the availability of the projectile and target materials.

4.5.5 Master Equations for Nucleon Transfer

Another description of the fusion and quasifission reactions is the possibility to apply master equations for the transfer of nucleons between the clusters of the DNS [111]. As coordinates we choose the charge, neutron and mass numbers of both clusters of the dinuclear system, respectively, namely $Z_P = Z$, $N_P = N$ and $A_P = Z + N$ for the projectile (light) cluster and $Z_T = Z_{tot} - Z_P$, $N_T = N_{tot} - N_P$ and $A_T = A_{tot} - A_P$ for the target (heavy) cluster. For the derivation of the master equations, we start with the shell model Hamiltonian of the DNS written in second quantization where n_P and n_T are the single-particle quantum numbers of the clusters P and T, respectively:

$$H = H_0 + V_{int}, \tag{4.44}$$

$$H_0 = \sum_{n_P} \varepsilon_{n_P} a_{n_P}^+ a_{n_P} + \sum_{n_T} \varepsilon_{n_T} a_{n_T}^+ a_{n_T}, \quad V_{int} = \sum_{n_P, n_T} \left(g_{n_P n_T}(R_m) a_{n_P}^+ a_{n_T} + h.c. \right).$$

$$\text{(4.45)}$$

For the matrix elements we simply set $g_{n_P n_T}(R_m) = g_{PT}(R_m) = \frac{1}{2}\langle n_P | U_P + U_T | n_T \rangle$ where R_m is the distance of the nuclei at the minimum of the internuclear potential and U_P and U_T are the single-particle potentials of the projectile (light) and target (heavy) nuclei, respectively. Since we assume a thermal equilibrium in the DNS, we disregard the excitation of the light fragment by the heavy one and vice versa. With Z, N (light fragment, projectile numbers) and $Z_{tot} - Z$, $N_{tot} - N$ (heavy fragment, target numbers) we are solving the eigenvalue problem

$$H_0 | Z, N, n \rangle = E_n^{Z,N} | Z, N, n \rangle \tag{4.46}$$

with the eigenvalues $E_n^{Z,N}$. Then we can formulate a master equation for the probability $P_{Z,N}(n, t)$ to find the system in the state (Z, N, n) at time t.

$$\frac{d}{dt} P_{Z,N}(n, t) = \sum_{Z',N',n'} \lambda(Z, N, n | Z', N', n') \left(P_{Z',N'}(n', t) - P_{Z,N}(n, t) \right)$$

$$- \left(\Lambda_{Z,N}^{qf}(n) + \Lambda_{Z,N}^{fis}(n) \right) P_{Z,N}(n, t). \quad \text{(4.47)}$$

The rates $\Lambda_{Z,N}^{qf}(n)$ and $\Lambda_{Z,N}^{fis}(n)$ regard the quasifission and the fission of the heavy nucleus in the DNS. The transition rates $\lambda(Z, N, n | Z', N', n') = \lambda(Z', N', n' | Z, N, n)$ are calculated in time-dependent first order perturbation theory:

$$\lambda(Z, N, n | Z', N', n') = |\langle Z, N, n | V_{int} | Z', N', n' \rangle|^2 \frac{\sin^2 \left(\Delta t (E_n^{Z,N} - E_{n'}^{Z',N'})/2\hbar \right)}{\Delta t (E_n^{Z,N} - E_{n'}^{Z',N'})^2/4}.$$

$$\text{(4.48)}$$

The interaction energy V_{int} induces only transitions between states which differ by an 1-particle-1-hole pair. The system of differential equations can be simplified by assuming the DNS in thermal equilibrium and, therefore, we factorize $P_{Z,N}(n, t)$ as

$$P_{Z,N}(n, t) = P_{Z,N}(t) \Phi_{Z,N}(n, T), \tag{4.49}$$

where $\Phi_{Z,N}(n, T)$ is the probability to find the DNS in the states at a local temperature $T(N, Z)$ and is normalized to unity. With temperature-depending Fermi occupation numbers for the single particle states and summing over the DNS states, we finally obtain the master equations used in the further calculations:

$$\frac{d}{dt}P_{Z,N}(t) = \Delta_{Z+1,N}^{(-,0)}P_{Z+1,N}(t) + \Delta_{Z-1,N}^{(+,0)}P_{Z-1,N}(t)$$

$$+ \Delta_{Z,N+1}^{(0,-)}P_{Z,N+1}(t) + \Delta_{Z,N-1}^{(0,+)}P_{Z,N-1}(t)$$

$$- \left(\Delta_{Z,N}^{(-,0)} + \Delta_{Z,N}^{(+,0)} + \Delta_{Z,N}^{(0,-)} + \Delta_{Z,N}^{(0,+)} \right) P_{Z,N}(t)$$

$$- (\Lambda_{Z,N}^{qf} + \Lambda_{Z,N}^{fis})P_{Z,N}(t) \tag{4.50}$$

with the transition rates

$$\Delta_{Z,N}^{(\pm,0)}(T) = \frac{1}{\Delta t}\sum_{P,T}^{Z}|g_{PT}|^2 n_{T}(T)(1 - n_{P}(T))\frac{\sin^2(\Delta t(\varepsilon_P - \varepsilon_T)/2\hbar)}{(\varepsilon_P - \varepsilon_T)^2/4},$$

$$\Delta_{Z,N}^{(0,\pm)}(T) = \frac{1}{\Delta t}\sum_{P,T}^{N}|g_{PT}|^2 n_{T}(T)(1 - n_{P}(T))\frac{\sin^2(\Delta t(\varepsilon_P - \varepsilon_T)/2\hbar)}{(\varepsilon_P - \varepsilon_T)^2/4},$$

$$\Lambda_{Z,N}^{qf}(T) = \sum_{n}\Lambda_{Z,N}^{qf}(n)\Phi_{Z,N}(n,T), \qquad \Lambda_{Z,N}^{fis}(T) = \sum_{n}\Lambda_{Z,N}^{fis}(n)\Phi_{Z,N}(n,T).$$

The initial condition for the master equations is $P_{Z,N} = \delta_{Z,Z_i}\delta_{N,N_i}$. The above rates dependent on temperature-dependent Fermi occupation numbers of the single particle states which were calculated with spherical Woods–Saxon potentials with spin-orbit force and Coulomb interaction. Also we took phenomenologically into account the rotation of the DNS in the single-particle energies.

The decay rates of the DNS for quasifission are treated with the one-dimensional Kramers rate

$$\Lambda_{Z,N}^{qf}(T) = \frac{\omega}{2\pi\omega^{B_{qf}^R}}\left(\sqrt{\left(\frac{\Gamma}{2\hbar}\right)^2 + (\omega^{B_{qf}^R})^2} - \frac{\Gamma}{2\hbar} \right)\exp\left(-\frac{B_{qf}^R(Z,N)}{T(Z,N)} \right). \tag{4.51}$$

They depend on the height B_{qf}^R of the outer potential barrier at the internuclear distance $R_b \approx R_P(1 + \beta_P\sqrt{5/4\pi}) + R_T(1 + \beta_T\sqrt{5/4\pi}) + 2.0$ fm which is nearly independent of the angular momentum for $J < 70$ since the DNS has a large moment of inertia. The height of the barrier is about 4.5 MeV for $Z = 20$ and less than 0.5 MeV for $Z = Z_{tot}/2 \pm 10$. The temperature is obtained with the Fermi-gas expression $T = (E^*/a)^{1/2}$ MeV by using the excitation energy $E^*(Z,N)$ of the DNS and $a = A_{tot}/12\,\text{MeV}^{-1}$. For a nearly symmetric DNS we have about $T = 1.5$ MeV. The potential is approximated by an inverted harmonic oscillator with the frequency $\omega^{B_{qf}^R}$ around the top of the barrier and by a harmonic oscillator with frequency ω at the pocket. We use constant values for these quantities: $\hbar\omega^{B_{qf}^R} = 1.0$ MeV, $\hbar\omega = 2.0$ MeV and set the width $\Gamma = 2.8\,MeV$.

The mass and charge yields are then obtained as

$$Y_{Z,N}(t_0) = \Lambda_{Z,N}^{qf}\int_0^{t_0} P_{Z,N}(t)dt \tag{4.52}$$

with the reaction time $t_0 \approx (3-4) \times 10^{-20}$ s which is about ten times longer than the time of deep-inelastic collisions. We determine the time t_0 by the balance equation of the probabilities:

$$\sum_{Z,N} \left(\Lambda_{Z,N}^{qf} + \Lambda_{Z_{tot}-Z,N_{tot}-N}^{fis} \right) \int_0^{t_0} P_{Z,N}(t)dt = 1 - P_{CN} \tag{4.53}$$

with the fusion probability P_{CN} where Z_{BG} and N_{BG} are determined by the barrier for fusion in the asymmetry coordinates:

$$P_{CN} = \sum_{Z>Z_{BG},N>N_{BG}} P_{Z,N}(t_0). \tag{4.54}$$

The DNS with $Z > Z_{BG}$ evolves to the compound nucleus in a time of 10^{-20} s which is much shorter than the decay time of the compound nucleus. The mass and charge yields of quasifission products are given by

$$Y(A_P) = \sum_Z Y_{Z,A_P-Z}(t_0), \qquad Y(Z_P) = \sum_N Y_{Z,N}(t_0). \tag{4.55}$$

The partial and total cross section for quasifission can be calculated as:

$$\sigma_{qf}(E_{c.m.}, A_P) = Y(A_P)\sigma_{cap}(E_{c.m.}),$$
$$\sigma_{qf}(E_{c.m.}) = \sum_{A_P} \sigma_{qf}(E_{c.m.}, A_P), \tag{4.56}$$

where P_f denotes the fission probability of the heavier nucleus

$$P_f = \sum_{Z,N} \Lambda_{Z_{tot}-Z,N_{tot}-N}^{fis} \int_0^{t_0} P_{Z,N}(t)dt. \tag{4.57}$$

The capture cross section, given as

$$\sigma_{cap}(E_{c.m.}) = \frac{\pi \hbar^2}{2\mu E_{c.m.}} J_{cap}(J_{cap} + 1), \tag{4.58}$$

depends on $J_{cap} \leq (2\mu R_b^2(E_{c.m.} - V_b)/\hbar^2)^{1/2}$ which is smaller than the critical angular momentum J_{crit}. Trajectories with $J \geq J_{crit}$ contribute to deep-inelastic and quasi-elastic collisions.

To explain the experimental total kinetic energy (TKE) of the quasifission products, one has to regard the large polarizations of the DNS nuclei. For nearly symmetric dinuclear systems with $(A_P + A_T)/2 - 20 \leq A_P \leq (A_P + A_T)/2 + 20$ we found deformations which are about 3–4 times larger than the deformations of the nuclei in their ground states. Let us assume the distribution of the fragments in charge, mass and deformation as

$$W(Z, N, \beta_P, \beta_T) = Y_{Z,N} w_{\beta_P}(Z, N) w_{\beta_T}(Z_{tot} - Z, N_{tot} - N), \quad (4.59)$$

where we set the distributions of the deformations β_P and β_T as Gaussian distributions

$$w_\beta(Z, N) = \frac{1}{\sqrt{2\pi\sigma_\beta^2}} \exp(-(\beta - \langle\beta\rangle)^2/(2\sigma_\beta^2)). \quad (4.60)$$

Here, $\sigma_\beta^2 = (\hbar\omega_{vib}/(2C_{vib})) \coth(\hbar\omega_{vib}/(2kT))$ with the frequency $\omega_{vib}(Z, N)$ and the stiffness parameter $C_{vib}(Z, N)$ of the quadrupole vibrations are determined from experimental spectra. The average TKE is obtained as a function of the mass number A_P of the light fragment

$$\langle TKE(A_P)\rangle = \frac{\int\int d\beta_P d\beta_T \sum_{\substack{Z,N \\ Z+N=A_P}} TKE \times W}{\int\int d\beta_P d\beta_T \sum_{\substack{Z,N \\ Z+N=A_P}} W} \quad (4.61)$$

with $TKE = V_N(R_b) + V_C(R_b)$. The variance of the TKE is

$$\sigma_{TKE}^2 \approx \sum_Z TKE^2|_{\substack{\beta_P = \langle\beta_P\rangle \\ \beta_T = \langle\beta_T\rangle}} \frac{Y_{Z,A_P-Z}(t_0)}{\sum_Z Y_{Z,A_P-Z}(t_0)} - \langle TKE(A_P)\rangle^2$$

$$+ \left(\sigma_{TKE}^{def}(A_P)\right)_P^2 + \left(\sigma_{TKE}^{def}(A_P)\right)_T^2 \quad (4.62)$$

with $(j = P, T)$

$$\left(\sigma_{TKE}^{def}(A_P)\right)_j^2 = \sum_Z \left(\frac{\partial TKE}{\partial\beta_j}\right)^2 |_{\substack{\beta_P = \langle\beta_P\rangle \\ \beta_T = \langle\beta_T\rangle}} \frac{\sigma_{\beta_j}^2 Y_{Z,A_P-Z}(t_0)}{\sum_Z Y_{Z,A_P-Z}(t_0)}. \quad (4.63)$$

4.5.6 Results for Quasifission

Quasifission was for example investigated by Itkis et al. in Dubna in reactions with ^{48}Ca projectiles incident on U, Pu, Cm, and Cf producing the elements 112, 114, 116, and 118. With the above formalism of master equations we calculated a large quantity of observable data like mass and charge distributions, distributions of total kinetic energies (TKE), variances of total kinetic energies and neutron multiplicities for cold and hot fusion reactions [111–113]. Therefore, the comparison of the theoretical description with experimental data provides sensitive information about the applicability and correctness of the used model.

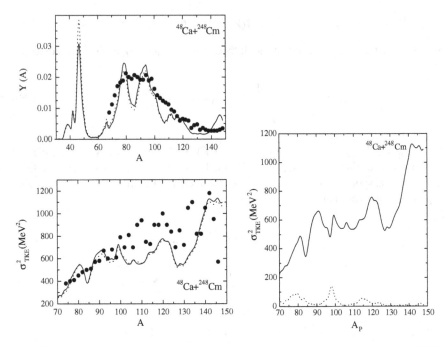

Fig. 4.28 *Left* The calculated and experimental mass yields (*upper part*) and variances of the TKE (*lower part*) of the quasifission products as a function of the mass number of the light fragment for the hot fusion reaction ^{48}Ca $+^{248}$Cm $\rightarrow ^{296}$116 at a bombarding energy corresponding to an excitation energy of the compound nucleus of 37 MeV. The results calculated for $J = 0$ and 70 are presented by *solid* and *dotted curves*, respectively. The experimental data are shown by *solid points*. *Right* The contributions of fluctuations in deformation (*solid line*) and of nucleon exchange (*dotted line*) to the variance of the TKE of quasifission products in the same reaction as on the *right hand side* as functions of the mass number of the light fragment

In Fig. 4.28 we show the calculated mass yield $Y(A_P)$ and the variances of the TKE of the fragments as functions of the light fragment mass number for the hot fusion reaction ^{48}Ca$+^{248}$Cm $\rightarrow ^{296}$116 in comparison with experimental data. At the chosen incident energy the compound nucleus has an excitation energy of 37 MeV. The small oscillations in the experimental data correspond to the accuracy of the measurements. Around the initial mass number of $A_P = 48$ the quasifission data were taken out from the experimental analysis since it is difficult to separate them from deep-inelastic events. The calculated peak around the initial mass number contains only quasifission events. The calculations disregard angular momenta which are larger than the critical one with which deep-inelastic and quasi-elastic collisions can happen. Since the quasifission process starts from the entrance channel, the peak around the initial mass number is pronounced.

Maxima in the mass and charge yields and the minima in the variances arise from the minima in the driving potential $U(R_m, Z_P, N_P, \beta_P^{gs}, \beta_T^{gs}, J)$ which are caused by shell effects in the dinuclear system. For $A_P > 48$, the maxima yields

Table 4.4 The calculated variances σ^2_{TKE} of the TKE for nearly symmetric quasifission products with $A_{tot}/2 - 20 \leq A_P \leq A_{tot}/2$, fractions of the fusion–fission events with respect to the quasifission events in the mass region $A_{tot}/2 - 20 \leq A_P \leq A_{tot}/2$, and the calculated total numbers ($M_n^{tot-sym}$) of emitted neutrons for nearly symmetric quasifission splitting with $A_{tot}/2 - 20 \leq A_P \leq A_{tot}/2$

Reaction	E^*_{CN} (MeV)	σ^2_{TKE} (MeV^2)	$P_{CN}/\sum_{A_{tot}/2-20}^{A_{tot}/2} Y(A_P)$	$M_n^{tot-sym}$
^{40}Ar + ^{165}Ho	89	119	11	5.5
	120	143	0.7	7.3
^{48}Ca + ^{244}Pu	34.8	805	1.4×10^{-2}	7.5
	50	893	1.1×10^{-1}	8.5
^{86}Kr + ^{208}Pb	17	738	2.1×10^{-7}	4.8
	30	813	2.0×10^{-5}	7.0

The reactions and the energies of the corresponding compound nuclei are indicated

of the quasifission products and the minima in the variance of the TKE appear if one of the fragments is a Pb, Zr, or Sn nucleus. Beside the maximum in $Y(A_P)$ for $A_P = 88$-95 corresponding to the Pb nucleus as heavy fragment, the maximum corresponding to about the neutron number $N = 50$ in the light fragment is also pronounced in the calculation of $Y(A_P)$ given in Fig. 4.28. Since the DNS has a large moment of inertia, the data calculated for $J = 0$ and 70 are very similar as shown in Fig. 4.28 . Therefore, the dependence of $Y(A_P)$ and $\sigma^2_{TKE}(A_P)$ on angular momentum is rather weak that confirms the applicability of the above equations. For $A_P > 100$, the variance $(\sigma^{def}_{TKE}(A_P))^2$, originating from the deformation, mainly contributes to $\sigma^2_{TKE}(A_P)$; the same is true for all the reactions which we considered. The contribution to the variance of TKE due to nucleon exchange is more important in the decay of more asymmetric DNS (see Fig. 4.28). With increasing excitation energy, the variance of the TKE of quasifission products with $A_{tot}/2 - 20 \leq A_P \leq A_{tot}/2 + 20$ smoothly increases, mainly due to the increase of σ^2_β with T.

The relative contributions of the fusion-fission with respect to the quasifission yields of symmetric fragmentations are listed in Table 4.4 for various reactions and excitation energies. The contributions of the fusion-fission products are mainly determined by the fusion probability P_{CN}, since the survival probability of the excited compound nucleus is much less than unity.

If the compound nucleus is quite stable to be detected, the quasifission process is the main factor suppressing the complete fusion of heavy nuclei. In fusion reactions the fusion-fission events are much smaller than the events of the quasifission products. The main contribution to symmetric and near symmetric fragmentations comes from quasifission. For the cold fusion reactions leading to superheavy nuclei, the quasifission products are almost associated with fragmentations near the initial (entrance) DNS. However, an increase of the neutron number in the DNS results in a larger fraction of nearly symmetric splitting. With our calculations we also predict mass yields for $A_P < A_P(initial)$. Complementary to the heavier fragments with $A_P > A_P(initial)$, the lighter fragments can give significant information about the dynamics and evolution of the DNS on its way in the mass asymmetry coordinate η to the fused system. The yields of light products are known to be larger for higher beam energies. It would be a challenge for experimentalists to measure also the very asymmetric region of the quasifission mass yield.

Beside the calculations for hot fusion reactions we also carried out calculations of quasifission and TKE variances for reactions with a ^{58}Fe beam, for cold fusion reactions with Pb targets, and for reactions with lighter nuclei, e.g. ^{40}Ar + ^{165}Ho [111].

4.6 Multinucleon Transfer Reactions

4.6.1 Production of Heaviest Nuclei in Transfer Reactions

The master equations also describe configurations of dinuclear systems which are more asymmetric than the DNS in the entrance channel. The processes of formation and decay are ruled by the same mechanism of diffusion in the same relevant collective coordinates: mass and charge asymmetries and relative distance.

With asymmetric-exit-channel quasifission (AECQ) reactions leading to nuclei with charge numbers larger than the charge number of the target, one can produce isotopes that can not be synthesized in complete fusion reactions. The direct production of transactinides in AECQ reactions would give nuclei with $101 \leq Z \leq 108$ in the reactions ^{48}Ca + ^{238}U, ^{243}Am, ^{244,246,248}Cm. The production of heavy actinides has been studied in the transfer-type reactions by bombarding of actinide targets with ^{16,18}O, ^{20,22}Ne and ^{40,44,48}Ca [114–116]. Nuclei with $Z > 102$ have not been observed because of the small cross sections or short lifetimes in the radiochemical identification of the nuclei.

The cross section $\sigma_{Z,N}$ of the production of a primary heavy nucleus with $Z = Z_H$ and $N = N_H$ (H = heavy) in the AECQ reaction is written as the product of the capture cross section σ_{cap} in the entrance reaction channel and the formation-decay probability Y_{Z_L,N_L} (L = light) of the DNS configuration: $\sigma_{Z,N} = \sigma_{cap} Y_{Z_L,N_L}$. The primary heavy nucleus is excited and evaporates x neutrons in the de-excitation process. The evaporation residue cross section for the heavy nucleus with charge number Z is obtained as

$$\sigma_{ER}(Z, N - x) = \sigma_{Z,N} W_{sur}(xn). \tag{4.64}$$

The actinide targets proposed for such reactions are deformed. Therefore, the minimum value of the incident energy $E_{c.m.}^{min}$, at which the collisions of nuclei at all orientations become possible, is larger than the Coulomb barrier calculated for spherical nuclei. In the AECQ reactions which occur slightly above the Coulomb barrier, only partial waves with $J \leq J_{cap} = 20$ contribute to the production of superheavy nuclei. For $J_{cap} = 20$, the primary heavy nucleus has an angular momentum of about 10.

Fig. 4.29 The DNS potential energies at R_m and $J = 0$ as functions of Z of the heavy nucleus are presented by *dotted*, *dashed* and *solid curves* for the reactions ^{48}Ca + ^{244,246,248}Cm, respectively. The arrow indicates the initial DNS. For the ^{48}Ca + ^{248}Cm reaction, the barriers $B_\eta^{sym}(Z_i = 20, N_i = 48)$ and $B_R(Z = 102, N = 160)$ are indicated. The $\times$ denotes $U(R_b, Z = 102, N = 160, J = 0)$. The potential energies refer to the energies of the corresponding compound nuclei. The mass table [79] is used

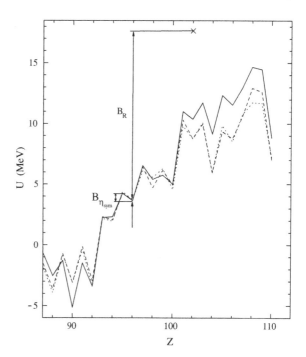

The primary mass and charge yield Y_{Z_L,N_L} of the decay fragments can be calculated as (see Eq. (4.52))

$$Y_{Z_L,N_L} = \Lambda_{Z_L,N_L}^{qf} \int_0^{t_0} P_{Z_L,N_L} dt. \tag{4.65}$$

For $J \leq 20$, the value of the probability of formation of the corresponding DNS configuration is weakly dependent on J and the factorization (4.64) is justified. The reactions with the transfer of many nucleons occur during a quite long time up to $t_0 \approx 10^{-20}$ s at $J \leq 20$. The DNS potential energies at R_m as a function of $Z = Z_H$ of the heavy fragment are shown in Fig. 4.29 for the reaction ^{48}Ca + ^{244,246,248}Cm [117]. A minimization with respect to the N/Z ratio is made for each Z and the deformations of the DNS nuclei are taken into consideration.

For $102 < Z < 110$, the potential energy decreases with the total number of neutrons of the DNS and a larger primary yield of superheavy nuclei is expected in the reactions with ^{244,246}Cm rather than with ^{248}Cm. This is demonstrated in Fig. 4.30, where the primary yields of the most probable isotopes of heavy nuclei are calculated with the master equations and with the statistical formula

$$Y_{Z,N} \approx 0.5 \exp\left(-\frac{B_R(Z, N) - B_\eta^{sym}(Z_i, N_i)}{T(Z_i, N_i)}\right), \tag{4.66}$$

Fig. 4.30 The calculated evaporation residue cross sections σ_{ER} are shown by *triangles, circles* and *squares* for the reactions ^{48}Ca + ^{244,246,248}Cm ($E_{c.m.} = 207$, 205.5 and 204 MeV, respectively). The heavy fragments after $1n$ evaporation are indicated in the *upper part* of the figure. The results obtained with (4.66) and (4.50) are indicated by *closed* and *open* symbols, respectively

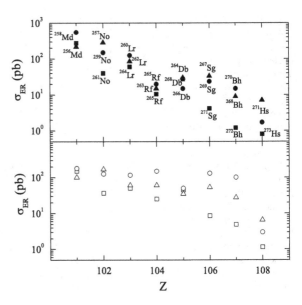

where B_R is given by $B_R(Z, N) = U(R_b, Z, N, J) - U(R_m, Z_i, N_i, J)$ (see Fig. 4.29) and T is the temperature calculated by using the Fermi-gas expression $T = \sqrt{E^*/a}$ with the excitation energy E^* of the initial DNS and the level-density parameter $a = A_{tot}/12\,\text{MeV}^{-1}$. Similar results are obtained with both methods.

In Fig. 4.30 the excitation energies of primary heavy nuclei correspond to $E_{c.m.} = 204 - 207\,\text{MeV}$. The excitation energy of the primary heavy nucleus is defined proportionally to its mass $A_H : E_H^*(Z, N) = (E^*(Z_i, N_i) - B_R(Z, N))A_H/(A_L + A_H)$. In this case $E_H^*(Z, N)$ is related to the maxima or to the right hand sides of excitation functions for one-neutron emission. For example, for ^{262}No we find $E_H^* = 16\,\text{MeV}$ and $W_{sur}(1n) = 2.4 \times 10^{-4}$. The experimental data as well as our treatment indicate the preference of a smaller number of evaporated neutrons to produce superheavy nuclei. So one can see that with the AECQ reactions on actinide targets, unknown isotopes of superheavy nuclei can be produced with suitable cross sections.

4.6.2 Transfer Products in Cold Fusion Reactions

In Fig. 4.31 the cross sections are drawn for producing isotopes with $Z = 84-94$ in the reactions ^{74}Ge + ^{208}Pb at $E_{c.m.} = 271.3\,\text{MeV}$ and ^{76}Ge + ^{208}Pb at $E_{c.m.} = 272.3\,\text{MeV}$ [118]. The products are among Po, At, Rn, Fr and Ra. The isotopes with short lifetimes decay during their flight from the target to the detector ($t \approx 2\,\mu\text{s}$). The cross sections increase with decreasing neutron number of the projectile. The measurement of the yields of transfer-type products in cold fusion reactions and their

Fig. 4.31 Calculated cross sections of the isotopes of nuclei with $Z = 84 - 94$ produced in the reactions ^{74,76}Ge $+ ^{208}$Pb at $E_{c.m.} = 271.3$ and 272.3 MeV, respectively. The order of the indicated mass numbers of the isotopes is in accordance with the decrease of their cross sections

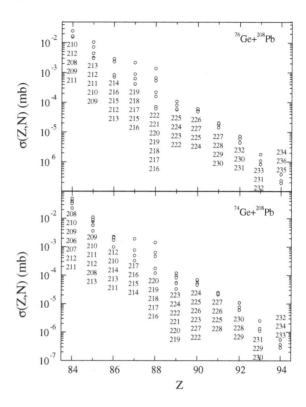

comparison with theoretical predictions are important for the understanding of the mechanism of fusion.

4.6.3 Production of Neutron-Rich Isotopes in Transfer Reactions

A recent study deals with the production of neutron-rich Zn and Ge isotopes with neutron numbers $N > 50$ (here $N = N_L$ and $Z = Z_L$) which are the products of multinucleon transfer channels of the reactions ^{48}Ca $+ ^{238}$U and ^{244}Pu at low energies [119]. In Fig. 4.32 we present calculated production cross sections of neutron-rich isotopes in the reactions ^{48}Ca $+ ^{238}$U and ^{244}Pu at incident energies near the Coulomb barrier. In both graphs the values of $E_{c.m.}$ correspond to the condition $E_L^*(Z, N, J) = B_n(Z, N)$ where the neutron separation energy B_n for unknown nuclei is taken from the finite range liquid drop model [82]. If we have $E_L^*(Z, N, J) > B_n(Z, N)$, then the primary neutron-rich nuclei are transformed into secondary nuclei with a less number of neutrons because of the deexcitation by nucleon emission. The DNS evolution in this reactions can be schematically presented in the following way: ^{48}Ca $+ ^{238}$U $\rightarrow ^{78,80}$Zn $+ ^{208,206}$Pb $\rightarrow ^{82,84,86}$Zn $+ ^{204,202,200}$Pb and ^{48}Ca $+ ^{244}$Pb $\rightarrow ^{82,84}$Ge $+$

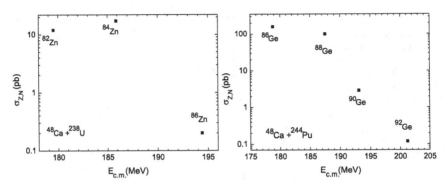

Fig. 4.32 *Left* Calculated cross sections for the indicated neutron-rich isotopes of Zn produced in the ^{48}Ca $+^{238}$U reaction at values of $E_{c.m.}$ providing the excitations of these isotopes to be equal to the corresponding thresholds for the neutron emission. *Right* The same as in the left figure, but for the indicated neutron-rich isotopes of Ge produced in the ^{48}Ca $+ ^{244}$Pu reaction

^{210,208}Pb $\rightarrow$ ^{86,88,90,92}Ca $+^{206,204,202,200}$ Pb. The system initially moves to the deep minimum of the potential energy surface which is caused by shell effects around the DNS with the magic heavy ^{208}Pb and light ^{78}Zn or ^{84}Ge nuclei. Then from this minimum it reaches the DNS with the exotic light nucleus by fluctuations in the mass asymmetry. For low excitation energy, the evolution of the DNS towards symmetry is hindered by the driving potential minimum. Since the predicted production cross sections for the new exotic isotopes ^{84,86}Zn and ^{90,92}Ge are larger than 0.1 pb, they can be synthesized with the present experimental possibilities (see also [120, 121]).

Within the DNS formalism, we studied for the future experiments the possibilities for producing new neutron-rich isotopes of nuclei with $Z = 64$–79 (here $N = N_H$ and $Z = Z_H$) as complementary to light fragments in the ^{48}Ca$+^{238}$U multinucleon transfer reaction at an incident energy $E_{c.m.}$ = 189 MeV taken as the height of the Coulomb barrier for the spherical nuclei. The primary neutron-rich nuclei of interest are excited and transformed into the secondary nuclei with less number of neutrons with the same cross section because the neutron emission is dominant over other deexcitation channels. The neutron emission channels are indicated in Fig. 4.33 for primary neutron-rich Os and Re isotopes. The calculated results demonstrate that the multinucleon transfer reactions provide a very efficient tool for producing new neutron-rich nuclei with $Z = 64$–79. It is apparent that with the use of a heavier actinide target as ^{244}Pu or ^{248}Cm one can reach more neutron-rich nuclei.

4.7 Binary and Ternary Fission in the Scission-Point Model

The measured fission properties of ^{258,259}Fm, ^{259,260}Md and ^{258,262}No show total kinetic energy (TKE) distributions of the fragments to be composed of two Gaussians [122–124]. The highest TKE is associated with a sharply symmetrical mass

Fig. 4.33 Calculated
production cross sections of
the primary Os and Re
isotopes versus the mass
number in the multinucleon
transfer reaction ^{48}Ca($E_{c.m.}$
$= 189$ MeV)$+^{238}$U. Neutron
evaporation channels for
neutron-rich primary
isotopes are indicated. The
heaviest known isotopes are
marked by *arrows*

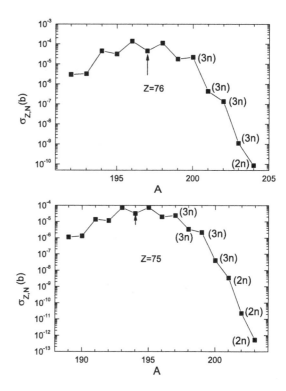

distribution while the fragments with lower energy result in a sharply symmetrical
or asymmetrical mass distribution. This phenomena is called the bimodal fission.
Here, we explain the bimodal fission with the dependence of the potential energy
surface of the fissioning system on the deformation parameters of the clusters in the
framework of the dinuclear system model. After the penetration through the fission
barrier the fissioning system moves along a trajectory to a compact dinuclear configuration belonging to the high TKE mode or along a trajectory to a highly deformed
and elongated dinuclear configuration belonging to the low TKE mode. In order to
describe the dynamics of these modes, we treat the fission process at the scission
point with the statistical scission-point model by using mass and charge asymmetry
coordinates, the deformation parameters of the fragments and the excitation energy
of the system [125–128].

4.7.1 Fission Potential With the Dinuclear System Model

The fissioning nucleus is assumed to form a dinuclear system at the scission point
with the two fragments in contact. The scission point is taken at a nuclear distance
$R = R_b$ with R_b as the position of the outer potential barrier. Then the following

parameters describe the system: the mass numbers A_i and charge numbers Z_i of the fragments $i = 1, 2$, their deformation parameters $\beta_i = c_i/a_i$ with the major (c_i) and minor (a_i) semiaxes of the ellipsoidal shapes of the fragments, the distance R between the fragments and the excitation energy E^* of the system. The negative binding energy of the system is obtained as the sum of the negative binding energies of the individual fragments, represented by the liquid drop energy U_{LD} and shell model correction energy U_{shell}, and by the Coulomb and nuclear interaction between the fragments (L = light, H = heavy):

$$U(\{A_i, Z_i, \beta_i\}, R, E^*) = U_{LD}^L(A_L, Z_L, \beta_L) + U_{LD}^H(A_H, Z_H, \beta_H)$$
$$+ \delta U_{shell}^L(A_L, Z_L, \beta_L, E^*) + \delta U_{shell}^H(A_H, Z_H, \beta_H, E^*)$$
$$+ V_C(\{A_i, Z_i, \beta_i\}, R) + V_N(\{A_i, Z_i, \beta_i\}, R).$$

$$(4.67)$$

The shell model correction energy δU_{shell} depends on the excitation energy. The excitation energy is given by

$$E^* = Q - \text{TKE}(\{A_i, Z_i, \beta_i\}) + B_n - E_{def}(\{A_i, Z_i, \beta_i\}) \qquad (4.68)$$

with the Q-value and the total kinetic energy of the fragments

$$\text{TKE}(\{A_i, Z_i, \beta_i\}) = V_C(\{A_i, Z_i, \beta_i\}, R_b) + V_N(\{A_i, Z_i, \beta_i\}, R_b). \qquad (4.69)$$

Here, we set $B_n = 0$ for spontaneous fission and $B_n = 8$ Mev for neutron induced fission. On the basis of the two-center shell model we first calculate the shell corrections δU_{shell} at zero excitation energy. In order to take into account the dependence of the shell corrections on the excitation energy E^* of the dinuclear system, we use a frequently applied phenomenological expression

$$\delta U_{shell}^i(A_i, Z_i, \beta_i, E^*) = \delta U_{shell}^i(A_i, Z_i, \beta_i, E^* = 0) \exp(-E_i^*/E_D), \qquad (4.70)$$

where the damping constant is chosen $E_D = 18.5$ MeV and the excitation energy in the pre-scission configuration is taken to be $E_i^* = A_i E^*/(A_L + A_H)$. The calculations have shown that for the excitation energies considered, a variation of the decay constant E_D over the range from 15 to 25 MeV has only slight effects on the results of the calculations. The deformation energy E_{def} is measured with respect to the ground states (g.s.) of the fragments of the dinuclear system and is calculated for $R = R_b$ as

$$E_{def}(\{A_i, Z_i, \beta_i\}) = U(\{A_i, Z_i, \beta_i\}, R_b, E^*) - U(\{A_i, Z_i, \beta_i^{g.s.}\}, R_b, E^*).$$

$$(4.71)$$

Fig. 4.34 shows the contour plots of the neutron-induced fission of ^{236}U for the fragmentations ^{104}Mo$+^{132}$Sn (upper part) and ^{104}Zr$+^{132}$Te (lower part) as functions of the deformations of the light and heavy fragments at the scission point. The potential energy surface of ^{104}Zr $+ ^{132}$Te has two minima which lead to bimodal fission with

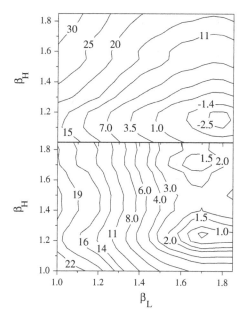

Fig. 4.34 Potential energy of scission configurations as a function of β_L and β_H for the neutron induced fission of ^{236}U leading to ^{104}Mo + ^{132}Sn (*upper part*) and ^{104}Zr + ^{132}Te (*lower part*)

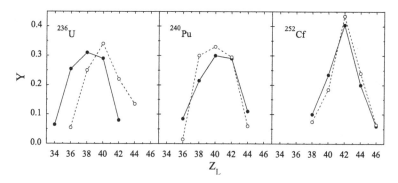

Fig. 4.35 Comparison of calculated and experimental charge number distributions for neutron induced fission of ^{236}U and ^{240}Pu, and spontaneous fission of ^{252}Cf. The yields are normalized to unity. The calculated and experimental points are shown as *open circles* and *solid points*, respectively, connected by *straight lines*

$\langle TKE \rangle$ = 181 and 168 MeV. At low values of E^*, there can be a few minima in the potential energy surface, but their number decreases with increasing E^*. Since the excitation energy of pre-scission configurations does not exceed 30 MeV in the cases of spontaneous and neutron induced fission, shell effects play an important role for the formation of the mass-energy distribution of fragments.

The bimodality or multimodality of fission can be also related to the variations of the charge and mass splittings in the small interval of charge and mass numbers [126]. In order to obtain high- and low-<TKE> regimes in the fissioning nucleus, one should not always require two fission paths, short and long, at the fixed charge and mass asymmetries. For example, for the fragmentations $^{235}U(n_{th}, f) \rightarrow {}^{104}Mo + {}^{132}Sn(< TKE > = 184 MeV)$ and $^{104}Zr + {}^{132}Te(< TKE > = 168 MeV)$ or $^{258}Fm(sf) \rightarrow {}^{126}Sn + {}^{132}Sn(< TKE > = 230 MeV)$ and $^{124}Cd + {}^{134}Te(< TKE > = 209 MeV)$. The suggested explanation of bimodal fission is rather simple and allows us to describe well the available experimental data. This explanation can be experimentally checked by measuring the masses and charges of the fission products along with their kinetic energies [126]. The identification of charge and mass numbers after the de-excitation of the fragments with γ-spectroscopy is very promising. The neutron evaporation from the fragments can be estimated with the calculated average values of excitation energies of the fragments. The suggested observation of bimodality requires the precise determination of the charges and masses of primary fragments. Additionally, one can expect different angular momenta of the fission fragments for different modes to be measured.

4.7.2 Binary Fission

The barrier at $R = R_b$ is assumed to keep the fragments in contact for a short time so that the fragments come into a thermodynamic equilibrium state and are distributed statistically in the potential energy surface. Then the relative primary yields of binary fission fragments before evaporation of neutrons can be calculated according to the statistical model as

$$Y(\{A_i, Z_i, \beta_i\}, E^*) = Y_0 \exp(-U(\{A_i, Z_i, \beta_i\}, R_b, E^*)/T). \qquad (4.72)$$

Here, the scission configuration has a certain distribution in β_L and β_H in thermodynamic equilibrium. The temperature T is related to the excitation energy E^*. For fixed A_i and Z_i the excitation energy is defined by the deepest minimum of the potential energy surface as a function of β_L and β_H with $1 \leq \beta_L, \beta_H \leq 2.1$. The corresponding effective temperature is not a free parameter, but obtained from the excitation energy: $T = (E^*/a)^{1/2}$ with $a = (A_L + A_H)/12$ MeV^{-1}. For fixed A_i the distribution of fission fragments in Z_i is very narrow. We found the most probable charge numbers Z_i^{min} for each A_i in good agreement with experimental data. If one needs to calculate the relative yield of a certain system regardless of deformation parameters, the probability is integrated over the deformations:

$$Y(A_L) = \sum_{Z_L} \int Y(A_L, Z_L, \beta_L, A_H, Z_H, \beta_H, E^*) d\beta_L d\beta_H. \qquad (4.73)$$

It should be noted that the statistical approach does not allow to calculate the absolute values of the yields; one just obtains relative yields. Since the temperatures of the

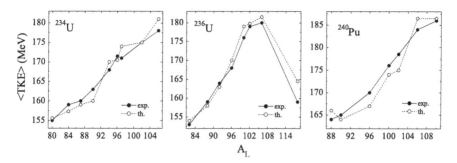

Fig. 4.36 Calculated and experimental ⟨TKE⟩ as a function of A_L for neutron induced fission of ^{234,236}U and ^{240}Pu. For fixed A_L, the most probable charge splitting was found by minimization of the total energy with respect to Z_L

systems considered here are about 1 MeV, only deformations near to the minima essentially contribute to the probabilities (yields). In order to get charge distributions one has to sum the probability over the mass number:

$$Y(Z_L) = \sum_{A_L} \int Y(A_L, Z_L, \beta_L, A_H, Z_H, \beta_H, E^*) d\beta_L d\beta_H. \qquad (4.74)$$

In Fig. 4.35 we compare experimental and calculated charge number distributions for spontaneous fission of ^{252}Cf and neutron induced fission of ^{236}U and ^{240}Pu. In the case of ^{236}U the calculated charge number distribution is shifted by two units relative to the experiment towards the symmetrical division. The mean total kinetic energy at fixed A_i, Z_i is calculated as

$$\langle TKE \rangle (\{A_i, Z_i\}) = \int TKE(\{A_i, Z_i, \beta_i\}) Y(\{A_i, Z_i, \beta_i\}, E^*) d\beta_L d\beta_H. \qquad (4.75)$$

To obtain the TKE—mass distribution, we sum over the charge number as follows

$$\langle TKE \rangle (A_L) = \sum_{Z_L} \langle TKE \rangle (\{A_i, Z_i\}). \qquad (4.76)$$

In Fig. 4.36 we show mean total kinetic energies as a function of A_L for ^{234,236}U and ^{240}Pu fission induced by thermal neutrons. For each fixed value of A_L the calculations were performed at the most probable value of Z_L. Since our quantities are the primary mass and energy distributions of the fission fragments before the prompt neutron emission, the experimental values of ⟨TKE⟩ have to be enlarged by 1–3 MeV. They are measured after the post - neutron emission. In Table 4.5 we listed the various calculated quantities for the spontaneous fission of ^{252}Cf in comparison with experimental data. Here, σ_{TKE}^2 are the variances of the TKE (see also [129]).

Table 4.5 Mean TKEs, variances of TKEs and mass yields in spontaneous fission of ^{252}Cf

Fragmentation	$< TKE >$ (MeV) exp	$< TKE >$ (MeV) theor	$\sigma^2_{<TKE>}$ (MeV^2) exp	$\sigma^2_{<TKE>}$ (MeV^2) theor	Y exp	Y theor
^{102}Zr $+^{150}$Ce	183.3	179.3	99(9)	106	0.21	0.18
^{106}Mo $+^{146}$Ba	189.3	190	89.5(4)	89	0.44	0.54
^{112}Ru $+^{140}$Xe	193.3	194.8	95(7)	90	0.25	0.21
^{118}Pd $+^{134}$Te	200	198		65	0.1	0.07

4.7.3 Ternary Fission

For the description of the ternary fission, we extended the model of binary fission. We assume that the ternary system consists of two prolate ellipsoidal fragments and a light charged particle (LCP) in between. The LCP can consist of one or several α-particles and neutrons originating from one or both binary fragments. The third light fragment (LCP) is assumed as stiff and spherical and lying between the heavier fragments. The potential depends on the deformations β_1 and β_2 of the outer fragments, but not so strongly as in the binary system.

Within the statistical method we calculate the relative yields of the ternary system for a given LCP, characterized by (A_3, Z_3). First, the relative probability of the binary system, $Y_{binary}(A_1^b, Z_1^b)$ with $A = A_1^b + A_2^b$ and $Z = Z_1^b + Z_2^b$, is determined. Then the ternary system is built up with the LCP between the heavier fragments. For each binary system and a certain LCP, we calculate the relative probabilities for ternary systems as

$$Y_{ternary}(A_1, Z_1, A_3, Z_3, A_1^b, Z_1^b) =$$
$$Y_t^0 \int \int \exp(-U(A_1, Z_1, \beta_1, A_2, Z_2, \beta_2, A_3, Z_3, E^*)/T) d\beta_1 d\beta_2. \quad (4.77)$$

Finally, the yields are summed with the same charge asymmetries and the same LCP. The following charge distribution is obtained:

$$Y_{ternary}(Z_1, A_3, Z_3) = \sum_{A_1^b, Z_1^b, A_1} Y_{binary}(A_1^b, Z_1^b) \times Y_{ternary}(A_1, Z_1, A_3, Z_3, A_1^b, Z_1^b).$$
$$(4.78)$$

The normalization factors are chosen so that the binary yields and the yields of ternary systems from a certain binary system are normalized to unity. The probability of formation of the α-particles from the heavy fragments is almost the same for all the heavy fragments and does not influence the relative yields.

In Fig. 4.37 we compare our calculations for ternary fission of ^{252}Cf and induced ternary fission of ^{56}Ni produced in the reaction ^{32}S $+^{24}$Mg [30]. In Table 4.6 we present spectroscopic factors S for the correlations of the formation probabilities of different LCP. The value of S and Y_{exp} are related to the value of ^{4}He which is $S(^4$He$)$

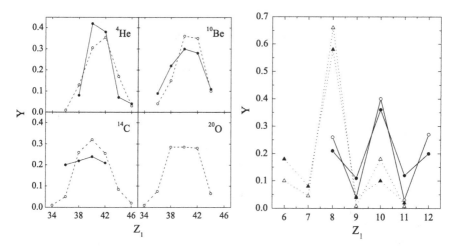

Fig. 4.37 *Left hand side* Charge distributions in spontaneous ternary fission of ^{252}Cf with different LCPs. The calculated and experimental points are depicted by *empty* and *filled circles*, respectively, connected by *straight lines*. *Right hand side* The experimental (*solid symbols*) and calculated (*open symbols*) charge distributions in induced ternary fission of ^{56}Ni with the middle particle ^{8}Be (*circles*) and ^{12}C (*triangles*), respectively

Table 4.6 Spectroscopic factors for the formation of the LCP with respect to $S(^{4}\text{He}) = 5 \times 10^{-2}$

LCP	Y_{exp}	$S_{\alpha+\alpha+...}$	S_{LCP}
^{4}He	1	1	1
^{10}Be	1.3×10^{-2}	5×10^{-2}	$\sim 5 \times 10^{-4}$
^{14}C	5×10^{-3}	2.5×10^{-3}	$\sim 10^{-6}$
^{20}O		1.25×10^{-4}	$\sim 10^{-9}$

$= 5 \times 10^{-2}$. For ^{10}Be we found that two sequentially formed ^{4}He are correlated to form the LCP ^{10}Be. The neutrons do not change the probability of formation. The spectroscopic factor for a direct formation of ^{10}Be, $S_{\text{LCP} = ^{10}\text{Be}}$, is by two orders of magnitude smaller.

4.8 Selected Summarizing and Concluding Remarks

Because of the richness of applications of the dinuclear model described in this article we think to renounce a detailed summary of the described methods. As one can recognize, the dinuclear model is a very basic model for the description of nuclear structure and heavy ion collisions, for example, not mentioned in this review deep-inelastic collisions [130–132]. In principle it explains the physics of a nuclear molecule, consisting of two touching clusters. The potential between the clusters is of diabatic nature, with a pocket around the touching distance and with a repulsive part at smaller relative distances, avoiding a melting of the two nuclei as it may happen with adiabatic potentials. Therefore, the main degrees of freedom in this model are the mass and charge asymmetries related to the transfer of nucleons (or clusters) between the clusters.

As it depends on the complexity of the methods the dinuclear model can be a phenomenological or a quantum theoretical description of the two touching clusters. The phenomenological description of diffusion in the collective mass asymmetry coordinate can be improved by microscopical master equations or by treatments of quantum and open quantum systems. The applied constants like single particle widths and cranking masses should be attributed to more microscopical calculations. Also further collective degrees of freedom, like vibrations, rotations and neck dynamics of the clusters should be adequately taken into consideration.

The basic assumption of a repulsive diabatic potential at smaller internuclear distances inherent in the dinuclear model does not contradict the experimental data. All applications of the dinuclear model give a qualitatively and quantitatively satisfying description of the present data. Therefore, we believe on the correctness of the internuclear potential of the dinuclear model. However, this is an important point which needs further microscopical proofs and experimental signs for its existence, especially as a function of the mass numbers of the clusters.

Acknowledgments We thank Prof. V.V. Volkov and the late Prof. S.P. Ivanova for introducing us into the ideas of the dinuclear system concept. Further, we are thankful to Prof. R.V. Jolos, Prof. Enguang Zhao and Prof. Junqing Li and all their and our coworkers for their very fruitful and pleasant collaboration with us in this actual field of nuclear physics. We thank the VW-Stiftung (Hannover), the DFG (Bonn) and the RFBR (Moscow) for their continuous financial support of this work.

References

1. Greiner, W., Park, J.Y., Scheid, W.: Nuclear Molecules. World Scientific, Singapore (1995)
2. Volkov, V.V.: Phys. Rep. **44**, 93 (1978)
3. Volkov, V.V.: Nuclear Reactions of Deep Inelastic Transfers. Energoizdat, Moscow (1982)
4. Volkov, V.V.: Izv. AN SSSR ser. fiz. **50**, 1879 (1986)
5. Volkov, V.V.: in Treatise on Heavy-Ion Science, Bromley, D.A. (ed.) Vol. 8 (Plenum Press, New York, 1989) p.255.
6. Bromley, D.A., Kuehner, J.A., Almqvist, E.: Phys. Rev. Lett. **4**, 365 (1960)
7. Maruhn, J., Greiner, W.: Z. Phys. **251**, 431 (1972)
8. Diaz Torres, A.: Phys. Rev. Lett. **101**, 122501 (2008)
9. Lukasiak, A., Cassing, W., Nörenberg, W.: Nucl. Phys. A **426**, 181 (1984)
10. Cassing, W., Nörenberg, W.: Nucl. Phys. A **433**, 467 (1985)
11. Diaz Torres, A., Antonenko, N.V., Scheid, W.: Nucl. Phys. A **652**, 61 (1999)
12. Adamian, G.G., Antonenko, N.V., Tchuvil'sky, Yu.M.: Phys. Lett. B **451**, 289 (1999)
13. Adamian, G.G., Antonenko, N.V., Jolos, R.V., Scheid, W.: Nucl. Phys. A **619**, 241 (1997)
14. Adamian, G.G., Antonenko, N.V., Ivanova, S.P., Scheid, W.: Nucl. Phys. A **646**, 29 (1999)
15. Adamian, G.G., Antonenko, N.V., Diaz Torres, A., Scheid, W.: Nucl. Phys. A **671**, 233 (2000)
16. Fink, H.J., Scheid, W., Greiner, W.: J. Phys. G **1**, 685 (1975)
17. Sargsyan, V.V., Kanokov, Z., Adamian, G.G., Antonenko, N.V., Scheid, W.: Phys. Rev. C **80**, 034606 (2009)
18. Krasznahorkay, A. et al.: Phys. Rev. Lett. **80**, 2073 (1998)

19. Krasznahorkay, A. et al.: Phys. Lett. B **461**, 15 (1999)
20. Blons, J.: Nucl. Phys. A **502**, 121c (1989)
21. Baumann, F.F., Brinkmann, K.Th.: Nucl. Phys. A **502**, 271c (1989)
22. Thirolf, P.G., Habs, D.: Prog. Part. Nucl. Phys. **49**, 325 (2002)
23. Cseh, J., Scheid, W.: J. Phys. G **18**, 1418 (1992)
24. Cseh, J., Lévai, G., Scheid, W.: Phys. Rev. C **48**, 1724 (1993)
25. Sanders, S.J., Szanto de Toledo, A., Beck, C.: Phys. Rep. **311**, 487 (1999)
26. Freer, M.: Rep. Prog. Phys. **70**, 2149 (2007)
27. Sciani, W. et al.: Phys. Rev. C **80**, 034319 (2009)
28. Cseh, J. et al.: Phys. Rev. C **80**, 034320 (2009)
29. Beck, C. et al.: Phys. Rev. C **80**, 034604 (2009)
30. von Oertzen, W. et al.: Phys. Rev. C **78**, 044615 (2008)
31. Adamian, G.G., Antonenko, N.V., Nenoff, N., Scheid, W.: Phys. Rev. C **64**, 014306 (2001)
32. Adamian, G.G., Antonenko, N.V., Jolos, R.V., Ivanova, S.P., Melnikova, O.I.: Int. J. Mod. Phys. E **5**, 191 (1996)
33. Migdal, A.B.: Theory of Finite Fermi Systems and Applications to Atomic Nuclei. Nauka, Moscow (1982)
34. Adamian, G.G., Andreev, A.V., Antonenko, N.V., Nenoff, N., Scheid, W., Shneidman, T.M.: Heavy Ion. Phys. **19**, 87 (2003)
35. Adamian, G.G., Andreev, A.V., Antonenko, N.V., Nenoff, N., Scheid, W., Shneidman, T.M.: Acta Phys. Pol. B **34**, 2147 (2003)
36. Kuklin, S.N., Adamian, G.G., Antonenko, N.V., Scheid, W.: Int. J. Mod. Phys. E **17**, 2019 (2008)
37. Zubov, A.S., Sargsyan, V.V., Adamian, G.G., Antonenko, N.V., Scheid, W.: Phys. Rev. C **81**, 024607 (2010)
38. Shneidman, T.M., Adamian, G.G., Antonenko, N.V., Ivanova, S.P., Scheid, W.: Nucl. Phys. A **671**, 119 (2000)
39. Shneidman, T.M., Adamian, G.G., Antonenko, N.V., Jolos, R.V., Scheid, W.: Phys. Lett. B **526**, 322 (2002)
40. Shneidman, T.M., Adamian, G.G., Antonenko, N.V., Jolos, R.V., Scheid, W.: Phys. Rev. C **67**, 014313 (2003)
41. Adamian, G.G., Antonenko, N.V., Jolos, R.V., Shneidman, T.M.: Phys. Rev. C **70**, 064318 (2004)
42. Shneidman, T.M., Adamian, G.G., Antonenko, N.V., Jolos, R.V.: Phys. Rev. C **74**, 034316 (2006)
43. Adamian, G.G., Antonenko, N.V., Jolos, R.V.: Nucl. Phys. A **584**, 205 (1995)
44. http://www.nndc.bnl.gov/nndc/ensdf/
45. Adamian, G.G., Antonenko, N.V., Jolos, R.V., Palchikov, Yu.V., Scheid, W.: Phys. Rev. C **67**, 054303 (2003)
46. Svensson, C.E. et al.: Phys. Rev. Lett. **82**, 3400 (1999)
47. Henry, R.G. et al.: Phys. Rev. Lett. **73**, 777 (1994)
48. Korichi, A. et al.: Phys. Lett. B **345**, 403 (1995)
49. Khoo, T.L. et al.: Phys. Rev. Lett. **76**, 1583 (1996)
50. Lopez-Martens, A. et al.: Phys. Lett. B **380**, 18 (1996)
51. Hackman, G. et al.: Phys. Rev. Lett. **79**, 4100 (1997)
52. Wilson, A.N. et al.: Phys. Rev. Lett. **90**, 142501 (2003)
53. Adamian, G.G., Antonenko, N.V., Jolos, R.V., Palchikov, Yu.V., Scheid, W., Shneidman, T.M.: Phys. Rev. C **69**, 054310 (2004)
54. Adamian, G.G., Antonenko, N.V., Scheid, W.: Nucl. Phys. A **678**, 24 (2000)
55. Oganessian, Yu.Ts. et al.: Phys. Rev. C **79**, 024608 (2009)
56. Diaz Torres, A., Adamian, G.G., Antonenko, N.V., Scheid, W.: Phys. Lett. B **481**, 228 (2000)
57. Antonenko, N.V., Cherepanov, E.A., Nasirov, A.K., Permjakov, V.B., Volkov, V.V.: Phys. Lett. B **319**, 425 (1993)

58. Antonenko, N.V., Cherepanov, E.A., Nasirov, A.K., Permjakov, V.B., Volkov, V.V.: Phys. Rev. C **51**, 2635 (1995)
59. Adamian, G.G., Antonenko, N.V., Scheid, W., Volkov, V.V.: Nuov. Cim. A **110**, 1143 (1997)
60. Adamian , G.G., Antonenko , N.V., Scheid , W., Volkov, V.V.: Nucl. Phys. A **627**, 361 (1997)
61. Adamian, G.G., Antonenko, N.V., Scheid, W., Volkov, V.V.: Nucl. Phys. A **633**, 409 (1998)
62. Giardina, G., Hofmann, S., Muminov, M.I., Nasirov, A.K.: Eur. Phys. J. A **8**, 205 (2000)
63. Volkov, V.V.: Part. Nucl. **35**, 797 (2004)
64. Zhao, E.G., Wang, N., Feng, Z.Q., Li, J.Q., Zhou, S.G., Scheid, W.: Int. J. Mod. Phys. E **17**, 1937 (2008)
65. Gäggeler, H. et al.: Z. Phys. A **316**, 291 (1984)
66. Adamian, G.G., Antonenko, N.V., Scheid, W.: Nucl. Phys. A **618**, 176 (1997)
67. Adamian, G.G., Antonenko, N.V., Ivanova, S.P., Scheid, W.: Phys. Rev. C **62**, 064303 (2000)
68. Zubov, A.S., Adamian, G.G., Antonenko, N.V., Ivanova, S.P., Scheid, W.: Phys. Rev. C **65**, 024308 (2002)
69. Zubov, A.S., Adamian, G.G., Antonenko, N.V., Ivanova, S.P., Scheid , W.: Eur. Phys. A **33**, 223 (2007)
70. Zubov, A.S., Adamian, G.G., Antonenko, N.V., Ivanova, S.P., Scheid, W.: Eur. Phys. J. A **23**, 249 (2005)
71. Zubov, A.S., Adamian, G.G., Antonenko, N.V., Ivanova, S.P., Scheid, W.: Phys. Rev. C **68**, 014616 (2003)
72. Hofmann, S.: Rep. Prog. Phys. **61**, 636 (1998)
73. Hofmann, S.: Eur. Phys. J. A **15**, 195 (2002)
74. Hofmann, S., Münzenberg, G.: Rev. Mod. Phys. **72**, 733 (2000)
75. Armbruster, P.: Ann. Rev. Nucl. Part. Sci. **50**, 411 (2000)
76. Hofmann, S.: Lect. Notes Phys. **764**, 203 (2009)
77. Morita, K.: Nucl. Phys. A **834**, 338c (2010)
78. Oganessian, Yu.Ts.: J. Phys. G **34**, R165 (2007)
79. Möller, P., Nix, R.: At. Data Nucl. Data Tables **39**, 213 (1988)
80. Cherepanov, E.A.: preprint JINR, E7-99-27 (1999)
81. Adamian , G.G., Antonenko, N.V., Scheid , W.: Phys. Rev. C **69**, 011601(R) (2004)
82. Möller, P. et al.: At. Data Nucl. Data Tables **59**, 185 (1995)
83. Myers , W.D., Swiatecki, W.J.: Nucl. Phys. A **601**, 141 (1996)
84. Adamian, G.G., Antonenko, N.V., Scheid, W.: Phys. Rev. C **69**, 014607 (2004)
85. Adamian , G.G., Antonenko , N.V., Scheid, W.: Phys. Rev. C **69**, 044601 (2004)
86. Adamian, G.G., Antonenko , N.V., Scheid, W., Zubov, A.S.: Phys. Rev. C **78**, 044605 (2008)
87. Adamian , G.G., Antonenko, N.V., Scheid, W.: Eur. Phys. J. A **41**, 235 (2009)
88. Adamian , G.G., Antonenko, N.V., Sargsyan, V.V.: Phys. Rev. C **79**, 054608 (2009)
89. Adamian, G.G., Antonenko, N.V., Sargsyan, V.V., Scheid, W.: Nucl. Phys. A **834**, 345c (2010)
90. Zubov , A.S., Adamian , G.G., Antonenko, N.V.: Phys. Part. Nucl. **40**, 847 (2009)
91. Sargsyan , V.V., Zubov, A.S., Kanokov, Z., Adamian, G.G., Antonenko, N.V.: Phys. Atom. Nucl. **72**, 425 (2009)
92. Sargsyan, V.V., Kanokov, Z., Adamian, G.G., Antonenko, N.V.: Phys. Part. Nucl. **41**, 175 (2010)
93. Yeremin, A.V.: Phys. Part. Nucl. **38**, 492 (2007)
94. Loveland, W. et al.: Phys. Rev. C **66**, 044617 (2002)
95. Gregorich, K.E. et al.: Phys. Rev. C **72**, 014605 (2005)
96. Stavsetra, L. et al.: Phys. Rev. Lett. **103**, 132502 (2009)
97. Yakushev, A.B. et al.: Radiochim. Acta **91**, 443 (2002)
98. Eichler, R. et al.: Radiochim. Acta **94**, 181 (2006)
99. Eichler, R. et al.: Nature **447**, 72 (2007)
100. Parkhomenko, O., Muntian, I., Patyk, Z., Sobiczewski, A.: Acta. Phys. Pol. B **34**, 2153 (2003)
101. Muntian, I., Hofmann, S., Patyk, Z., Sobiczewski, A.: Acta. Phys. Pol. B **34**, 2073 (2003)
102. Muntian, I., Patyk, Z., Sobiczewski, A.: Acta. Phys. Pol. B **32**, 691 (2001)

103. Muntian, I., Patyk, Z., Sobiczewski, A.: Acta. Phys. Pol. B **34**, 2141 (2003)
104. Sobiczewski, A., Gareev, F.A., Kalinkin, B.N.: Phys. Lett. B **22**, 500 (1966)
105. Möller , P., Nix, R.: J. Phys. G **20**, 1681 (1994)
106. Sobiczewski , A., Pomorski, K.: Prog. Part. Nucl. Phys. **58**, 292 (2007)
107. Bender, M., Heenen, P.H., Reinhard, P.G.: Rev. Mod. Phys. **75**, 121 (2003)
108. Liran , S., Marinov, A., Zeldes, N.: Phys. Rev. C **63**, 017302 (2001)
109. Liran, S., Marinov, A., Zeldes, N.: Phys. Rev. C **66**, 024303 (2002)
110. Oganessian, Yu.Ts. et al.: Phys. Rev. C **79**, 024603 (2009)
111. Adamian , G.G., Antonenko, N.V., Scheid, W.: Phys. Rev. C **68**, 034601 (2003)
112. Diaz Torres , A., Adamian , G.G., Antonenko, N.V., Scheid, W.: Phys. Rev. C **64**, 024604 (2001)
113. Diaz Torres , A., Adamian, G.G., Antonenko, N.V., Scheid, W.: Nucl. Phys. A **679**, 410 (2001)
114. Gäggeler, H. et al.: Phys. Rev. C **33**, 1983 (1986)
115. Hoffman, D.C. et al.: Phys. Rev. C **31**, 1763 (1985)
116. Türler, A. et al.: Phys. Rev. C **46**, 1364 (1992)
117. Adamian , G.G., Antonenko, N.V., Zubov, A.S.: Phys. Rev. C **71**, 034603 (2005)
118. Adamian , G.G., Antonenko , N.V.: Phys. Rev. C **72**, 064617 (2005)
119. Adamian , G.G., Antonenko, N.V., Sargsyan, V.V., Scheid, W.: Phys. Rev. C **81**, 024604 (2010)
120. Penionzhkevich , Yu.E., Adamian, G.G., Antonenko, N.V.: Phys. Lett. B **621**, 119 (2005)
121. Penionzhkevich , Yu.E., Adamian, G.G., Antonenko, N.V.: Eur. Phys. J. A **27**, 187 (2006)
122. Hulet, E.K. et al.: Phys. Rev. Lett. **56**, 313 (1986)
123. Hulet, E.K. et al.: Phys. Rev. C **40**, 770 (1989)
124. Britt, H.C. et al.: Phys. Rev. C **30**, 559 (1984)
125. Andreev , A.V., Adamian, G.G., Antonenko, N.V., Ivanova, S.P., Scheid, W.: Eur. Phys. J. A **22**, 51 (2004)
126. Andreev, A.V., Adamian, G.G., Antonenko, N.V., Ivanova, S.P.: Eur. Phys. J.A **26**, 327 (2005)
127. Andreev , A.V., Adamian, G.G., Antonenko, N.V., Ivanova, S.P., Kuklin, S.N., Scheid, W.: Eur. Phys. J. A **30**, 579 (2006)
128. Andreev , A.V., Adamian, G.G., Antonenko, N.V., Ivanova, S.P., Scheid, W.: Phys. Atom. Nucl. **69**, 197 (2006)
129. Bolgova , O.N., Adamian, G.G., Antonenko, N.V., Zubov, A.S., Ivanova, S.P., Scheid, W.: Phys. Atom. Nucl. **72**, 928 (2009)
130. Adamian , G.G., Nasirov, A.K., Antonenko, N.V., Jolos, R.V.: Phys. Part. Nucl. **25**, 583 (1994)
131. Heinz , S. et al.: Eur. Phys. J. A **38**, 227 (2008)
132. Heinz, S. et al.: Eur. Phys. J. A **43**, 181 (2010)
133. Belozerov, A.V. et al.: Eur. Phys. J. A **16**, 447 (2003)
134. Oganessian, Yu.Ts. et al.: Phys. Rev. C **64**, 054606 (2001)
135. Gäggeler, H.W. et al.: Nucl. Phys. A **502**, 561 (1989)

Chapter 5
Nuclear Alpha-Particle Condensates

T. Yamada, Y. Funaki, H. Horiuchi, G. Röpke, P. Schuck and A. Tohsaki

Abstract The α-particle condensate in nuclei is a novel state described by a product state of α's, all with their c.o.m. in the lowest $0S$ orbit. We demonstrate that a typical α-particle condensate is the Hoyle state ($E_x = 7.65\,\text{MeV}$, 0_2^+ state in ^{12}C),

T. Yamada (✉)
Laboratory of Physics, Kanto Gakuin University,
Yokohama 236-8501, Japan
e-mail: yamada@kanto-gakuin.ac.jp

Y. Funaki
Nishina Center for Accelerator-based Science,
The Institute of Physical and Chemical Research (RIKEN),
Wako 351-0098, Japan

H. Horiuchi · A. Tohsaki
Research Center for Nuclear Physics (RCNP), Osaka University,
Ibaraki, Osaka 567-0047, Japan

H. Horiuchi
International Institute for Advanced Studies,
Kizugawa, Kyoto 619-0225, Japan

G. Röpke
Institut für Physik, Universität Rostock,
Rostock 18051, Germany

P. Schuck
Institut de Physique Nucléaire, CNRS,
UMR 8608, Orsay F-91406, France

P. Schuck
Université Paris-Sud,
Orsay F-91505, France

P. Schuck
Labratoire de Physique et Modélisation des Milieux Condensés,
CNRS et, Université Joseph Fourier, 25 Av. des Martyrs, BP 166,
F-38042 Grenoble Cedex 9, France

C. Beck (ed.), *Clusters in Nuclei, Vol.2*, Lecture Notes in Physics 848,
DOI: 10.1007/978-3-642-24707-1_5, © Springer-Verlag Berlin Heidelberg 2012

which plays a crucial role for the synthesis of ^{12}C in the universe. The influence of antisymmentrization in the Hoyle state on the bosonic character of the α particle is discussed in detail. It is shown to be weak. The bosonic aspects in the Hoyle state, therefore, are predominant. It is conjectured that α-particle condensate states also exist in heavier $n\alpha$ nuclei, like ^{16}O, ^{20}Ne, etc. For instance the 0_6^+ state of ^{16}O at $E_x = 15.1$ MeV is identified from a theoretical analysis as being a strong candidate of a 4α condensate. The calculated small width (140 keV) of 0_6^+, consistent with data, lends credit to the existence of heavier Hoyle-analogue states. In non-self-conjugated nuclei such as ^{11}B and ^{13}C, we discuss candidates for the product states of clusters, composed of α's, triton's, and neutrons etc. The relationship of α-particle condensation in finite nuclei to quartetting in symmetric nuclear matter is investigated with the help of an in-medium modified four-nucleon equation. A nonlinear order parameter equation for quartet condensation is derived and solved for α particle condensation in infinite nuclear matter. The strong qualitative difference with the pairing case is pointed out.

5.1 Introduction

Cluster as well as mean-field pictures are crucial to understand the structure of light nuclei [1, 2]. It is well known that many states in light nuclei as well as neutron rich nuclei [3] and hypernuclei [4] have cluster structures. Recently, it was found that certain states in self-conjugate nuclei around the α-particle disintegration threshold can be described dominantly as product states of α particles, all in the lowest $0S$ orbit. They are called "α-particle condensate states". Considerable theoretical and experimental work has been devoted to this since this idea was first put forward in 2001 [5].

The ground state of ^{8}Be has a pronounced α-cluster structure [6, 7]. Its average density in the 0^+ ground state is, therefore, very low, only about a third of usual nuclear saturation density. The two α particles are held together only by the Coulomb barrier and ^{8}Be is, therefore, unstable but with a very long life time (10^{-17}s). No other atomic nucleus is known to have such a structure in its ground state. However, it is demonstrated with a purely microscopic approach that, e.g. ^{12}C also has such a structure but as an excited state [8–16]: the famous "Hoyle" state [17, 18], i.e. the 0_2^+ state at 7.65 MeV [19]. It is formed by three almost independent α particles, only held together by the Coulomb barrier. It is located about 300 keV above the disintegration threshold into 3α particles and has a similar life time as ^{8}Be, i.e. also very long. A new-type of antisymmetrized α-particle product state wave function, or THSR α-cluster wave function proposed by Tohsaki, Horiuchi, Schuck, and Röpke [5, 20–23] describes well the structure of the Hoyle state. The THSR wave function is analogous to the (number-projected) BCS wave function [24], replacing, however, Cooper pairs by α particles (quartets). The 3α particles, to good approximation, can be viewed to move in their own bosonic mean field where they occupy the lowest $0S$ level. We, therefore, talk about an alpha particle condensate. A more accurate theory reveals that there exist residual correlations, mostly of the Pauli type, among the alpha particles and, in reality, their occupation of the $0S$ level is reduced but still amounts to over 70% [25–27]. This number is typical for nuclear mean field approaches. The theory [5, 14] reproduces almost all measured data of the Hoyle state, as for instance the inelastic form factor from (e,e'), very accurately. It is predicted that the Hoyle state has about triple to quadruple volume compared with the one of the ^{12}C ground state. Excitations of one alpha out of the condensate into $0D$ and $1S$ states of the mean field can be formed and the 2_2^+ [26, 28, 29] and 0_3^+ [30, 31] states in ^{12}C are reproduced in this way (the latter, so far only tentatively). This triplet of states are precisely the ones which, even with the most modern no-core shell model codes [32–35], cannot be reproduced at all. The new interpretation of the Hoyle state as an α condensate has stimulated a lot of theoretical and experimental works on α-particle condensation phenomena in light nuclei [28–31, 36, 37, 38–50].

The establishment of the novel aspects of the Hoyle state incited us to conjecture 4α condensation in ^{16}O. The theoretical calculation [51] of the OCM (orthogonality condition model) type [52–54] succeeded in describing the structure of the first six 0^+ states up to about 16 MeV, including the ground state with its closed-shell structure, and showed that the 0_6^+ state at 15.1 MeV around the 4α threshold is a strong candidate for a 4α-particle condensate, having a large α condensate fraction of 60%. Similar gas-like states of α clusters have been predicted around their α cluster disintegration thresholds in self-conjugate $A=4n$ nuclei with the THSR wave function [5, 55] and the Gross–Pitaevskii-equation approach [56]. Besides the $4n$ nuclei, one can also expect cluster-gas states composed of alpha and triton clusters (including valence neutrons, etc.) around their cluster disintegration thresholds in $A \neq 4n$ nuclei, in which all clusters are in their respective $0S$ orbits, similar to the Hoyle state with its $(0S_\alpha)^3$ configuration. The states, thus, can be called "Hoyle-analogues" in non-self-conjugated nuclei. It is an intriguing subject to investigate whether or not Hoyle-analogue states exist in $A \neq 4n$ nuclei, for example, ^{11}B, composed of 2α and a t cluster [57–59] or ^{13}C, composed of 3α and $1n$ [60–62]. The $2\alpha + t$ ($3\alpha + n$) OCM [59, 61] calculation indicates that the $1/2_2^+$ ($1/2_3^+$) state at $E_x = 11.95$ (12.14) MeV just above the $2\alpha + t$ ($3\alpha + n$) threshold is a candidate for the Hoyle-analogue.

It has been pointed out that in homogeneous nuclear matter and asymmetric matter α condensation is a possible phase [63–67] at low densities. Therefore, the above mentioned α-particle product states in finite nuclei is related to Bose–Einstein condensation (BEC) of α particles in infinite matter. The infinite matter study used a four particle (quartet) generalization of the well known Thouless criterion for the onset of pairing as a function of density and temperature. The particular finding in the four nucleon case was that α-particle condensation can only occur at very low densities where the quartets do not overlap appreciably. This result is consistent with the structure of the Hoyle state as well as the 0_6^+ state of ^{16}O, in which the average density is about one third or one fourth of the saturation density. It is interesting to note that the low density condition for quartetting was in the meanwhile confirmed in Refs. [68–70] with a theoretical study in cold atom physics.

At this point it may be worthwhile to remark that nuclear physics is predestinated for cluster physics. This stems from the fact that in nuclear physics there are four

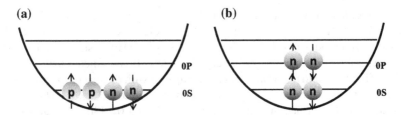

Fig. 5.1 (Color online) Sketch of **a** α-particle configuration with the two protons and two neutrons occupying the lowest $0S$ level in the mean field potential of harmonic oscillator shape, and **b** the energetically lowest configuration in the case of four neutrons with two neutrons in the $0S$ orbit and the other two in the $0P$ orbit

different fermions (proton–neutron, spin up–down), all attracting one another with about equal strength. Such a situation is very rare in interacting fermion systems. Most of the time there are only two species of fermions, as e.g. electrons, spin up–down. However, four different fermions are needed to form a quartet. This is easily understood in a mean field picture where the four nucleons can be put into the lowest $0S$ level of a harmonic potential, whereas were there only neutrons two of four neutrons would have to be put into the p-orbit which is energetically very penalizing, see Fig. 5.1. This is the reason why there is no bound state of four neutrons, while the α-particle is very strongly bound. However, recently experiments in cold atom physics try to trap more than one species of fermions [71, 72] which then also may open up interesting cluster physics in that field.

The purpose of this lecture is to demonstrate the novel aspects of nuclear α-particle condensates, in particular, emphasizing the structure study of ^{12}C and ^{16}O with the THSR wave function and the OCM approach.

The paper is organized as follows. In Sect. 5.2 we first review briefly the RGM framework to describe $n\alpha$ nuclear states [73, 74], which is basic for the THSR wave function and OCM. Then, we formulate the THSR wave function and OCM. Before discussing the Hoyle state, we study the structure of ^{8}Be with the THSR wave function, and discuss the difference between the THSR-type wave function and Brink-type wave function [75, 76] in Sect. 5.3. The latter type of wave function is based on a geometrical, crystal-like viewpoint of the cluster structure. Section 5.4 is dedicated to a discussion of the structure of the Hoyle state, studying the antisymmetrization effect among the 3α clusters, occupation probability and momentum distribution of α particles, and the de Broglie wave length, etc. Then, we discuss the Hoyle-analogue states in ^{16}O with the 4α OCM and THSR wave function, together with ^{11}B and ^{13}C. The Gross–Pitaevskii-equation approach is devoted to investigate α-particle condensation in heavier $4n$ nuclei. In Sect. 5.5, we focus on the α-particle condensation in nuclear matter and its relation with that in finite nuclei. The density dependence of the α condensation fraction is discussed and a 'gap' equation for the α particle order parameter is established and solved. The strong qualitative difference with the pairing case is discussed. Finally, in Sect. 5.6 we present the summary and conclusions.

5.2 Formulation of Alpha-Condensation: THSR Wave Function and OCM Approach

5.2.1 Resonating Group Method (RGM)

The microscopic $n\alpha$ wave function $\Psi_{n\alpha}$ incorporating α-cluster substructures can in general be expressed in the following RGM form [73, 74]:

$$\Psi_{n\alpha} = \mathcal{A} \left\{ \chi(\xi) \prod_{i=1}^{n} \phi_{\alpha_i} \right\} = \int d\mathbf{a} \Psi_{n\alpha}(\mathbf{a}) \chi(\mathbf{a}), \tag{5.1}$$

$$\Psi_{n\alpha}(\mathbf{a}) \equiv \mathcal{A} \left\{ \prod_{j=1}^{n-1} \delta(\xi_j - \mathbf{a}_j) \prod_{i=1}^{n} \phi_{\alpha_i} \right\}, \tag{5.2}$$

with $\mathcal{A}$ the antisymmetrizer of $4n$ nucleons. The intrinsic wave function of the ith α cluster, ϕ_{α_i}, is taken as a Gaussian (with size parameter b),

$$\phi_{\alpha_i} \propto \exp\left[-\sum_{1 \leq k < l \leq 4} (\mathbf{r}_{i,k} - \mathbf{r}_{i,l})^2 / (8b^2) \right], \tag{5.3}$$

representing the intrinsic spatial part of the $(0s)^4$ shell-model configuration, where $\{\mathbf{r}_{i,1}, \ldots, \mathbf{r}_{i,4}\}$ denote the coordinates of the four nucleons in the i-th cluster. The spin-isospin part in Eq. (5.3) is not explicitly written out but supposed to be of scalar-isoscalar form. We will not mention it henceforth. The wave function χ for the c.o.m. motion of the α's is chosen translationally invariant and depends only on the corresponding Jacobi coordinates $\xi = \{\xi_1, \xi_2, \ldots, \xi_{n-1}\}$. The function $\Psi_{n\alpha}(\mathbf{a})$ in Eq. (5.2) describes the α-cluster state located at the relative positions specified by a set of the Jacobi parameter coordinates $\mathbf{a} = \{\mathbf{a}_1, \mathbf{a}_2, \ldots, \mathbf{a}_{n-1}\}$.

The internal part of the Hamiltonian for the relevant $A = 4n$ nucleus is composed of kinetic energy $-\frac{\hbar^2}{2M} \nabla_i^2$, with nucleon mass M, the Coulomb force (V_{ij}^C), the effective two-nucleon ($V_{ij}^{(2)}$) and three-nucleon ($V_{ijk}^{(3)}$) interactions:

$$H = -\sum_{i=1}^{4n} \frac{\hbar^2}{2M} \nabla_i^2 - T_G + \sum_{i<j}^{4n} V_{ij}^C + \sum_{i<j}^{4n} V_{ij}^{(2)} + \sum_{i<j<k}^{4n} V_{ijk}^{(3)}, \tag{5.4}$$

where the c.o.m. kinetic energy of the total system T_G is subtracted.

The Schrödinger equation for the fermionic $n\alpha$ system is

$$H\Psi_{n\alpha} = E\Psi_{n\alpha}. \tag{5.5}$$

Substituting the total wave function of Eq. (5.2) into Eq. (5.5), we obtain the equation of motion for the relative wave function χ,

$$\int d\mathbf{a}' \left\{ H(\mathbf{a}, \mathbf{a}') - EN(\mathbf{a}, \mathbf{a}') \right\} \chi(\mathbf{a}') = 0, \quad \text{or} \quad (\mathcal{H} - E\mathcal{N}) \chi = 0, \qquad (5.6)$$

where the Hamiltonian and norm kernels, $H(\mathbf{a}, \mathbf{a}')$ and $N(\mathbf{a}, \mathbf{a}')$, are defined as

$$\left\{ \begin{array}{c} H(\mathbf{a}, \mathbf{a}') \\ N(\mathbf{a}, \mathbf{a}') \end{array} \right\} = \langle \Psi_{n\alpha}(\mathbf{a}) \mid \left\{ \begin{array}{c} H \\ 1 \end{array} \right\} \mid \Psi_{n\alpha}(\mathbf{a}') \rangle. \qquad (5.7)$$

Equation (5.6) is called the RGM equation [74]. One also can formulate the RGM framework for non $4n$ nuclei such as ^{11}B and ^{13}C with the microscopic $2\alpha + t$ and $3\alpha + n$ cluster model, respectively.

5.2.2 THSR Wave Function

In the THSR description [5, 22], the relative wave function χ in Eq. (5.1) is expressed in the following $n\alpha$ condensation form,

$$\chi_{n\alpha}^{\text{THSR}}(B : \mathbf{R}_1, \mathbf{R}_2, \ldots, \mathbf{R}_n) = \prod_{i=1}^{n} \varphi_0(B : \mathbf{R}_i - \mathbf{X}_G), \qquad (5.8)$$

$$\varphi_0(B : \mathbf{R}) = \exp(-2\mathbf{R}^2/B^2), \qquad (5.9)$$

where $\mathbf{R}_i = (\mathbf{r}_{i,1} + \cdots + \mathbf{r}_{i,4})/4$ denotes the c.o.m. coordinate of the i-th α particle, $\mathbf{X}_G = (\mathbf{R}_1 + \cdots + \mathbf{R}_n)/n$ is the total c.o.m. coordinate of the $n\alpha$ system and $\varphi_0(B : \mathbf{R})$ represents a Gaussian with a large width parameter B which is of the nucleus' dimension. Usually, one uses Jacobi coordinates $\{\boldsymbol{\xi}_i\}$ splitting off the total c.o.m. part of the wave function. Then the THSR ansatz for χ in Eq. (5.8) is given by

$$\chi_{n\alpha}^{\text{THSR}} (B : \mathbf{R}_1, \mathbf{R}_2, \ldots, \mathbf{R}_n) = \exp\left(-2 \sum_{i=1}^{n-1} \mu_i \frac{\boldsymbol{\xi}_i^2}{B^2} \right), \qquad (5.10)$$

with $\mu_i = i/(i + 1)$. A slight generalization of Eq. (5.10) is possible, taking into account nuclear deformation (see Sect. 5.3). With Eqs. (5.8) and (5.1), one can write the THSR wave function in the following $n\alpha$ product form,

$$\Psi_{n\alpha} \to \langle \mathbf{r}_{1,1}, \ldots, \mathbf{r}_{n,4} | \text{THSR} \rangle = \mathcal{A}[\psi_{\alpha_1} \psi_{\alpha_2} \cdots \psi_{\alpha_n}], \qquad (5.11)$$

where

$$|\text{THSR}\rangle = |\text{THSR}(B)\rangle \equiv \mathcal{A}|B\rangle, \qquad (5.12)$$

$$\langle \mathbf{r}_{1,1}, \ldots, \mathbf{r}_{n,4} | B \rangle = \psi_{\alpha_1} \psi_{\alpha_2} \cdots \psi_{\alpha_n}, \qquad (5.13)$$

where $\psi_{\alpha_i} = \varphi_0(B : \mathbf{R}_i - \mathbf{X}_G)\phi_{\alpha_i}$ and definitions of Eqs. (5.12) and (5.13) will be useful later. Equations (5.11)~(5.13) show the analogy of the THSR wave function with the number-projected BCS wave functions for pairing

$$\langle \mathbf{r}_{1,1}, \ldots, \mathbf{r}_{n,2}|\text{BCS}\rangle = \mathcal{A}\left[\phi_{\text{pair}}(\mathbf{r}_{1,1}, \mathbf{r}_{1,2})\phi_{\text{pair}}(\mathbf{r}_{2,1}, \mathbf{r}_{2,2})\cdots\phi_{\text{pair}}(\mathbf{r}_{n,1}, \mathbf{r}_{n,2})\right],$$

(5.14)

where $\phi_{\text{pair}}(\mathbf{r}_{i,1}, \mathbf{r}_{i,2})$ denotes the Cooper pair wave function.

The product of n identical $0S$ wave functions in Eq. (5.11) reflects the boson condensate character. This feature is realized as long as the action of the antisymmetrizer in Eq. (5.1) is sufficiently weak. On the other hand, in the limit where B is taken to be $B = b$, the normalized THSR wave function is equivalent to an SU(3) shell model wave function with the lowest harmonic oscillator quanta [77, 78]; for example, in the case of the 2α, 3α and 4α systems, they respectively are given by

$$\lim_{B \to b} N_{2\alpha}(B)\Psi_{2\alpha}(B) = |(0s)^4(0p)^4; (\lambda\mu) = (4, 0), J^\pi = 0^+\rangle,$$

(5.15)

$$\lim_{B \to b} N_{3\alpha}(B)\Psi_{3\alpha}(B) = |(0s)^4(0p)^8; (\lambda\mu) = (0, 4), J^\pi = 0^+\rangle,$$

(5.16)

$$\lim_{B \to b} N_{4\alpha}(B)\Psi_{4\alpha}(B) = |(0s)^4(0p)^{12}; (\lambda\mu) = (0, 0), J^\pi = 0^+\rangle,$$

(5.17)

where the $N_{n\alpha}(B)$ are the normalization factors. The shell model wave functions in Eqs. (5.15), (5.16) and (5.17) are the dominant configurations of the ground-state wave functions of ^{8}Be, ^{12}C and ^{16}O, respectively. In fact, the components of Eqs. (5.16) and (5.17) in the ground states of ^{12}C and ^{16}O have weights over 60 and 90%, respectively, because both of the states have shell-model-like compact structures. On the other hand, in the case of the ground state of ^{8}Be, the component of Eq. (5.15) is as small as about 20% but still the largest, and the remaining components distribute monotonously in a lot of higher SU(3) configurations, when one expands the wave function of the ^{8}Be ground state in terms of the SU(3) basis. This characteristic comes from a pronounced 2α cluster structure in the ground state (see Sect. 5.3).

The wave functions of the quantum states in $A = 4n$ nucleus can be expanded using the $n\alpha$ THSR wave function, like

$$\Psi_k = \sum_m f_k(B^{(m)})\Psi_{n\alpha}(B^{(m)}),$$

(5.18)

where $\Psi_{n\alpha}(B^{(m)})$ is the $n\alpha$ THSR wave function which has the form of

$$\Psi_{n\alpha}(B^{(m)}) = \mathcal{A}\left[\chi_{n\alpha}^{\text{THSR}}(B^{(m)}; \mathbf{R}_1, \mathbf{R}_2, \ldots, \mathbf{R}_n)\phi_{\alpha_1}\phi_{\alpha_2}\cdots\phi_{\alpha_n}\right].$$

(5.19)

The discrete variational parameters $B^{(m)}$ represent the generator coordinate of the Hill–Wheeler ansatz. The expansion coefficients $f_k(B^{(m)})$ and the corresponding eigenenergy E_k for the k-th eigenstate are obtained by solving the following Hill–Wheeler equation [79, 80],

$$\sum_{m'} \left\langle \Psi_{n\alpha}(B^{(m)}) \middle| H - E_k \middle| \Psi_{n\alpha}(B^{(m')}) \right\rangle f_k \left(B^{(m')} \right) = 0. \tag{5.20}$$

This equation has the same structure as the RGM equation in Eq. (5.6) but reducing it to a one parameter equation. The superposition of THSR wave functions fine tunes the results but a single THSR wave function with an optimized B-value already yields excellent results as will be demonstrated below.

5.2.3 $n\alpha$ Boson Wave Function and OCM

In order to study the bosonic properties of the $n\alpha$ system, one needs to map the microscopic (fermionic) $n\alpha$ cluster model wave function $\Psi_{n\alpha}$ in Eq. (5.1) onto an $n\alpha$ boson wave function $\Phi_{n\alpha}^{(B)}$. The RGM framework given in Sect. 5.2.1 is useful and appropriate for the mapping. Taking into account the normalization of $\Psi_{n\alpha}$, $1 = \langle \Psi_{n\alpha} | \Psi_{n\alpha} \rangle = \langle \chi(\xi) | N(\xi, \xi') | \chi(\xi') \rangle$, the $n\alpha$ bosonic wave function is provided in the following form [74],

$$\Phi_{n\alpha}^{(B)}(\xi) \equiv \mathcal{N}^{1/2} \chi = \int d\xi' N^{1/2}(\xi, \xi') \chi(\xi'), \tag{5.21}$$

where χ represents the relative wave function with the set of Jacobi coordinates, $\xi = \{\xi_1, \xi_2, \ldots, \xi_{n-1}\}$, with respect to the c.o.m. of α clusters. The square-root matrix $N^{1/2}(\xi, \xi')$ is related to the norm kernel of the $n\alpha$ RGM wave function in Eq. (5.7). It is noted that $\Phi_{n\alpha}^{(B)}$ depends only on the Jacobi coordinates ξ, and all of the internal coordinates of $n\alpha$ particles are integrated out in $\Phi_{n\alpha}^{(B)}$.

From the RGM equation (5.6), the equation of motion for $\Phi_{n\alpha}^{(B)}(\xi)$ is obtained in the form

$$\left(\mathcal{N}^{-1/2} \mathcal{H} \mathcal{N}^{-1/2} - E \right) \Phi_{n\alpha}^{(B)} = 0, \tag{5.22}$$

where $\mathcal{H}$ denotes the Hamiltonian kernel defined in Eq. (5.6). Then, one can interpret $\mathcal{N}^{-1/2} \mathcal{H} \mathcal{N}^{-1/2}$ as the nonlocal $n\alpha$ boson Hamiltonian. In Eq. (5.22) care should be taken that before inversion all zero eigenvalues of the norm $\mathcal{N}$ are properly eliminated. The eigenfunctions belonging to the zero eigenvalues are the so-called Pauli forbidden states $u_F(\mathbf{r})$ which satisfy the condition $\mathcal{N} u_F = \mathcal{A} \{ u_F(\xi) \prod_{i=1}^{n} \phi_{\alpha_i} \} = 0$.

The boson wave function has the following properties: (1) $\Phi_{n\alpha}^{(B)}$ is totally symmetric for any 2α-particle exchange, (2) $\Phi_{n\alpha}^{(B)}$ satisfies the equation motion (5.22), and (3) $\Phi_{n\alpha}^{(B)}$ is orthogonal to the Pauli forbidden states $u_F(\mathbf{r})$. In order to obtain the boson wave function $\Phi_{n\alpha}^{(B)}$, we need to solve the equation of motion of the bosons in Eq. (5.22). Solving the boson equation, however, is difficult in general even for the 3α case. Thus, it is requested to use more feasible frameworks for the study of the bosonic properties and the amount of α condensation for the $N\alpha$ system. One such framework is OCM (orthogonality condition model) [52–54]. The OCM scheme, which is an approximation to RGM, is known to describe nicely the structure of low-lying states in light nuclei [26, 30, 31, 51, 52–54, 81–86]. The essential properties of the $n\alpha$ boson wave function $\Phi_{n\alpha}^{(B)}$, as mentioned above, can be taken into account in OCM in a simple manner. We will demonstrate this below.

In OCM, the α cluster is treated as a point-like particle. We approximate the nonlocal $n\alpha$ boson Hamiltonian in Eq. (5.22) by an effective (local) one, that is $H^{(\mathrm{OCM})}$,

$$\mathcal{N}^{-1/2}\mathcal{H}\mathcal{N}^{-1/2} \sim H^{(\mathrm{OCM})} \tag{5.23}$$

$$H^{(\mathrm{OCM})} \equiv \sum_{i=1}^{n} T_i - T_G + \sum_{i<j=1}^{n} V_{2\alpha}^{\mathrm{eff}}(i, j) + \sum_{i<j<k=1}^{n} V_{3\alpha}^{\mathrm{eff}}(i, j, k), \tag{5.24}$$

where T_i denotes the kinetic energy of the i-th α cluster, and the center-of-mass kinetic energy T_G is subtracted from the Hamiltonian. The effective local 2α and 3α potentials are presented as $V_{2\alpha}^{\mathrm{eff}}$ (including the Coulomb potential) and $V_{3\alpha}^{\mathrm{eff}}$, respectively. Then, the equation of the relative motion of the $n\alpha$ particles with $H^{(\mathrm{OCM})}$, called the OCM equation, is written as

$$\left\{ H^{(\mathrm{OCM})} - E \right\} \Phi_{n\alpha}^{(\mathrm{OCM})} = 0, \tag{5.25}$$

$$\langle u_F \mid \Phi_{n\alpha}^{(\mathrm{OCM})} \rangle = 0, \tag{5.26}$$

where u_F denotes the Pauli-forbidden state of the $n\alpha$ system as mentioned above. In the case of 2α system, the Pauli-forbidden states between the two α-particles are $0S$, $0D$ and $1S$ states with the total oscillator quanta Q less than 4. It is pointed out that the Pauli-forbidden states in the $n\alpha$ system can be constructed from those of the 2α system [87, 88].

The bosonic property of the wave function Φ can be taken into account by symmetrizing the wave function with respect to any 2α-particle exchange,

$$\Phi_{n\alpha}^{(\mathrm{OCM})} = \mathcal{S}\Phi_{n\alpha}^{(\mathrm{OCM})}(1, 2, \ldots, n), \tag{5.27}$$

where $\mathcal{S}$ denotes the symmetrization operator, $\mathcal{S} = (1/\sqrt{n!}) \sum_k \mathcal{P}_k$, where the sum runs over all permutations $\mathcal{P}$ of the n α-particles. It is noted that the completely collapsed state of the $n\alpha$ particles is forbidden within the present framework because of the Pauli-blocking effect in Eq. (5.26).

The OCM equation (5.25) with the condition (5.26) is solved with the help of the Gaussian expansion method (GEM) [89, 90]. Combining OCM and GEM provides a powerful tool to study the structure of light nuclei [4, 26, 51, 59, 61] as well as light hypernuclei [4, 91], because the Pauli-blocking effect among the clusters is properly taken into account and GEM covers an approximately complete model space [89, 90].

It is also useful to apply Kukulin's method [92] for removing the Pauli-forbidden states u_F's from the wave function $\Phi_{n\alpha}^{(OCM)}$. The present OCM-GEM framework, for example, in the case of ^{16}O, can cover a model space large enough to describe the dilute α gas-like configuration, as well as $\alpha + {}^{12}C$ cluster and shell-model-like ground state structures.

5.2.4 Single α-Particle Density Matrix and Occupation Probabilities

A literal interpretation of an α condensate in a finite system is that all the α particles occupy the lowest $0S$-wave orbit of an α mean field potential. Due to residual interactions and the action of the Pauli principle, the occupation probabilities may spread out over several orbits, but a particular orbit should be occupied with a significant probability if a state is called a condensate. The occupation probability can be calculated by solving the eigenvalue problem of a single α-particle density matrix [25–27, 93–96].

The single α-particle density matrix for the $n\alpha$ boson system can be defined with the use of the $n\alpha$ Boson wave function $\Phi_{n\alpha}^{(B)}(\xi)$ in Eq. (5.21) mapped from the translationally invariant normalized microscopic $n\alpha$ wave function in Eq. (5.1),

$$\rho_{\text{int}}^{(1)}(\mathbf{q}_1, \mathbf{q}_1{}') = \left(\frac{n}{n-1}\right)^3 \rho_{\text{int,J}}^{(1)}(\xi_1, \xi_1{}'), \tag{5.28}$$

$$\rho_{\text{int,J}}^{(1)}(\xi_1, \xi_1{}') = \int \prod_{i=2}^{n-1} d\xi_i\, \Phi_{n\alpha}^{(B)*}(\xi_1, \xi_2, \ldots, \xi_{n-1})\Phi_{n\alpha}^{(B)}(\xi_1{}', \xi_2, \ldots, \xi_{n-1}), \tag{5.29}$$

where $\mathbf{q}_1 = \frac{n-1}{n}\xi_1 = \mathbf{R}_1 - \mathbf{X}_G$ is the 1st particle coordinate ($\mathbf{R}_1$) with respect to the c.o.m coordinate of the system ($\mathbf{X}_G$) and ξ_1 denotes the relative coordinate between the 1st particle and the remaining $(n-1)$ ones. The factor in Eq. (5.28) is the Jacobian $\partial\xi_1/\partial\mathbf{q}_1$. Since the wave function $\Phi_{n\alpha}^{(B)}(\xi)$ is totally symmetric with respect to particle permutation, the choice of the 1st particle is arbitrary. The definition (5.29) is called the Jacobi-type one-particle density matrix. The diagonal density matrix $\rho_{\text{int}}^{(1)}(\mathbf{q}, \mathbf{q})$ stands for the density distribution of α particles with respect to the c.o.m. coordinate of the $n\alpha$ system. The eigenvalue problem of the density matrix $\rho_{\text{int}}^{(1)}$,

$$\int d\mathbf{q}'\rho_{\text{int}}^{(1)}(\mathbf{q}, \mathbf{q}')\varphi(\mathbf{q}') = \lambda\varphi(\mathbf{q}), \tag{5.30}$$

gives the single α-particle orbit $\varphi(\mathbf{q})$ and its occupation probability λ, where $\mathbf{q}$ is measured from the c.o.m. coordinate of the system. The spectrum of eigenvalues of the density matrix $\rho_{\text{int}}^{(1)}$ gives information on the occupancy of the orbits of the system and it is obviously equal to that of $\rho_{\text{int},J}^{(1)}$. The occupation probability is labeled with the angular momentum L and the quantum number of a positive integer n_L, like L_{n_L}. In this article, for a single-α orbit with an angular momentum L, we denote the largest occupation probability as L_1 ($n_L = 1$), the second largest as L_2 ($n_L = 2$), the third largest as L_3 ($n_3 = 3$), etc. Please notice that the positive number n_L is different from the number of nodes for the radial part of the corresponding single-α orbit $\varphi(\mathbf{q})$ (for instance, in Fig. 5.13, the single-α orbits labeled as $L = 0$ and $n_L = 1$ (S_1) have $2S$ and $0S$ nodal behaviors for the ground state (0_1^+) and the Hoyle state (0_2^+), respectively).

Let us remind that one should use the Jacobi coordinate system for the choice of the internal coordinates of the density matrix. If an internal coordinate system other than the Jacobi coordinate system is adopted (for example, that adopted by Pethick and Pitaevskii (PP) [97]), an unphysical result is obtained even for condensation of a finite number of ideal bosons in a harmonic trap, contrary to what PP expected. Two physically motivated criteria for the choice of the adequate coordinate system lead to a unique answer for the internal one-particle density matrix, i.e. the Jacobi-type internal density matrix, while the PP-type one-body density matrix does not satisfy the criteria (see Refs. [95, 96] for details).

In general even for the 3α system, one encounters numerical difficulties to obtain the boson wave function mapped from the microscopic $n\alpha$ wave function, $\Phi_{n\alpha}^{(B)} = \mathcal{N}^{1/2}\chi$ in Eq. (5.21), by solving the boson equation in Eq. (5.22), as mentioned above. Thus, it is hard in general to calculate the one-body density matrix for the α particle $\rho_{\text{int}}^{(1)}$ in Eq. (5.28). To overcome this difficulty, the following two approximate ways have so far been proposed to evaluate the density matrix. One is, as for the boson wave function, to use the $n\alpha$ OCM wave function (5.27) obtained by solving the OCM equations (5.25) and (5.26), i.e. $\Phi_{n\alpha}^{(B)} \simeq \Phi_{n\alpha}^{(\text{OCM})}$ [26, 51]. The application of this method was done for the 3α and 4α systems, the results of which are presented in Sects. 5.4.1 and 5.4.2. The other is to make the following approximation for the boson wave function as proposed in Refs. [25, 27], $\Phi_{n\alpha}^{(B)}(\xi) \simeq \mathcal{N}\chi/\sqrt{\langle \mathcal{N}\chi | \mathcal{N}\chi \rangle}$. This method was used for the 3α and 4α THSR wave functions. The results are discussed in Sect. 5.4.2. The two approximations give quantitative similar results for the occupation probabilities and for the single-α orbits in the 3α and 4α systems, which are obtained by solving the eigenvalue problem of Eq. (5.30).

5.3 THSR Wave Function Versus Brink Wave Function for ^{8}Be

Before discussing the Hoyle state, it is instructive to study ^{8}Be in some detail because even this nucleus which is known to have intrinsically a two-alpha dumbbell structure [6, 7] can very well be described in the laboratory frame with the THSR wave function. Let us repeat Eq. (5.1) for this particular case

$$\Psi_{2\alpha} = \mathcal{A}\left[\chi(\mathbf{r})\phi_{\alpha_1}\phi_{\alpha_2}\right], \tag{5.31}$$

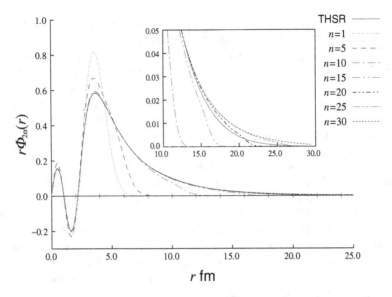

Fig. 5.2 (Color online) Comparison of THSR wave function with a single component "Brink" wave function with $D = 3.45$ fm (denoted by $n = 1$). The convergence rate with the superposition of several (n) "Brink" wave functions is also shown. The line denoted by $n = 30$ corresponds to the full RGM solution. The Volkov No. 1 force is taken with Majorana parameter value $M = 0.56$. Figure is taken from Ref. [22]

with the relative coordinate between the 2α particles, $\mathbf{r} = \mathbf{R}_1 - \mathbf{R}_2$. Note that Eq. (5.31) is a fully antisymmetric and translationally invariant wave function in $8 - 1 = 7$ coordinates. Solving the RGM equation in Eq. (5.6) with a given Hamiltonian, one obtains the energy E of ^{8}Be and χ. The 2α boson wave function $\Phi_{2\alpha}^{(B)}(\mathbf{r})$ representing the relative motion of the two α-particles, mapped from the corresponding fermionic 2α wave function $\Psi_{2\alpha}$, is given in Eq. (5.21),

$$\Phi_{2\alpha}^{(B)}(\mathbf{r}) = \int d\mathbf{r}' N^{1/2}(\mathbf{r}, \mathbf{r}') \chi(\mathbf{r}'), \qquad (5.32)$$

Expressions (5.31) and (5.32) have been obtained with very high numerical accuracy since 50 years with excellent results for all low energy properties of ^{8}Be [6]. The radial part of the 2α boson wave function $r \Phi_{2\alpha}(\mathbf{r})$ in the ground state ($J^\pi = 0^+$) is shown in Fig. 5.2 denoted by $n = 30$. We see that there exist two nodes, an effect which stems from the Pauli principle.

Here we will discuss two approximate forms for $\chi(r)$: the THSR wave function and the Brink cluster wave function [75]. Let us start with the latter. In the Brink wave function, the α particles are placed at certain positions in space. In the case of ^{8}Be, placing the 2α particles at the positions of $\mathbf{D}/2$ and $-\mathbf{D}/2$, respectively, this leads to

$$\chi^{\text{Brink}}(r) = \widehat{P}^{J=0}\exp\left[-\frac{\left(\mathbf{R}_1 - \mathbf{D}/2 - (\mathbf{R}_2 + \mathbf{D}/2)\right)^2}{b^2}\right]$$

$$= \widehat{P}^{J=0}\exp\left[-\frac{1}{b^2}(\mathbf{r} - \mathbf{D})^2\right], \tag{5.33}$$

where $\widehat{P}^{J=0}$ denotes the projection operator onto spin $J=0$. Though this kind of geometrical, crystal-like viewpoint of the cluster structure works well for many cases, for instance, parity-violating $^{12}C+\alpha$, $^{16}O+\alpha$, and $^{40}Ca+\alpha$ structures in ^{16}O, ^{20}Ne and ^{44}Ti, respectively [2, 98–100], and also when additionally neutrons are involved [101], it is on the contrary known since several decades that this picture fails for the description of the famous Hoyle state, i.e. the 0_2^+ state in ^{12}C (see Sect. 5.4). The ansatz of the two α particles being placed at a distance $\mathbf{D}$ from one another seems reasonable, since the Quantum Monte Carlo calculation with realistic two- and three-nucleon potentials in Ref. [7] indeed indicates that the two α's are about 4 fm apart. Obviously, the parameter $\mathbf{D}$ can be varied to find the optimal position of the α-particles. The result of such a procedure is shown in Fig. 5.2 with the line denoted by $n=1$ taking the optimal value $D=3.45$ fm (b is kept fixed at its free space value, $b=1.36$ fm). Qualitatively such a "Brink" wave function follows the full variational solution (line denoted by $n=30$). However, in the outer part, for instance in the exponentially decaying tail quite strong differences appear. The squared overlap with the exact solution is 0.722. Of course, the Brink wave functions also can serve as a basis and it is interesting to study the convergence properties. We, therefore, write for the ^8Be wave function appearing in Eq. (5.6)

$$\Psi_{2\alpha} = \mathcal{A}\left[\chi(r)\phi_{\alpha_1}\phi_{\alpha_2}\right] = \sum_i f_i \Psi_{2\alpha}^{\text{Brink}}\left(r, D^{(i)}, b\right), \tag{5.34}$$

$$\Psi_{2\alpha}^{\text{Brink}}\left(r, D^{(i)}, b\right) = \mathcal{A}\left[\chi_{D^{(i)}}^{\text{Brink}}(r)\phi_{\alpha_1}\phi_{\alpha_2}\right] \tag{5.35}$$

where the $D^{(i)}$ indicate the various positions of the α-particles and f_i are the expansion coefficients. The convergence of the squared overlap with the exact solution is studied where we take for the positions $D^{(1)} = 1$ fm, $D^{(2)} = 2$ fm, $\ldots$, $D^{(n)} = n$ fm. We start with $n=5$. In Fig. 5.3 the convergence rate is shown as a function of n for the squared overlap and for the energy. The point of $n=1$ is with the optimized single Brink wave function ($D^{(1)} = 3.45$ fm). We see that the convergence is not extremely fast but for $n=20$ the squared overlap with the full RGM solution amounts to 0.9999. Also energy is converged to within 10^{-4}. In Fig. 5.2 we show the convergence of the 2α boson wave function $r\Phi_{2\alpha}(r)$. In the insert we see that there is still a slight change in the far tail going from $n=25$ to $n=30$.

Let us now investigate the THSR ansatz for $\chi(r)$. There it is assumed from the beginning that the α's are delocalised and a single Gaussian e^{-r^2/B^2} centered at the origin with, however, a large width $B^2 = b^2 + 2\beta^2$, with β a variational parameter, is taken. Very much improved results over the single component Brink wave function

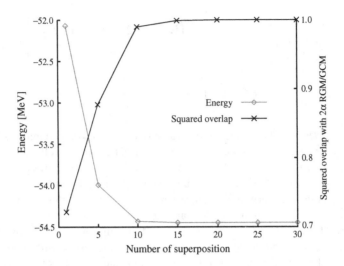

Fig. 5.3 (Color online) Binding energy corresponding to $\Psi_{2\alpha}$ (see Eq. (5.34)) with the superposition of n Brink wave functions and the squared overlap between the full RGM solution and $\Psi_{2\alpha}$. For $n=1$, a single Brink wave function with optimized $R=3.45$ fm is adopted. Figure is taken from Ref. [22]

are obtained. With $\beta = 3.24\,fm$ the squared overlap becomes 97.24%. However, practically 100% accuracy, compared with the exact solution, can be achieved starting with a slightly improved ansatz, i.e. with an axially symmetric deformed Gaussian which is then projected on the ground-state spin $J=0$ (projections on $J=2,4$ yield the rotational band of ^{8}Be) [102],

$$\chi^{\mathrm{THSR}}(r) = \widehat{P}^{J=0}\exp\left(-\frac{r_\perp^2}{b^2 + 2\beta_\perp^2} - \frac{r_z^2}{b^2 + 2\beta_z^2}\right)$$
$$\propto \frac{\exp(-r^2/B_\perp^2)}{ir}\mathrm{Erf}\left(i\frac{(B_z^2 - B_\perp^2)^{1/2}}{B_\perp B_z}r\right), \qquad (5.36)$$

with $B_i^2 = b^2 + 2\beta_i^2$ and $r_\perp^2 = r_x^2 + r_y^2$, and Erf($x$) the error function. The second line of Eq. (5.36) is obtained from a simple calculation.

Such an intrinsically deformed ansatz is, of course, physically motivated by the observation of the rotational spectrum of ^{8}Be indicating a large value of the corresponding moment of inertia. The minimization of the energy yields $\beta_\perp = \beta_x = \beta_y = 1.78$ fm and $\beta_z = 7.85$ fm. With these numbers, the squared overlap between the exact $\Psi_{2\alpha}$ and $\Psi_{2\alpha}^{\mathrm{THSR}}$ is with 0.9999 extremely precise. In Fig. 5.2 we also show that the THSR wave function agrees very well even far out in the tail with the "exact" solution with 30 "Brink" components.

As seen above, the single component, two parameter THSR ansatz, Eq. (5.36), for the relative wave function of two alpha's seems to grasp the physical situation extremely well. The most important part of this wave function is the outer one beyond some 3 fm. There, the two alpha's are in an S wave of essentially Gaussian shape. The corresponding harmonic oscillator frequency is estimated to $\hbar\omega \sim 2\,\text{MeV}$. Therefore, as long as the two alpha's do not overlap strongly, they swing in a very low frequency harmonic oscillator mode in a wide and delocalized fashion, reminiscent of a weakly bound gas like state. Inside the region $r < 2-3$ fm where the two alpha's heavily overlap, because of the strong action of the Pauli principle, the relative wave function has two nodes and small amplitude, as shown in Fig. 5.2. Contrary to the outer part of the wave function determined dynamically, the behavior of the relative wave function in this strongly overlapping region is determined kinematically, solely reflecting the r-dependence of the norm kernel in Eq. (5.7). This is clearly seen from the fact that both THSR and Brink wave functions have very nearly the same behavior in this region. Thus, we found that the alpha's in ^{8}Be move practically as pure bosons in a relative $0S$ state of very low frequency as long as they do not come into one another's way, that is as long as they do not overlap. One should stress that this picture holds after projection on good total momentum and good spin, that is in the laboratory frame. It is equally true, as already mentioned, that in the intrinsic frame ^{8}Be can be described as a strongly deformed two alpha structure, see ansatz (5.36), reminiscent of a dumbbell.

5.4 Alpha-Gas Like States in Light Nuclei

5.4.1 ^{12}C Case

The α cluster nature of ^{12}C has been studied by many authors using various approaches. Figure 5.4 shows the energy spectrum of ^{12}C [19]. The 0_2^+ state, located near the 3α breakup threshold, is called the Hoyle state [17, 18], which plays an astrophysically crucial role in the synthesis of ^{12}C in the universe. Its small excitation energy of 7.65 MeV is very difficult to explain by the shell model, even using the most modern non-core shell model approach [32–35]. The fully microscopic 3α cluster models [8–13], however, succeeded in the 1970s in explaining the observed data such as the small excitation energy and the inelastic form factor of the (e,e') reaction etc., together with the structures of the ground-band states $(0_1^+ - 2_1^+ - 4_1^+)$, 2_2^+, and negative-parity states $(3_1^- - 1_1^-)$. The cluster model studies with the 3α GCM (generator coordinate method) [8–10] and 3α RGM [11–13] showed that the Hoyle state has a weakly interacting gas like 3α-cluster structure with a very large radius (about 1/3 of the ground-state density), whereas the ground state has a shell-model-like compact structure.

This 3α gas-like nature of the Hoyle state is demonstrated in Fig. 5.5, in which the overlap between a Brink-type wave function and the full RGM solution obtained by solving the 3α RGM equation in Eq. (5.6) is shown. The overlap is quite poor and in the best case the squared overlap reaches only about 50%. This means that the 0_2^+ state has a distinct clustering and has no definite spacial or geometrical configuration. The

Fig. 5.4 Experimental
energy spectra of ^{12}C [19]
together with the calculated
ones using the 3α RGM
[11–13]

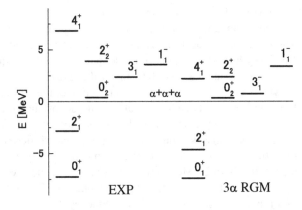

Fig. 5.5 Structure of the 0_2^+
state shown by the overlap
between the Brink-type
cluster wave function of the
isosceles configuration and
the exact 0_2^+ wave function.
Figure adopted from
Ref. [2, 8–10]

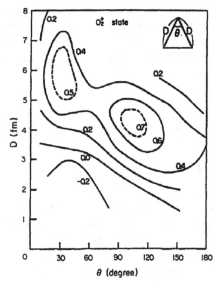

situation is also pointed out in a recent work [15, 16], in which about 55 components
of the Brink-type wave functions are needed to reproduce accurately the full RGM
solution for the Hoyle state. However, this Hoyle-state wave function is shown to be
almost completely equivalent to a "single THSR wave function" as discussed in next
section.

5.4.1.1 THSR Description of the Hoyle State

The total wave function for ^{12}C in the THSR description is obtained by solving the
Hill–Wheeler equation based on Eqs. (5.18), (5.19) and (5.20). Table 5.1 shows the
results of the energies, r.m.s. radii, and monopole strengths in the THSR description

Table 5.1 Comparison of the total energies, r.m.s. radii ($R_{\text{r.m.s.}}$), and monopole strengths ($M(0_2^+ \rightarrow 0_1^+)$) for ^{12}C given by solving Hill–Wheeler equation based on Eq. (5.3) and by Ref. [11–13]. The effective two-nucleon force Volkov No. 2 [103] was adopted in the two cases for which the 3α threshold energy is calculated to be -82.04 MeV

		THSR w.f. (Hill–Wheeler)	3α RGM [11–13]	Exp.
E (MeV)	0_1^+	-89.52	-89.4	-92.2
	0_2^+	-81.79	-81.7	-84.6
$R_{\text{r.m.s.}}$ (fm)	0_1^+	2.40	2.40	2.44
	0_2^+	3.83	3.47	
$M(0_2^+ \rightarrow 0_1^+)$ (fm^2)		6.45	6.7	5.4

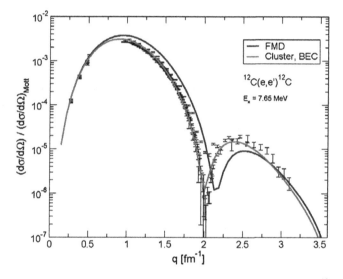

Fig. 5.6 (Color online) Comparison of the experimental inelastic form factor of ^{12}C(e, e') with the RGM (denoted by cluster), THSR (BEC) and FMD calculations. Figure is adopted from Ref. [15, 16]

together with those of the full 3α RGM calculation and the data. One can see that the THSR description succeeds to reproduce the properties of the two 0^+ states. Inspecting the r.m.s. radii, the Hoyle state has a volume 3–4 times larger than that of the ground state of ^{12}C. The inelastic form factor of ^{12}C from the ground state to the Hoyle state in the THSR description is displayed in Fig. 5.6. We reproduce very accurately the experimental data.

In order to study how good a single 3α THSR wave function reproduces the full RGM solutions, we use the THSR wave function with axially symmetric deformation, presented as

$$\Psi_{3\alpha}(\beta_\perp, \beta_z) = \mathcal{A} \left\{ \chi_{3\alpha}^{\text{THSR}}(\beta_\perp, \beta_z) \phi_\alpha \phi_\alpha \phi_\alpha \right\}, \tag{5.37}$$

$$\chi_{3\alpha}^{\text{THSR}}(\beta_\perp, \beta_z) = \exp\left[-2\sum_{i=1}^{2}\mu_i\left(\frac{\xi_{i\perp}^2}{b^2 + 2\beta_\perp^2} + \frac{\xi_{iz}^2}{b^2 + 2\beta_z^2}\right)\right], \qquad (5.38)$$

where $\boldsymbol{\xi}_{1,2}$ are the two Jacobi coordinates with $\mu_1 = 1/2$ and $\mu_2 = 2/3$, and $\beta_\perp$ and β_z are the deformation parameters with $\beta_\perp = \beta_x = \beta_y$. The wave function with good total spin $J=0$ is written as

$$\Psi_{3\alpha}^{J=0}(\beta_\perp, \beta_z) = \widehat{P}^{J=0}\Psi_{3\alpha}(\beta_\perp, \beta_z), \qquad (5.39)$$

where $\widehat{P}^J$ is the angular momentum projection operator. In what concerns the THSR wave function for the description of the Hoyle state, the situation is slightly more complicated than in the ^{8}Be case by the fact that the loosely bound 3α configuration is now no longer the ground state but the 0_2^+ state at 7.65 MeV excitation energy (as a side remark, let us mention that usually Bose–Einstein condensates of cold atoms also are not the ground states of the systems which are given by small crystals). As discussed in Sect. 5.2, the wave function (5.39) has the dominant configuration of the ground state in the limit of $\beta_\perp = \beta_z = 0$. Thus, in order to discuss the Hoyle state, we have to use the 3α wave function $\widetilde{\Psi}_{3\alpha}^{J=0}$ which is orthogonal to the ground state, expressed as

$$\widetilde{\Psi}_{3\alpha}^{J=0}(\beta_\perp, \beta_z) = \widehat{P}^{J=0}\widehat{P}_\perp^{\text{g.s}}\Psi_{3\alpha}(\beta_\perp, \beta_z), \qquad (5.40)$$

where $\widehat{P}_\perp^{\text{g.s}}$ keeps the wave function in Eq. (5.40) to be orthogonal to the ground-state wave function, i.e. $\widehat{P}_\perp^{\text{g.s}} = 1 - |0_1^+\rangle\langle 0_1^+|$.

On the left side of Fig. 5.7, we show the contour map of the energy surface corresponding to the state (5.39) in the two parameter space $(\beta_\perp, \beta_z)$, defined as

$$E(\beta_\perp, \beta_z) = \frac{\langle\Psi_{3\alpha}^{J=0}(\beta_\perp, \beta_z)|H|\Psi_{3\alpha}^{J=0}(\beta_\perp, \beta_z)\rangle}{\langle\Psi_{3\alpha}^{J=0}(\beta_\perp, \beta_z)|\Psi_{3\alpha}^{J=0}(\beta_\perp, \beta_z)\rangle}, \qquad (5.41)$$

where H is the microscopic Hamiltonian of ^{12}C used in the 3α RGM calculation. One sees a minimum at $\beta_\perp = 1.5$ fm and $\beta_z = 1.5$ fm, which means a spherical shape. The minimum energy of -87.68 MeV is about 1.7 MeV higher than the total energy of -89.4 MeV obtained by the full 3α RGM calculation (see Table 5.1). When the Hill–Wheeler equation in Eq. (5.20) is solved in the two-parameter space of $\beta_\perp$ and β_z, we can reproduce the total energy of the RGM result.

On the right side of Fig. 5.7, the contour map of the energy surface corresponding to the state (5.40) orthogonal to the ground state is displayed, where we use the ground-state solution of the Hill–Wheeler equation in the two-parameter space of $\beta_\perp$ and β_z. We see an energy minimum at $\beta_\perp = 5.2$ fm and $\beta_z = 1.5$ fm in the prolate region of the map and a second energy minimum at $\beta_\perp = 2.6$ fm and $\beta_z = 7.5$ fm in the oblate region. The minimum energy value is -81.75 MeV. This value is almost the same as the total energy of -81.67 MeV obtained by the full 3α RGM (see Table 5.1). The minimum energy of -81.75 MeV is close to the second minimum

(a) **(b)**

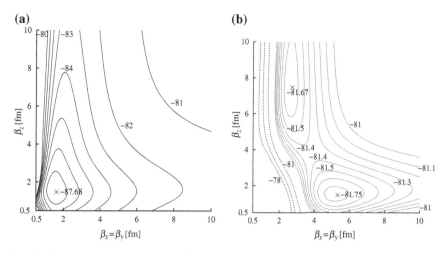

Fig. 5.7 (Color online) Contour map of the energy surface in the two parameter space $(\beta_\perp, \beta_z)$ for (*left*) $\Psi_{3\alpha}^{J=0}(\beta_\perp, \beta_z)$ in Eq. (5.39) and for (*right*) $\widetilde{\Psi}_{3\alpha}^{J=0}(\beta_\perp, \beta_z)$ in Eq. (5.40) orthogonal to the ground state

energy of -81.67 MeV and there is a valley with an almost flat bottom connecting these two minima. This means that the energy of the spherical configuration is only slightly higher than that of the deformed configuration, that is, the energy gain due to the deformation is small.

A very remarkable result from the right side of Fig. 5.7 is that the wave function at the minimum energy point ($\beta_\perp = 5.3$ fm and $\beta_z = 1.5$ fm) has 99.3% squared overlap with the full RGM solution (see Fig. 5.8), although the spherical wave function ($\beta_\perp = \beta_z = 4.0$ fm) gives already a squared overlap of 92%. The THSR wave function Eq. (5.37) is of Gaussian type with a wide extension, centered at the origin. It is completely different from a Brink type wave function with the three α-particles placed at definite values in space. A slight improvement of Eq. (5.40) can still be achieved in taking the β_i parameters as Hill–Wheeler coordinates and superpose a couple of wave functions of the type (5.40) with different width parameters. Practically 100% squared overlap with the wave function of the full RGM result is then achieved. It should be pointed out that the superposition of several Gaussians of the type (5.40) does not at all change the physical content of the THSR wave function as a wide extended distribution centered around the origin. Therefore, the Hoyle state can be seen as three almost inert α-particles moving in their own mean field potential, to good approximation given by a wide harmonic oscillator, whereas the α's are represented by four nucleons captured in narrow harmonic potentials. The situation is given as a cartoon in Fig. 5.9.

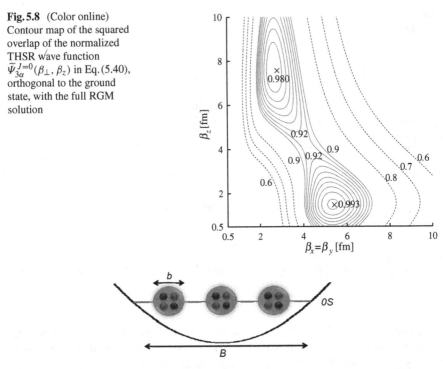

Fig. 5.8 (Color online) Contour map of the squared overlap of the normalized THSR wave function $\widetilde{\Psi}_{3\alpha}^{J=0}(\beta_\perp, \beta_z)$ in Eq. (5.40), orthogonal to the ground state, with the full RGM solution

Fig. 5.9 (Color online) Pictorial representation of the THSR wave function for $n=3$ (^{12}C). The three α-particles are trapped in the $0S$-state of a wide harmonic oscillator (B) and the four nucleons of each α are confined in the $0s$-state of a narrow one (b). All nucleons are antisymmetrized

5.4.1.2 Influence of Antisymmetrization and Orthogonalization

A crucial question is whether for the Hoyle state the THSR wave function (5.1) with (5.8) can be considered to good approximation as a product state of α particles condensed with their c.o.m. motion into the $0S$ orbital. For this, one has to quantify the influence of the antisymmetrizer $\mathcal{A}$ in Eq. (5.1). A direct way to measure the influence of antisymmetrization is to study the following expectation value of the antisymmetrizer $\mathcal{A}$,

$$N(B) = \frac{\langle B|\mathcal{A}|B\rangle}{\langle B|B\rangle}, \tag{5.42}$$

where $|B\rangle$ is the THSR wave function in Eq. (5.12) without the antisymmetrization, that is, just the product state $\psi_{\alpha_1}\psi_{\alpha_2}\psi_{\alpha_3}$ in Eq. (5.13). The normalization of the antisymmetrizer $\mathcal{A}$ is chosen so that $N(B)$ becomes unity in the limit where the intercluster overlap disappears, i.e. for the with parameter B $\rightarrow \infty$.

The result of $N(B)$ is shown in Fig. 5.10 as a function of the width parameter B. We chose, as optimal values of B for describing the ground and Hoyle states,

Fig. 5.10 (Color online) Expectation value of the antisymmetrization operator for the product state $|B\rangle$. The value at the optimal B values, B_g for the ground state and B_H for the Hoyle state, are denoted by a circle and a cross, respectively

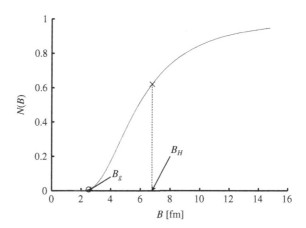

$B = B_g = 2.5$ fm and $B = B_H = 6.8$ fm, for which the normalized THSR wave functions give the best approximation of the ground state 0_1^+ and the Hoyle state 0_2^+, respectively, which are obtained by solving the Hill–Wheeler equation (5.20). The squared overlaps are 0.93 and 0.78, respectively. From Fig. 5.10 we find that $N(B_H) \sim 0.62$ and $N(B_g) \sim 0.007$. These results indicate that the influence of the antisymmetrization is strongly reduced in the Hoyle state compared with the influence in the ground state. An important point in the present consideration is that the THSR wave function at $B = B_H$ is not automatically orthogonal to the ground state. This is contrary to the situation with condensed cold bosonic atoms, for which the density is so low that the overlap of the electron clouds can, on average, be totally neglected. In the present case, the squared overlap of $|\text{THSR}(B = B_H)\rangle$ with $|\text{THSR}(B = B_g)\rangle$ (or with the ground state 0_1^+ obtained by solving the Hill–Wheeler equation) is less than 0.12. This small value indicates that the orthogonality with the ground state is nearly realized.

An explicit orthogonalization with $|\text{THSR}(B)\rangle$ to the ground state 0_1^+ obtained by solving the Hill–Wheeler equation gives non-negligible effects for a quantitative description of the Hoyle state with the THSR wave function. As mentioned in Sect. 5.4.1.1, the normalized THSR wave function orthogonal to the ground state 0_1^+ $\left(\sim \hat{P}_\perp^{(\text{g.s})}|\text{THSR}(B)\rangle\right)$ gives a squared overlap of 0.92 (for $B=6.1$ fm) with the 0_2^+ state obtained by solving the Hill–Wheeler equation, although the squared overlap using the normalized THSR wave function without the orthogonalization gives already a value of 0.78. In addition, as shown in Fig. 5.11, the energy curves for the THSR wave function,

$$E(B) = \frac{\langle \text{THSR}(B)|H|\text{THSR}(B)\rangle}{\langle \text{THSR}(B)|\text{THSR}(B)\rangle}, \tag{5.43}$$

indicates a minimum corresponding to the ground state at $B \sim B_g$, but the second minimum corresponding to the Hoyle state is not present. This is due to the fact that the THSR state with $B = B_H$, $|\text{THSR}(B = B_H)\rangle$, still includes the ground-state

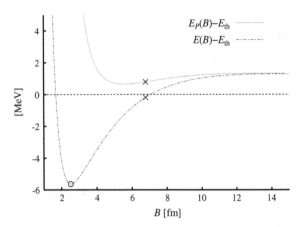

Fig. 5.11 (Color online) Energy curve in the orthogonal space to the ground state, denoted by $E_P(B)$, together with $E(B)$. The values at the optimal B values, B_g and B_H for the ground state and Hoyle state, respectively, are marked by a *circle* and a *cross*, respectively

component of about 10%, as mentioned above. In fact, if one calculates the energy taking into account the explicit orthogonalization to the ground state,

$$E_P(B) = \frac{\langle \hat{P}_\perp^{(g.s)} \text{THSR}(B) | H | \hat{P}_\perp^{(g.s)} \text{THSR}(B) \rangle}{\langle \hat{P}_\perp^{(g.s)} \text{THSR}(B) | \hat{P}_\perp^{(g.s)} \text{THSR}(B) \rangle}, \tag{5.44}$$

there appears the minimum corresponding to the Hoyle state at B $\sim B_H$, as shown in Fig. 5.11. Thus, the small admixture of the ground-state components to the Hoyle state is never negligible, and explicit elimination by $\hat{P}_\perp^{(g.s)}$ plays an essential role to describe the Hoyle state. It is true that the effect of the antisymmetrization is not negligible even for the Hoyle state in the sense that the projection operator $\hat{P}_\perp^{(g.s)}$ excludes the compact ground-state components which are strongly subject to the antisymmetrizer. Nevertheless, it is worth emphasizing that as a result of the explicit orthogonalization to the ground state, the Hoyle state can not have a compact structure but has a dilute density, for which, in the end, the effect of antisymmetrization is small.

5.4.1.3 Alpha-Particle Occupation Probabilities, Momentum Distribution, and the de Broglie Wave Length in the Hoyle State

Direct quantities indicating how well the Hoyle state is described by a product state of three α's are the α-particle occupation probabilities and single particle orbits, which are obtained by diagonalizing the internal single α-particle density matrix $\rho_{\text{int}}^{(1)}(q, q')$ defined in Eq. (5.28). The occupation of the single-α orbits of the Hoyle state is shown in Fig. 5.12. One finds that the α particles occupy the S_1 orbit to over 70%, and those for other orbits are very small. This means that each of the three α particles in the 0_2^+ state is in the S_1 orbit with occupation probability as large as about 70%. The radial behavior of the S_1 orbit is illustrated with the solid line

Fig. 5.12 (Color online) Occupation of the single-α orbitals of the Hoyle state of ^{12}C compared with the ground state

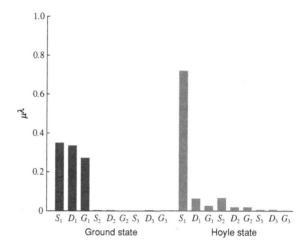

in Fig. 5.13b. We see no nodal behavior but small oscillations in the inner region ($r < 4$ fm) and a long tail up to $r \sim 10$ fm. For reference, the radial behavior of the S-wave Gaussian function, $\varphi_{0s}(r) = N_{0s}(B)\exp(-r^2/(2B^2))$, is drawn with the dashed line in Fig. 5.13b, where the size parameter B is chosen to be 3.6 fm, and $N_{0s}(B)$ denotes the normalization factor. The radial behavior of the S_1 orbit is similar to that of the S-wave Gaussian function, in particular, in the outer region ($r > 4$ fm), whereas a slight oscillation of the former around the latter can be seen in the inner region ($r < 4$ fm). Thus, the Hoyle state can be described as the product state of $(0S)_\alpha^3$ being realized with a probability of over 70%.

In the case of the ground state of ^{12}C, the α-particle occupations are equally shared between S_1, D_1 and G_1 orbits (see Fig. 5.12), thus invalidating a condensate picture for the ground state. These occupancies can be explained quite well from the following fact: The ground state has as main configuration the SU(3) shell model wave function $(\lambda, \mu) = (04)$. Figure 5.13a demonstrates the radial parts for the S_1-, D_1, and, G_1-orbits, the number of nodes of which are two, one and zero, respectively. Reflecting the SU(3) character, the radial behavior of the three orbits is similar to those of the harmonic oscillator wave functions (u_{NL}) with $Q = 4$, u_{02}, u_{21} and u_{40}, respectively, where $N(L)$ denotes the number of nodes (orbital angular momentum). We see that the radial parts of the single α-particle orbits oscillate strongly in the inside region ($r < 4$ fm). This is due to the important Pauli blocking effect for the ground state with its compact shell-model-like structure.

Another important quantity to demonstrate the 3α condensate nature of the Hoyle state is the momentum distribution of a single-α particle. It is defined as a double Fourier transformation of the internal single-α density matrix $\rho_{\text{int}}^{(1)}(q, q')$ defined in Eq. (5.28),

$$\rho(k) = \int d\mathbf{q}' d\mathbf{q} \frac{e^{i\mathbf{k}\cdot\mathbf{q}'}}{(2\pi)^{3/2}} \rho_{\text{int}}^{(1)}(\mathbf{q}, \mathbf{q}') \frac{e^{-i\mathbf{k}\cdot\mathbf{q}}}{(2\pi)^{3/2}}, \quad \int d\mathbf{k}\rho(k) = 1, \qquad (5.45)$$

Fig. 5.13 Radial parts of the single α orbits, **a** S_1 (*solid line*), D_1 (*dashed*) and G_1 (*dotted*), in the 0_1^+ state, and **b** the S_1 (*solid*) orbit in the 0_2^+ state compared with an S-wave Gaussian function (*dotted*), $r\varphi_{0s}$, with the size parameter $B = 3.6$ fm (see text) [26]. Note that all the radial parts in figures are multiplied by r

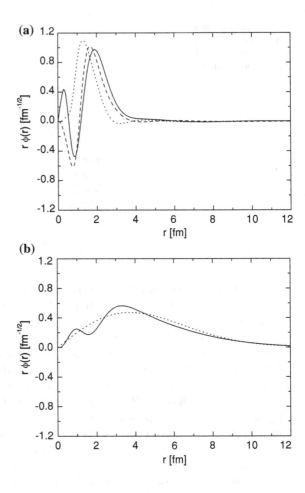

Let us remind that $\rho(k)$ would have a δ-function like peak around $k = 0$ for an ideal dilute condensed state in homogeneous infinite matter.

The momentum distributions of the α particle, $\rho(k)$ and $k^2 \times \rho(k)$, are shown for the 0_1^+ and 0_2^+ states in Fig. 5.14. Reflecting the dilute structure of the Hoyle, we see a strong concentration of the momentum distribution in the $k < 1$ fm^{-1} region, and the behavior of $\rho(k)$ is of the δ-function type, similar to the momentum distribution of the dilute neutral atomic condensate states at very low temperature trapped by an external magnetic field [104]. On the other hand, the ground state has higher momentum components up to $k \sim 6$ fm^{-1} as seen from the behavior of $k^2 \times \rho(k)$ reflecting the compact structure. The above results for the radial behavior of the S_1 orbit, occupation probability and momentum distribution for the 0_2^+ state again lead us to conclude that this state is of the 3α condensate character with as much as about 70% occupation probability.

Fig. 5.14 Momentum distribution of the α particle in ^{12}C, **a** $\rho(k)$ and **b** $k^2\rho(k)$, for the 0_1^+ (*solid line*) and 0_2^+ (*dotted*) states [26]

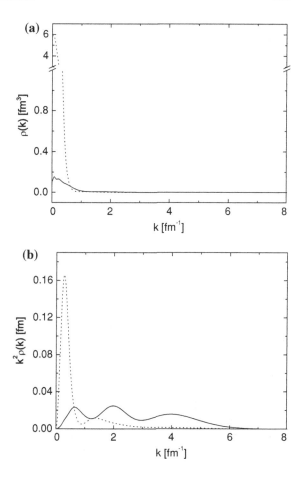

The de Broglie wave length of the α's moving in the Hoyle state is an interesting quantity. It can be estimated from the resonance energy of ^{8}Be being roughly 100 keV. Otherwise, one can estimate the kinetic energy of the α-particles from a bosonic mean field picture using the Gross–Pitaevskii equation [56] (see Sect. 5.4.3). The mean field potential of α-particles in the Hoyle state (see Fig. 5.23) indicates the position of the single α particle energy (180 keV). The kinetic energy of the single α particle is calculated to be 380 keV. From this, the de Broglie wave length $\lambda = 2\pi \left(\frac{2M_\alpha}{\hbar^2} E_\alpha\right)^{-1/2}$ is, therefore, estimated to be of a lower limit of approximately 20 fm. A more reliable estimate of the de Broglie wave length is to use the expectation value of k^2 for the wave number k of the α particle in the Hoyle state, evaluated from the momentum distribution of the alpha particle, $\rho(k)$, in Fig. 5.14, obtained by a 3α OCM calculation [26]. The result is $\lambda = 2\pi/\sqrt{\langle k^2 \rangle} \sim 20$ fm, consistent with the previous value. These estimates all indicate that the de Broglie wave length is much longer than the inter α-particle distance, contrary to what is claimed in Ref. [105], using qualitative arguments.

Fig. 5.15 (Color online)
Theoretical interpretation of
the 0_2^+, 2_2^+ and 0_3^+ states

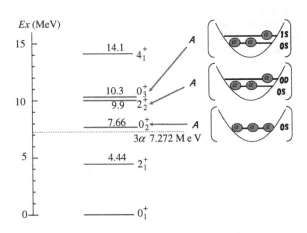

5.4.1.4 Family of the Hoyle State: 2_2^+ and 0_3^+

In the previous section, we found that the Hoyle state has a dilute 3α-condensate-like structure with a main configuration of $(0S)_\alpha^3$. Then, an excited state of the Hoyle state, for example, a 2^+ state with $(0S)_\alpha^2(0D)_\alpha$, may exist somewhat higher up in energy than the Hoyle state. Itoh et al. observed the 2_2^+ state at 2.6 ± 0.3 MeV above the 3α threshold with a width of 1.0 ± 0.3 MeV by measuring α particles decaying from excited ^{12}C states with inelastic α scattering [36]. This state was quite recently confirmed by experiment with a high-energy-resolution magnetic spectrometer [39].

A deformed calculation using the THSR wave function in Eqs. (5.37) and (5.38) was performed for the 2^+ state of ^{12}C. Projecting on good angular momentum with a treatment of resonances yields the position of the 2_2^+-state in ^{12}C (2.1 MeV above 3α threshold) which is in good agreement with the experimental value [28, 36]. Also the calculated width (0.64 MeV) gives a quite reasonable estimate of the data. Detailed investigation of the wave function of the 2_2^+-state shows that it can essentially be described in lifting out of the condensate state with the three α's in the $0S$-orbit, one α-particle in the next $0D$-orbit. It is tempting to imagine that the 0_3^+-state which, experimentally, is almost degenerate with the 2_2^+-state, is obtained by lifting one α-particle into the $1S$-orbit. Preliminary theoretical studies [30, 31] indicate that this scenario might indeed apply. However, the width of the 0_3^+ state is very broad (~ 3 MeV), rendering a theoretical treatment rather delicate. Further investigations are necessary to validate or reject this picture which is shown graphically in Fig. 5.15. Anyway, it would be quite satisfying, if the triplet of states, $(0_2^+, 2_2^+, 0_3^+)$ could all be explained from the α-particle perspective, since those three states are *precisely* the ones which cannot be explained within a (no core) shell model approach [32–35].

5.4.1.5 Precursors of a 3α Condensate State: 3_1^- and 1_1^-

The 3^- state at $E_{3\alpha} = 2.37\,\text{MeV}$ measured from the 3α threshold in Fig. 5.4 is interesting from the point of view of the dilute α condensation. If the state is a condensate with all of the 3α particles in the P orbit, there is the possibility of a superfuid with a vortex line, similar to the rotating dilute atomic condensate at very low temperature [104]. Thus, it is an intriguing problem to study the structure within the 3α OCM. The OCM [26] reproduces well the energy of the 3_1^- state as well as the 1_1^- state ($E_{3\alpha} = 3.57\,\text{MeV}$) with respect to the 3α threshold.

The calculated nuclear radius of the 3^- state is 2.95 fm, the value of which is larger than that for the ground state (0_1^+), while it is smaller than that for the 0_2^+ state. This suggests that the structure of the 3^- state is intermediate between the shell-model-like compact structure (0_1^+) and the dilute 3α structure (0_2^+). The occupation probabilities of the single-α orbits of the state are 44.7% for P_1-orbit and 27.9% for F_1-orbit. Although the concentration of the single P_1 orbit amounts to about 50%, the radial behavior of the single-α orbit has two nodes in the inner region. However, the amplitude of the inner oscillations is significantly smaller than that for the ground state in Fig. 5.13a [26]. The small oscillations indicate a weak Pauli-blocking effect, and thus, we can see the precursor of a 3α condensate state [26], although the 3^- state is not an ideal rotating dilute 3α condensate.

As for the 1_1^- state, the calculated nuclear radius, 3.32 fm, is larger than that of the ground state and the 3_1^- state but is still smaller than that of the 0_2^+ one. The occupation probabilities of the α particles in the 1_1^- state are 35% for P_1 orbit and 16% for F_1 orbit. Thus, there is no concentration of the occupation probability to a single orbit like in the 0_2^+ state. Since the α particles in the 1_1^- state are distributed over several orbits, the state is not of the dilute α-condensate type. On the other hand, the radial behavior of the P_1 orbit has two nodes in the inner region, the behavior of which is rather similar to the $2P$ harmonic oscillator wave function. However, the F_1 orbit has a F-wave Gaussian-type behavior. Also the oscillatory behavior of the F_1 orbit for $0 < r < 2\,\text{fm}$ is similar to the one of the S_1 orbit in the 0_2^+ state in Fig. 5.13. These interesting behaviors of the F_1 orbit indicate some signal of dilute α condensation, reflecting the relatively large nuclear radius (3.32 fm) of the 1_1^- state.

5.4.2 ^{16}O Case

In the previous section, we showed that the Hoyle state, which has about one third of saturation density, can be described, to good approximation, as a product state of three α-particles, condensed, with their c.o.m. motion, into the lowest α mean field $0S$-orbit [25, 26]. These novel aspects indicate that the Hoyle state has a 3α-condensate-like structure. The establishment of the novel aspect of the Hoyle state naturally leads us to speculate about the possibility of 4α-particle condensation in ^{16}O.

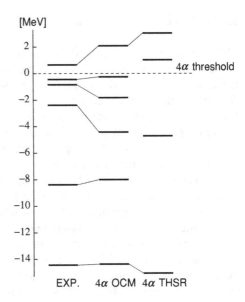

Fig. 5.16 Comparison of energy spectra among experiment, the 4α OCM calculation [51], and the THSR treatment [27]. *Dotted line* denotes the 4α threshold. Experimental data are taken from Ref. [19] and from Ref. [48] for the 0_4^+ state

The experimental 0^+ spectrum of ^{16}O up to about the 4α disintegration threshold is shown in Fig. 5.16. In the past, the 0_1^+ (g.s), 0_2^+ and 0_3^+ states up to about 13 MeV excitation energy has very well been reproduced with a semi-microscopic cluster model, i.e. the $\alpha + ^{12}$C OCM (Orthogonality Condition Model) [87, 88]. In particular, this model calculation, as well as that of an $\alpha + ^{12}$C GCM (Generator-Coordinate-Method) one [106], demonstrates that the 0_2^+ state at 6.05 MeV and the 0_3^+ state at 12.05 MeV have $\alpha + ^{12}$C structures [107] where the α-particle orbits around the ^{12}C(0_1^+)-core in an S-wave and around the ^{12}C(2_1^+)-core in a D-wave, respectively. Consistent results were later obtained by the 4α OCM calculation within the harmonic oscillator basis [86]. However, the model space adopted in Refs. [83, 84, 86, 106] is not sufficient to account simultaneously for the $\alpha + ^{12}$C and the 4α gas-like configurations. On the other hand, the 4α-particle condensate state was first investigated in Ref. [5] and its existence was predicted around the 4α threshold with the THSR wave function. While the THSR wave function can well describe the dilute α cluster states as well as shell model like ground states, other structures such as $\alpha + ^{12}$C clustering can not be treated and are only incorporated in an average way. Thus, it is important to explore the 4α condensate without any a priori assumption with respect to the structure of the 4α system. For this purpose, a full four-body 4α OCM calculation with Gaussian basis functions was performed [51]. This model space is large enough to cover the 4α gas, the $\alpha + ^{12}$C cluster, as well as the shell-model configurations. In this section, we first present the results of the 4α OCM calculation, and then discuss a recent analysis with the THSR wave function for the 4α system.

Table 5.2 Energies ($E - E_{4\alpha}^{\text{th}}$), r.m.s. radii ($R$), and monopole transition matrix elements to the ground state [$M(E0)$]. $R_{\text{exp.}}$ and $M(E0)_{\text{exp.}}$ are the corresponding experimental data. The finite size of α particle is taken into account in R and $M(E0)$ (see Ref. [26] for details)

	4α OCM				Experiment			
	$E - E_{4\alpha}^{\text{th}}$	R	$M(E0)$	Γ	$E - E_{4\alpha}^{\text{th}}$	$R_{\text{exp.}}$	$M(E0)_{\text{exp.}}$	$\Gamma_{\text{exp.}}$
0_1^+	−14.37	2.7			−14.44	2.71 ± 0.02		
0_2^+	−8.00	3.0	3.9		−8.39		3.55 ± 0.21	
0_3^+	−4.41	3.1	2.4		−2.39		4.03 ± 0.09	
0_4^+	−1.81	4.0	2.4	0.60	−0.84		No data	0.6
0_5^+	−0.248	3.1	2.6	0.20	−0.43		3.3 ± 0.7	0.185
0_6^+	2.08	5.6	1.0	0.14	0.66		No data	0.166

5.4.2.1 4α OCM Analysis

The 4α OCM Hamiltonian was presented in Eq. (5.24). The effective α–α interaction $V_{2\alpha}^{\text{eff}}$ is constructed by the folding procedure from an effective two-nucleon force [108, 109] including the Coulomb interaction. One should note that the folded $\alpha - \alpha$ potential reproduces the $\alpha - \alpha$ scattering phase shifts and energies of the ^{8}Be ground state and of the Hoyle state. The three-body force $V_{3\alpha}^{\text{eff}}$ was phenomenologically introduced so as to fit the ground state energy of ^{12}C. In addition, the phenomenological four-body force $V_{4\alpha}$ was adjusted to the ground state energy of ^{16}O. The origin of the three-body and four-body forces is considered to be deducible from the state dependence of the effective nucleon–nucleon interaction and the additional Pauli repulsion between more than two α-particles. However, they are short-range, and hence only act in compact configurations. The expectation values of those forces is less than 10% of the one of the corresponding two-body term.

The $J^\pi = 0^+$ energy spectrum obtained by the 4α OCM is shown in Fig. 5.16. We can reproduce the full spectrum of 0^+ states up to about the 4α disintegration threshold, and tentatively make a one-to-one correspondence of those states with the six lowest 0^+ states of the experimental spectrum. In view of the complexity of the situation, the agreement is considered to be very satisfactory.

We show in Table 5.2 the calculated r.m.s. radii and monopole transition matrix elements to the ground state, together with the corresponding experimental values. The r.m.s. radius of the ground state is reproduced well, and those for the other five 0^+ states are by about 10% or more larger than the ground state. The $M(E0)$ values for the 0_2^+ and 0_5^+ states are consistent with the corresponding experimental values. The $M(E0)$ value for the 0_3^+ state is accurate only within a factor of two.

In order to analyze the obtained wave functions, it is useful to study the overlap amplitude $\mathcal{Y}(r)$ and spectroscopic factor S^2, which are defined as follows:

$$\mathcal{Y}(r) = \sqrt{\frac{4!}{3!1!}} \left\langle \left[\frac{\delta(r' - r)}{r'^2} Y_L(\hat{\mathbf{r}}') \Phi_L(^{12}C) \right]_0 \middle| \Psi(0_6^+) \right\rangle, \qquad (5.46)$$

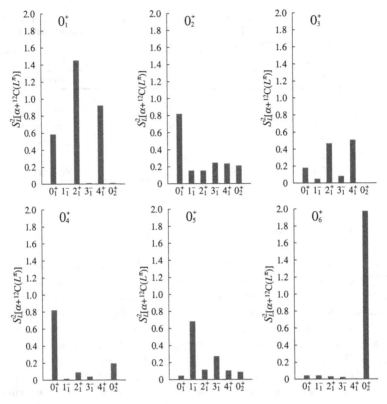

Fig. 5.17 Spectroscopic factors of the $\alpha + {}^{12}C(L_n^\pi)$ channels ($L_n^\pi = 0_1^+, 1_1^-, 2_1^+, 3_1^-, 4_1^+, 0_2^+$) in the six 0^+ sates of ${}^{16}O$

$$S^2 = \int_0^\infty dr\, [r\mathcal{Y}(r)]^2 , \qquad (5.47)$$

where $\Phi_L({}^{12}C)$ is the wave function of ${}^{12}C$, given by the 3α OCM calculation [26], and r is the relative distance between the center-of-mass of ${}^{12}C$ and the α particle. From this quantity we can see how large is the component in a certain $\alpha + {}^{12}C$ channel which is contained in the wave functions obtained by the 4α OCM. The results of S^2 factors are shown in Fig. 5.17. Since the ground state has a closed shell structure with the dominant component of SU(3)$(\lambda, \mu) = (0, 0)$, the values of the S^2 factors for 0_1^+ in Fig. 5.17 can be explained by the SU(3) nature of the state. As mentioned above, the structures of the 0_2^+ and 0_3^+ states are well established as having $\alpha + {}^{12}C(0_1^+)$ and $\alpha + {}^{12}C(2_1^+)$ cluster structures, respectively. These structures of the 0_2^+ and 0_3^+ states are confirmed by the 4α OCM calculation. In fact, one sees that the S^2 factors for the $\alpha + {}^{12}C(0_1^+)$ and $\alpha + {}^{12}C(2_1^+)$ channels are dominant in the 0_2^+ and 0_3^+ states, respectively.

Fig. 5.18 (Colors online) Overlap amplitudes multiplied by r, defined by Eq. (5.46), for the 0_6^+ state in ^{16}O

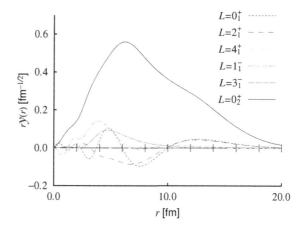

On the contrary, the structures of the observed 0_4^+, 0_5^+ and 0_6^+ states in Fig. 5.16 have, in the past, not clearly been understood, since they have never been discussed with the previous cluster model calculations [83, 84, 86, 106]. Although Ref. [5] predicts the 4α condensate state around the 4α threshold, it is not clear to which of those states it corresponds to. As shown in Fig. 5.16, the 4α OCM calculation succeeded, for the first time, to reproduce the 0_4^+, 0_5^+ and 0_6^+ states, together with the 0_1^+, 0_2^+ and 0_3^+ states. From the analyses of the overlap amplitudes and the S^2 factors (see Fig. 5.17), the 4α OCM showed that the 0_4^+ and 0_5^+ states mainly have $\alpha +^{12}$ C(0_1^+) structure with higher nodal behavior and an $\alpha +^{12}$ C(1^-) structure, respectively. The monopole strength of the 0_5^+ state is reproduced nicely within the experimental error.

In Table 5.2, the largest r.m.s. radius is about 5 fm for the 0_6^+ state. Comparing with the relatively smaller r.m.s. radii of the 0_4^+ and 0_5^+ states, this large size suggests that the 0_6^+ state may be composed of a weakly interacting gas of α particles of the condensate type. In addition, the 0_6^+ state has a large overlap amplitude with the $\alpha +^{12}$ C(0_2^+) channel with a S^2 factor of about 2.0 (see Fig. 5.17), whereas the amplitudes in the other channels are much suppressed (see Fig. 5.18). The amplitude in the Hoyle-state channel has no oscillations and a long tail stretches out to ∼20 fm.

While a large size is generally necessary for forming an α condensate, the best way for its identification is to investigate the single-α orbit and its occupation probability, which can be obtained by diagonalizing the one-body (α) density matrix as defined in Eq. (5.28) [25, 26, 93–96]. As a result of the calculation of the $L = 0$ case, a large occupation probability of 61% of the lowest $0S$-orbit is found for the 0_6^+ state, whereas the other five 0^+ states all have appreciably smaller values, at most 25% (0_2^+). The corresponding single-α S orbit is shown in Fig. 5.19. It has a strong spatially extended behavior without any node ($0S$). This behavior is very similar to that of the overlap amplitude of 0_6^+ for the $\alpha +^{12}$ C(0_2^+) channel shown in Fig. 5.18. These results indicate that α particles are condensed into a very dilute $0S$ single-α orbit, see also Refs. [26, 110]. In addition, Fig. 5.20 shows the momentum distribution $k^2\rho(k)$

Fig. 5.19 (Color online)
Radial parts of single-α
orbits with $L = 0$ belonging
to the largest occupation
number, for the ground and
0_6^+ states

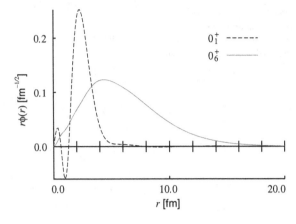

Fig. 5.20 (Color online)
Momentum distribution of α
particles multiplied by
k^2, $k^2\rho(k)$, in the six 0^+
states of ^{16}O

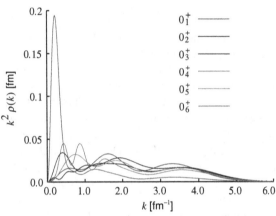

of α particles in the six 0^+ states defined in Eq. (5.45). One sees that the momentum distribution of the 0_6^+ state concentrates strongly in the narrow region $k < 1\,\mathrm{fm}^{-1}$, and the behavior is quite similar to that of the Hoyle state in Fig. 5.14. Thus, the 0_6^+ state clearly has, according to our calculation, a 4α condensate character.

Comparing the single-α orbit of the 0_6^+ state in Fig. 5.19 with that of the Hoyle state shown in Fig. 5.13, one can see an almost identical shape. This is also an important indication that the 0_6^+ state has α-particle condensate nature. Of course, the extension is slightly different because of the smallness of the system. The nodeless character of the wave function is very pronounced and only some oscillations with small amplitude are present in ^{12}C, reflecting the weak influence of the Pauli principle between the α's, as discussed in Sect. 5.4.1.2. On the contrary, due to the much reduced ground-state radii, the "α-like" clusters strongly overlap in ^{12}C and ^{16}O, producing strong amplitude oscillations which take care of antisymmetrization between clusters [26, 51]. In fact, on sees in Fig. 5.19 that the single-α orbit for the ground state has maximum amplitude at around 3 fm and oscillations in the interior with two nodal

Table 5.3 Partial α widths in $^{16}O^*$ decaying into possible channels and the total width. The reduced widths defined in Eq. (5.48) are also shown. a is the channel radius

	$^{12}C(0_1^+)+\alpha$ ($a = 8.0$ fm)	$^{12}C(2_1^+)+\alpha$ ($a = 7.4$ fm)	$^{12}C(0_2^+)+\alpha$ ($a = 8.0$ fm)	Total
Γ_L (keV)	104	32	8×10^{-7}	136
$\theta_L^2(a)$	0.024	0.016	0.60	

(2S) behavior, due to the Pauli principle and reflecting the shell-model configuration (also see Fig. 5.13 for the ground state of ^{12}C).

The α decay width constitutes a very important information to identify the 0_6^+ state from the experimental point of view. The width Γ_L can be estimated, based on the R-matrix theory [111],

$$\Gamma_L = 2P_L(a) \cdot \gamma_L^2(a),$$

$$P_L(a) = \frac{ka}{F_L^2(ka) + G_L^2(ka)},$$

$$\gamma_L^2(a) = \theta_L^2(a)\gamma_W^2(a),$$

$$\gamma_W^2(a) = \frac{3\hbar^2}{2\mu a^2}, \quad \theta_L^2(a) = \frac{a^3}{3}\mathcal{Y}_L^2(a), \tag{5.48}$$

where k, a and μ are the wave number of the relative motion, the channel radius, and the reduced mass, respectively, and F_L, G_L, and $P_L(a)$ are the regular and irregular Coulomb wave functions and the corresponding penetration factor, respectively. The reduced width of $\theta_L^2(a)$ is related with the overlap amplitude $\mathcal{Y}(r)$ defined in Eq. (5.46). In Table 5.3, we show the partial α decay widths Γ_L of the 0_6^+ state decaying into the $\alpha + {}^{12}C(0_1^+)$, $\alpha + {}^{12}C(2_1^+)$ and $\alpha + {}^{12}C(0_2^+)$ channels, and also the total α decay width which is obtained as a sum of the partial widths, and reduced widths $\theta_L^2(a)$ defined in Eq. (5.48). Experimental values are all taken as given by the decay energies. Thus the excitation energy of the calculated 0_6^+ state is assumed to be the experimental value, i.e. 15.1 MeV.

The obtained very small total α decay width of 136 keV, in reasonable agreement with the corresponding experimental value of 160 keV, indicates that this state is unusually long lived. The reason of this fact can be explained in terms of the present analysis as follows: Since this state has a very exotic structure composed of gas-like four alpha particles, the overlap between this state and $\alpha + {}^{12}C(0_1^+)$ or $\alpha + {}^{12}C(2_1^+)$ wave functions with a certain channel radius becomes very small, as this is, indeed, indicated by small $\theta_L^2(a)$ values, 0.024 and 0.016, respectively, and therefore by small $\gamma_L^2(a)$ values. These largely suppress the decay widths expressed by Eq. (5.48) in spite of the large values of the penetration factors caused by large decay energies 7.9 and 3.5 MeV into these two channels, $\alpha + {}^{12}C(0_1^+)$ and $\alpha + {}^{12}C(2_1^+)$, respectively. On the other hand, the decay into the $\alpha + {}^{12}C(0_2^+)$ channel is also suppressed due to the very small penetration caused by the very small decay energy 0.28 MeV, even though the corresponding reduced width takes a relatively large value $\theta_L^2(a) = 0.60$. This is natural since the 0_2^+ state of ^{12}C has a gas-like three-alpha-particle structure. It is very likely that the above mechanism holds generally for the alpha condensate states in heavier $n\alpha$ systems, and therefore the alpha condensate states can also be expected to exist in heavier systems as a relatively long lived resonance.

As for the decay widths of the 0_4^+ and 0_5^+ states, as evaluated by Eq. (5.48), they are shown in Table 5.2. The calculated width of the 0_4^+ state is 600 keV, which is quite a bit larger than that found for the 0_5^+ state 200 keV. Both are qualitatively consistent with the corresponding experimental data, 600 and 185 keV, respectively. We should note that our calculation consistently reproduces the ratio of the widths of the 0_4^+, 0_5^+, and 0_6^+ states, i.e. about 3:1:1, respectively, though the magnitudes of the widths are underestimated by about a factor of four with respect to the experimental values (see Table 5.2). The reason why the width of the 0_4^+ state is larger than that of the 0_5^+ state, though the 0_4^+ state has lower excitation energy, is due to the fact that the former has a much larger component of the $\alpha + ^{12}$C(0_1^+) decay channel, reflecting the characteristic structure of the 0_4^+ state. The 4α condensate state, thus, should not be assigned to the 0_4^+ or 0_5^+ state [27, 51] but very likely to the 0_6^+ state.

5.4.2.2 THSR Wave Function Analysis

As mentioned already, the first investigation of the 4α-particle condensate state was performed with the THSR wave function [5]. It was conjectured that 4α-particle condensation should occur around the 4α threshold. As mentioned in Sect. 5.2, the THSR wave function allows only for two limiting configurations, that is a pure Slater determinant for B $= b$ and a pure α-particle gas for B $\gg b$. Asymptotic configurations like $\alpha + ^{12}$C(0_1^+) are absent. Thus, the THSR wave function can well describe the dilute α cluster states as well as shell model like ground states, whereas other structures such as $\alpha + ^{12}$C(0_1^+) clustering may be only incorporated in an average way. In this section, we see whether or not the counterpart of the 0_6^+ state obtained by the 4α OCM calculation can also be found with the THSR wave function, and then we study how well the 4α condensate state is described with the THSR wave function.

The microscopic wave function of ^{16}O with the THSR ansatz is described in Eqs. (5.18) and (5.19), and the eigenenergies and eigenstates are obtained by solving the Hill–Wheeler equation Eq. (5.20). The Hamiltonian is given in Eq. (5.4), where the effective nucleon–nucleon interaction called F1 [112] was adopted. The resulting 0^+ spectrum is shown in Fig. 5.16. Hereafter, we assign the four 0^+ states as $(0_1^+)_{\text{THSR}}$–$(0_4^+)_{\text{THSR}}$. In Table 5.4, the energies, r.m.s. radii, monopole transition matrix elements M(E0) to the ground state, and α-decay widths of the $(0_1^+)_{\text{THSR}}$–$(0_4^+)_{\text{THSR}}$ states are displayed. The $(0_3^+)_{\text{THSR}}$ state has a large r.m.s. radius of 4.2 fm and the $(0_4^+)_{\text{THSR}}$ state has an even larger one of 6.1 fm. They are comparable to the values for the $(0_4^+)_{\text{OCM}}$ and $(0_6^+)_{\text{OCM}}$ states in the 4α OCM calculation, respectively, where the six 0^+ states obtained by the 4α OCM calcula-

Table 5.4 Energies $E - E_{4\alpha}^{th}$, r.m.s. radii R_{rms}, monopole matrix elements $M(E0)$, and α decay widths Γ, obtained within the 4α THSR framework, where $E_{4\alpha}^{th} = 4E_\alpha$ denotes the 4α threshold energy, with E_α the binding energy of the α particle [27]

	$E - E_{4\alpha}^{th}$[MeV]	R_{rms}[fm]	$M(E0)$[fm^2]	Γ[MeV]
$(0_1^+)_{THSR}$	-15.05	2.5		
$(0_2^+)_{THSR}$	-4.7	3.1	9.8	
$(0_3^+)_{THSR}$	1.03	4.2	2.5	1.6
$(0_4^+)_{THSR}$	3.04	6.1	1.2	0.14

Fig. 5.21 (Color online) Overlap amplitude multiplied by r for the $(0_4^+)_{THSR}$ state in the $\alpha + ^{12}C(0_1^+, 0_2^+)$ channels, defined in Eq. (5.46)

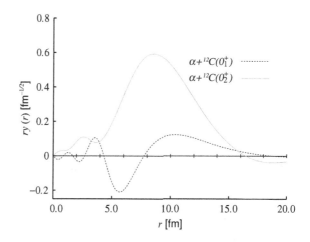

In Table 5.2 are labeled as $(0_1^+)_{OCM}$–$(0_6^+)_{OCM}$. On the other hand, the $M(E0)$ values of the $(0_3^+)_{THSR}$ and $(0_4^+)_{THSR}$ states well agree with those of the $(0_4^+)_{OCM}$ and $(0_6^+)_{OCM}$ states, respectively. This suggests that the $(0_4^+)_{THSR}$ state corresponds to the $(0_6^+)_{OCM}$ state, and hence to the 15.1 MeV state.

More quantitative evidences for the $(0_4^+)_{THSR}$ state being the counterpart of the $(0_6^+)_{OCM}$ state are presented from the analyses of the overlap amplitudes of the $\alpha + ^{12}C(0_1^+, 0_2^+)$ channels and the one-body density matrix for the α particle etc. with the wave function of the $(0_4^+)_{THSR}$ state. In Fig. 5.21, we show the overlap amplitudes of the $\alpha + ^{12}C(0_1^+, 0_2^+)$ channels for the $(0_4^+)_{THSR}$ state. The radial behaviors of the $\alpha + ^{12}C(0_1^+, 0_2^+)$ channels are very similar to those of the $(0_6^+)_{OCM}$ case in Fig. 5.18. In addition, we found that the single-α S orbit occupancy in the $(0_4^+)_{THSR}$ state is as large as 64%, which is comparable to that in the $(0_6^+)_{OCM}$ state, and the radial behavior of the single-α S orbit in the former state is illustrated in Fig. 5.22, and is similar to that in the latter in Fig. 5.19.

In conclusion we found that the $(0_4^+)_{THSR}$ state corresponds to the $(0_6^+)_{OCM}$ state and is most appropriately considered to be the 4α condensate state. This further

Fig. 5.22 (Color online) Radial parts of single-α-particle orbits with $L = 0$ and $n_L = 1$ for the $(0_1^+)_{\text{THSR}}$ (*dotted curve*) and $(0_4^+)_{\text{THSR}}$ (*solid*) states

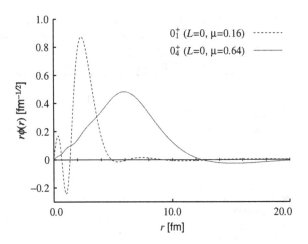

gives us a strong support that the 4α condensate state exists around the 4α breakup threshold and is very likely to correspond to the observed 0_6^+ state at 15.1 MeV.

5.4.3 Heavier 4n Nuclei: Gross–Pitaevskii Equation

In principle, one could go on, increasing the number of α-particles, as for ^{20}Ne, ^{24}Mg, etc. and study their structure with use of the THSR wave function or within the OCM framework. However, one easily imagines that the complexity of the calculations quickly becomes prohibitive. In order to get a rough idea what happens for more α-particles, drastic approximations have to be performed. One such approximation is to consider the α-particles as ideal inert bosons and to treat them in mean field approximation. This then leads to the Gross–Pitaevskii Equation (GPE) [113–116] which is widely employed in the physics of cold atoms [104]. One interesting question that can be asked in this connection is: How many α's can maximally exist in a self-bound α-gas state? Seeking an answer, it is interesting to investigate it schematically using an effective $\alpha - \alpha$ interaction within an α-gas mean-field calculation of the Gross–Pitaevskii type [56].

In the mean field approach, the total wave function of the condensate $n\alpha$-boson system is represented as

$$\Phi(n\alpha) = \prod_{i=1}^{n} \varphi(\mathbf{r}_i), \tag{5.49}$$

where φ is the normalized single-α wave function of the i-th α boson. Then, the equation of motion for the α boson, called the Gross–Pitaevskii equation, is of the non-linear Schrödinger-equation type,

$$-\frac{\hbar^2}{2m}\left(1-\frac{1}{n}\right)\nabla^2\varphi(\mathbf{r}) + U(\mathbf{r})\varphi(\mathbf{r}) = \varepsilon\varphi(\mathbf{r}), \tag{5.50}$$

$$U(\mathbf{r}) = (n-1)\int d\mathbf{r}'\,|\varphi(\mathbf{r}')|^2\,V_{\alpha\alpha}(\mathbf{r}',\mathbf{r}), \tag{5.51}$$

where m stands for the mass of the α particle and U is the mean-field potential of α-particles. The center-of-mass kinetic energy correction, $1 - \frac{1}{n}$, is taken into account together with the finite number corrections, $n - 1$. In the present study, only the S-wave state is solved self-consistently with the iterative method. The effective 2α interaction v_2 is taken of Gaussian-type including a repulsive density-dependent term, to account for the Pauli repulsion at short distances, which is of similar form as the Gogny interaction (known as an effective NN potential) [117] used in nuclear mean-field calculations

$$V_{\alpha\alpha}(\mathbf{r},\mathbf{r}') = V_0\exp\left[-0.7^2(\mathbf{r}-\mathbf{r}')^2\right] - 130\exp\left[-0.475^2(\mathbf{r}-\mathbf{r}')^2\right]$$
$$+ (4\pi)^2 g\delta(\mathbf{r}-\mathbf{r}')\rho\left(\frac{\mathbf{r}+\mathbf{r}'}{2}\right) + V_{\text{Coul}}(\mathbf{r}-\mathbf{r}'), \tag{5.52}$$

where the units of v_2 and r are MeV and fm, respectively, and ρ denotes the density of the $n\alpha$ system. The folded Coulomb potential V_{Coul} is presented as

$$V_{\text{Coul}}(\mathbf{r}-\mathbf{r}') = \frac{4e^2}{|\mathbf{r}-\mathbf{r}'|}\text{erf}(a|\mathbf{r}-\mathbf{r}'|), \tag{5.53}$$

The Gaussian-potential part in Eq. (5.52) is based on the Ali-Bodmer potential [118], which is known to reproduce well the elastic α-α scattering phase shift up to about 60 MeV for $V_0 = 500$ MeV. The two parameters of the force, V_0 and g, were adjusted to reproduce the energy (measured from the 3α threshold) and r.m.s. radius of the Hoyle state obtained from the THSR analysis.

The corresponding α mean-field potential for three α's of ^{12}C is shown in Fig. 5.23. One sees the 0S-state lying slightly above threshold but below the Coulomb barrier. As more α-particles are added, the Coulomb repulsion drives the loosely bound system of α-particles farther and farther apart. For example, in the case of six α's in ^{24}Mg, the Coulomb barrier is lower and its position is moved outwards. Thus, eventually the Coulomb barrier should fade away in some limiting nucleus. According to our estimate [56], a maximum of eight to ten α-particles can be held together in a condensate. However, there may be ways to lend additional stability to such systems. We know that in the case of ^{8}Be, adding one or two neutrons produces extra binding without seriously disturbing the pronounced α-cluster structure. Therefore, one has reason to speculate that adding a few neutrons to a many-α state may stabilize the condensate. But again, state-of-the-art microscopic investigations are necessary before anything definite can be said about how extra neutrons will influence an α-particle condensate.

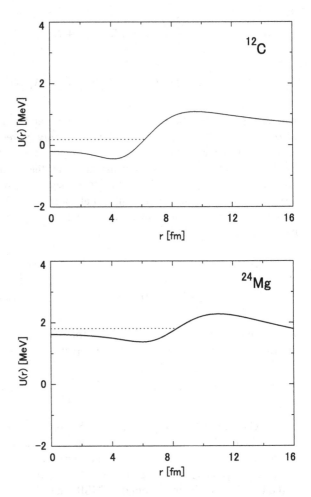

Fig. 5.23 Alpha-particle mean-field potential for three $\alpha's$ in ^{12}C and six $\alpha's$ in ^{24}Mg. The *dotted line* denotes the energy of the single-α 0S orbit (from Ref. [56])

Concerning excitation of condensate states with many α particles, heavy ion reactions and Coulomb excitation may be appropriate tools. As an ideal case let us imagine that ^{40}Ca has been excited by Coulomb excitation to a state of about 60 MeV. Coulomb excitation favors 0$^+$-states and 60 MeV is the threshold for disintegration into 10 α-particles. Since the Coulomb barrier is absent for ten α's, this state may perform a Coulomb explosion of a 10 α particle coherent state. A cartoon of such a scenario is sketched in Fig. 5.24. With heavy ion reactions, experiments with coincident measurements are being analyzed to detect multi α events by Borderie et al. [49]. W. v. Oertzen et al. seem to have detected an enhancement of multi α decay out of α-condensates in compound states of heavier $N=Z$ nuclei, see Ref. [119].

Fig. 5.24 (Color online) Cartoon of a Coulomb explosion of 10 α-particles from ^{40}Ca

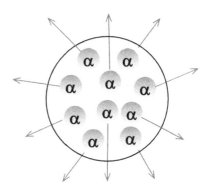

5.4.4 Hoyle-Analogue States in Non-4n Nuclei: ^{11}B and ^{13}C

In the previous sections, we discussed the α-gas-like states in $4n$ nuclei. On the other hand, one can also expect cluster-gas states composed of alpha and triton clusters (including valence neutrons etc.) around their cluster disintegrated thresholds in $A \neq 4n$ nuclei, in which all clusters are in their respective $0S$ orbits, similar to the Hoyle state with $(0S_\alpha)^3$. The states, thus, can be called *Hoyle-analogue* states in non-self-conjugated nuclei. It is an intriguing subject to investigate whether or not the Hoyle-analogue states exist in $A \neq 4n$ nuclei, as for example, ^{11}B, composed of 2α and $1t$ clusters as well as ^{13}C, composed of 3α and $1n$.

The structure of $3/2^-$ and $1/2^+$ ($1/2^-$ and $1/2^+$) states in ^{11}B (^{13}C) up to around the $2\alpha + t$ ($3\alpha + n$) threshold were investigated by the $2\alpha + t$ OCM [59] $3\alpha + n$ OCM [61]) for ^{11}B(^{13}C) combined with the Gaussian expansion method. The model space for the $2\alpha + t$ ($3\alpha + n$) OCM is large enough to cover the $2\alpha + t$ ($3\alpha + n$) gas, the ^{7}Li $+ \alpha$ and $^8Be + t$ (^{9}Be $+ \alpha$ and ^{12}C $+ n$) clusters, as well as the shell-model configurations. As well known, the $\alpha - t$ and $\alpha - n$ potentials have a strong parity dependence [2]. In the odd waves they are strongly attractive to produce the bound states (for $\alpha - t$) and resonant states ($\alpha - t$ and $\alpha - n$), while the even ones are weakly attractive and have no ability to produce any resonant states up to $E_x \sim 15$ MeV [120]. Thus, the gas-like states in ^{11}B and ^{13}C might appear in their even-parity states.

The energy levels of $3/2^-$ and $1/2^+$ states in ^{11}B are shown in Fig. 5.25. The $3/2^-_1$ state is the ground state with a shell-model-like structure. The calculated nuclear radius is $R = 2.22$ fm($R^{exp} = 2.43 \pm 0.11$ fm). The dominant configuration of this state is SU(3)$[f](\lambda, \mu)_L = [443](1, 3)_1$ with $Q = 7$ harmonic oscillator quanta (95%), having the main angular momentum channel of $(L, S)_J = (1, \frac{1}{2})_{\frac{3}{2}}$. On the other hand, also the $3/2^-_2$ state has a shell-model-like structure.

The $3/2^-_3$ state appears at $E_x = 8.2$ MeV ($E = -2.9$ MeV referring to the $2\alpha + t$ threshold). The radius of $3/2^-_3$ is 3.00 fm. This value is by about 30% larger than that of the ground state of ^{11}B, and the $\alpha - \alpha$ r.m.s. distance (distance between ^{8}Be(2α) and t) is 4.47 fm (3.49 fm). Thus, $3/2^-_3$ has a $2\alpha + t$ cluster structure. A characteristic

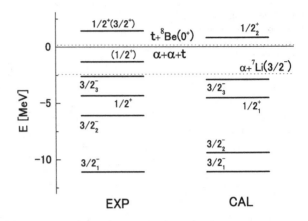

Fig. 5.25 (Color online)
Calculated energy levels of
$3/2^-$ and $1/2^+$ states in ^{11}B
with respect to the $2\alpha + t$
threshold, together with the
experimental data

feature of $3/2_3^-$ is that the isoscalar monopole transition rate $B(E0:IS)$ is as large as $96 \pm 16\,\mathrm{fm}^4$, comparable to that of the Hoyle state ($120 \pm 9\,\mathrm{fm}^4$). The present model ($92\,\mathrm{fm}^4$) reproduces well the data. It is interesting to study whether $3/2_3^-$ possesses an α gas nature like the Hoyle state. To this purpose, we study the single-cluster orbits and their occupation probabilities in the $3/2_3^-$ state by solving the eigenvalue equation of the single-cluster density matrices.

Figure 5.26 shows the occupation probabilities of the n-th L-wave single-α-particle (single-t-particle) orbit in the $3/2_1^-$ and $3/2_3^-$ states. In the $3/2_1^-$ state, the occupation probabilities of α particles spread out in several orbits, and those of t orbits concentrate mainly on two orbits. These results originate from the SU(3) nature of the $3/2_1^-$ state as mentioned above. On the other hand, in the $3/2_3^-$ state, there also is no concentration of α occupation probability on a single orbit. This result is in contrast with those of the Hoyle state (see Fig. 5.12). Consequently the $3/2_3^-$ state can not be identified as the analogue of the Hoyle state. The reason why the $3/2_3^-$ state is not of Hoyle-type is as follows: The $3/2_3^-$ state is bound by 2.9 MeV with respect to the $2\alpha + t$ threshold, while the Hoyle state is located by 0.38 MeV above the 3α threshold. This extra binding energy of $3/2_3^-$ with respect to the $2\alpha + t$ threshold suppresses strongly the development of the gas-like $2\alpha + t$ structure.

As for the $1/2^+$ states, the $1/2_1^+$ state appears as a bound state at $E_x^{\mathrm{exp}} = 6.79$ MeV around the ^{7}Li + α threshold. This low excitation energy indicates that α-type correlations should play an important role in this state. In fact, we found that the $1/2_1^+$ state with the radius of 3.14 fm has a ^{7}Li(g.s) + α structure with P-wave relative motion, although the ^{7}Li($\alpha + t$) part is rather distorted in comparison with the ground state of ^{7}Li. Since the $3/2_3^-$ state has the largest S^2 factor for the ^{7}Li(g.s) + α channel with S-wave relative motion, the $1/2_1^+$ and $3/2_3^-$ states of ^{11}B can be interpreted as parity-doublet partners.

In addition to the $1/2_1^+$ state, the $1/2_2^+$ state appears as a resonant state at $E_x = 11.95$ MeV ($\Gamma = 190$ keV) around the $2\alpha + t$ threshold using the complex-scaling method [121–126]. The large radius ($R_N = 5.98$ fm) indicates that the state has a dilute cluster structure. The analysis of the single-cluster properties showed that

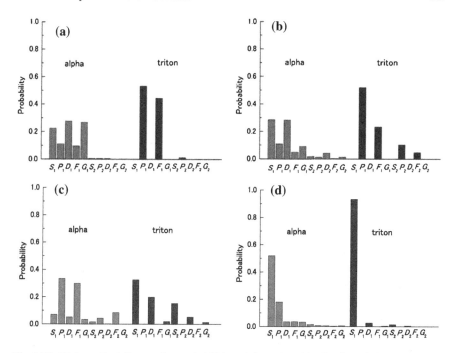

Fig. 5.26 (Color online) Occupation probabilities of the α (t) orbits for the **a** $3/2_1^-$ and **b** $3/2_3^-$ states together with the **c** $1/2_1^+$ and **d** $1/2_2^+$ ones in ^{11}B

this state has as main configuration $(0S_\alpha)^2(0S_t)$ orbital occupation with about 65% probability (see Fig. 5.26). Thus, the $1/2_2^+$ state can be called the Hoyle-analogue. Recently, the $1/2^+(3/2^+)$ state at $E_x = 12.56$ MeV with $\Gamma = 210 \pm 20$ keV (located at 1.4 MeV above the $2\alpha + t$ threshold) was observed in the $\alpha + {}^7$Li decay channel [127–129]. The energy and width of the 12.56-MeV state are in good correspondence to the present study. The Hoyle-analogue state in ^{11}B, thus, could be assigned as the 12.56-MeV state. It should be reminded that the $1/2_2^+$ state is located by 0.75 MeV above the $\alpha + \alpha + t$ threshold, while $1/2_1^+$ is bound by 4.2 MeV with respect to the three cluster threshold. The latter binding energy leads to a suppression of the development of the gas-like $\alpha + \alpha + t$ structure in $1/2_1^+$, whereas it is generated with a large nuclear radius in the $1/2_2^+$ state because of its appearance above the three-body threshold.

The calculated energy spectrum of $1/2^-$ states in ^{13}C is shown in Fig. 5.27. The four $1/2^-$ energy levels are in good correspondence with the experimental data. The ground state ($1/2_1^-$) is described as having a shell-model configuration. The calculated nuclear radius (2.39 fm) agrees with the data (2.44 fm). The three isoscalar monopole transition strengths, $M(1/2_1^- - 1/2_2^-) = 4.2$, $M(1/2_1^- - 1/2_3^-) = 5.6$, and $M(1/2_1^- - 1/2_4^-) = 8.2$, are also consistent with experiment [60], 6.1 ± 0.5, 4.2 ± 0.4, and 4.9 ± 0.4, respectively, in units of fm^2. The nuclear radii for the three excited $1/2^-$ states are 3.36, 2.96, and 3.19 fm, respectively.

Fig. 5.27 (Color online) Calculated energy levels of $1/2^-$ and $1/2^+$ states in ^{13}C with respect to the $3\alpha + t$ threshold, together with the experimental data

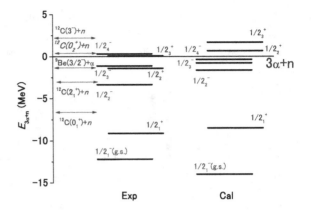

From the analysis of the radial behavior of the overlap amplitudes referring to the ^{12}C $+ n$ and ^{9}Be$+ \alpha$ channels, the $1/2_2^-$ and $1/2_3^-$ states are characterized as having large components of ^{12}C(g.s, 2^+) $+ n$ and ^{12}C(3^-) $+ n$ with ^{9}Be(g.s)$+ \alpha$, respectively. On the other hand, the $1/2_4^-$ state contains a somewhat large component of the ^{12}C(Hoyle)$+n$ channel together with ^{12}C(2^+) $+ n$ and ^{9}Be(g.s) $+ \alpha$. However, this state does not have as large an α condensate component as the Hoyle state in ^{12}C. This is due to strong attraction of the odd-wave $\alpha - n$ potentials which induces the coupling of ^{12}C(2^+) $+ n$ and ^{9}Be(g.s)$+ \alpha$ structures with the ^{12}C(Hoyle) $+ n$ configuration, and disturbs significantly the structure of the 3α condensate in ^{13}C. In the mirror nucleus ^{13}N, the $3\alpha + p$ OCM analysis gives qualitatively similar results to those of the present ^{13}C case.

As for the $1/2^+$ states of ^{13}C, the energy spectrum of the first three $1/2^+$ states correspond well with the data (see Fig. 5.27). We found that the $1/2_1^+$ state has a main configuration $[^{12}$C(g.s) $\otimes 2s_{1/2}]$. Reflecting the fact that the neutron binding energy of the $1/2_1^+$ state with respect to the ^{12}C(g.s.) $+ n$ threshold is as small as 1.9 MeV, this state has a neutron-halo-like structure. In fact the calculated nuclear radius of this state (2.68 fm) is larger than that of the ground state (2.39 fm), and this enhancement of the radius comes from the neutron-halo-like structure.

On the other hand, the $1/2_2^+$ state has a dominant configuration of the extra neutron coupled with the Hoyle state, with non-negligible mixing of ^{9}Be(g.s, $1/2_1^-$) $+ \alpha$ channels. The nuclear radius is about 4.0 fm, which is smaller than that of the Hoyle state in the 3α OCM calculation [26]. We found that the size of the 3α part in this state is reduced by about 15 % in comparison with that of the Hoyle state. The occupation probability of α particle in $0S$ orbit in this state is less than 30%, which is much smaller than that for the Hoyle state.

The $1/2_3^+$ state around the $3\alpha + n$ threshold has the nuclear radius of 5.40 fm with a dilute α condensate feature, in which 3α particles occupy an identical $0S$ orbit with 55% probability. This state has a rather large overlap with the ^{9}Be($1/2_1^+$) $+ \alpha$ channel as well as with the ^{12}C(Hoyle) $+ n$ one. It is noted that the ^{9}Be($1/2_1^+$) state is known to have a neutron-halo-like structure (or $2\alpha + n$ gas-like structure). Thus, these results suggest that the $1/2_3^+$ state is a candidate for the Hoyle-analogue state.

With this we terminate our consideration of cluster and condensate aspects in finite nuclei. An important connection with the finite systems is given by clustering, for instance α particle clustering and condensation in infinite matter. In the next section we turn to these issues.

5.5 Clusters in Nuclear Matter and α-Particle Condensation

5.5.1 Nuclear Clusters in the Medium

Of course, it is also interesting and important to study how α-clusters behave and actually condensate in infinite symmetric and asymmetric nuclear matters. This not only in regard to better understand what finally happens in a finite nucleus but in collapsing and compact stars one may speculate about the existence of a macroscopic α-particle condensate. So let us first consider the modification an α particle undergoes when it is embedded in a nuclear medium.

Medium modifications of single-particle states as well as of few-nucleon states become of importance with increasing density of nuclear matter. The self-energy of an A-particle cluster can in principle be deduced from contributions describing the single-particle self-energies as well as medium modifications of the interaction and the vertices. A guiding principle in incorporating medium effects is the construction of consistent ("conserving") approximations, which treat medium corrections in the self-energy and in the interaction vertex at the same level of accuracy. This can be achieved in a systematic way using the Green functions formalism [130]. At the mean-field level, we have only the Hartree–Fock self-energy $\Gamma^{HF}(1) = \sum_2 \bar{V}(12, 12) f(2)$ together with the Pauli blocking factors, which modify the two-nucleon interaction from $V(12, 1'2')$ to $V(12, 1'2')[1 - f(1) - f(2)]$, with $f(1) = [1 + \exp(\varepsilon^{HF}(1) - \mu)/T]^{-1}$ and $\bar{V}(12, 12) = V(12, 12) - V(12, 21)$. In the case of the two-nucleon system ($A = 2$), the effective wave equation which includes those corrections is presented in the following form [131, 132],

$$\left[\varepsilon^{HF}(1) + \varepsilon^{HF}(2) - E_{2,P}\right] \psi_{2,P}(12)$$
$$+ \frac{1}{2} \sum_{1'2'} [1 - f(1) - f(2)] \bar{V}(12, 1'2') \psi_{2,P}(1'2') = 0. \tag{5.54}$$

This effective wave equation describes bound states as well as scattering states. The onset of pair condensation is achieved when the binding energy $E_{2,P=0}$ coincides with 2μ, where P denotes the total momentum of the two-nucleon system. It is noted that the Gor'kov equation in BCS theory of superfluidity is a special case of Eq. (5.54).

Fig. 5.28 Shift of binding energy of the light clusters (*d—dash dotted, t/h—dotted,* and *α—dashed:* perturbation theory, full line: non-perturbative Faddeev–Yakubovski equation) in symmetric nuclear matter as a function of density for given temperature $T = 10$ MeV [64]

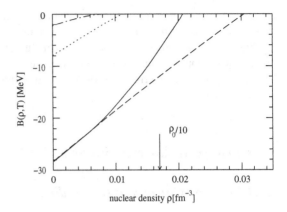

Similar equations have been derived from the Green function approach for the case $A = 3$ and $A = 4$, describing triton/helion (^{3}He) nuclei as well as α-particles in nuclear matter. The effective wave equation contains in mean field approximation the Hartree–Fock self-energy shift of the single-particle energies as well as the Pauli blocking of the interaction. We give the effective wave equation for $A = 4$,

$$\left[\varepsilon^{\mathrm{HF}}(1) + \varepsilon^{\mathrm{HF}}(2) + \varepsilon^{\mathrm{HF}}(3) + \varepsilon^{\mathrm{HF}}(4) - E_{4,P}\right]\psi_{4,P}(1234)$$
$$+ \frac{1}{2}\sum_{i<j}\sum_{1'2'3'4'}[1 - f(i) - f(j)]\bar{V}(ij, i'j')\prod_{k\neq i,j}\delta_{k,k'}\psi_{4,P}(1'2'3'4') = 0. \quad (5.55)$$

A similar equation is obtained for $A = 3$, which is an equation for a fermionic cluster.

The effective wave equation has been solved using separable potentials for $A = 2$ by integration. For $A = 3, 4$ we can use a *Faddeev approach* [64]. The shifts of binding energy can also be calculated approximately via perturbation theory. In Fig. 5.28 we show the shift of the binding energy of the light clusters (d, t/h and α) in symmetric nuclear matter as a function of density for temperature $T = 10$ MeV [64].

It is found that the cluster binding energy decreases with increasing density. Finally, at the *Mott density* $\rho_{A,P}^{\mathrm{Mott}}(T)$ the bound state is dissolved. The clusters are not present at higher densities, merging into the nucleonic medium. For a given cluster type characterized by A, n, we can also introduce the Mott momentum $P_A^{\mathrm{Mott}}(\rho, T)$ in terms of the ambient temperature T and nucleon density ρ, such that the bound states exist only for $P \geq P_A^{\mathrm{Mott}}(\rho, T)$. We do not present an example here, but it is intuitively clear that a cluster with high c.o.m. momentum with respect to the medium is less affected by the Pauli principle than a cluster at rest, because the overlap of the bound state wave function in momentum space and the Fermi distribution function becomes smaller.

5.5.2 Four-Particle Condensates and Quartetting in Nuclear Matter

In general, it is necessary to take into account *all bosonic clusters* to gain a complete picture of the onset of superfluidity. As is well known, the deuteron is weakly bound as compared to other nuclei. Higher A-clusters can arise that are more stable. In this section, we will consider the formation of α-particles, which are of special importance because of their large binding energy per nucleon (~ 7 MeV). We will not include tritons or helions, which are fermions and not so tightly bound. Moreover, we will not consider nuclei in the iron region, which have even larger binding energy per nucleon than the α-particle and thus constitute, in principle, the dominant component at low temperatures and densities. However, the latter are complex structures of many particles and are strongly affected by the medium as the density increases for given temperature, so that they are assumed not to be of relevance in the density region considered here.

The in-medium wave equation for the four-nucleon problem has been solved using the Faddeev–Yakubovski technique, with the inclusion of Pauli blocking, see also below. The binding energy of an α-like cluster with zero c.o.m. momentum vanishes at around $\rho_0/10$, where $\rho_0 \simeq 0.16$ nucleons/fm^3 denotes the saturation density of isospin-symmetric nuclear matter, see Fig. 5.28. Thus, the four-body bound states make no significant contribution to the composition of the system above this density. Given the medium-modified bound-state energy $E_{4,P}$, the bound-state contribution to the EOS is

$$\rho_4(\beta, \mu) = \sum_P \left[e^{\beta(E_{4,P} - 2\mu_p - 2\mu_n)} - 1 \right]^{-1}. \tag{5.56}$$

We will not include the contribution of the excited states nor that of scattering states. Because of the large specific binding energy of the α particle, low-density nuclear matter is predominantly composed of α particles. This observation underlies the concept of α matter and its relevance to diverse nuclear phenomena [133–139].

As exemplified by Eq. (5.55), the effect of the medium on the properties of an α particle in mean-field approximation (i.e., for an uncorrelated medium) is produced by the Hartree–Fock self-energy shift and Pauli blocking. The shift of the α-like bound state has been calculated using perturbation theory [131, 132] as well as by solution of the Faddeev–Yakubovski equation [64]. It is found that the bound states of clusters d, t, and h with $A < 4$ are already dissolved at a Mott density $\rho_\alpha^{\text{Mott}} \approx \rho_0/10$, see Fig. 5.28. Since Bose condensation only is of relevance for d and α, and the fraction of d, t and h becomes low compared with that of α with increasing density, we can neglect the contribution of them to an equation of state. Consequently, if we further neglect the contribution of the four-particle scattering phase shifts in the different channels, we can now construct an equation of state $\rho(T, \mu) = \rho^{\text{free}}(T, \mu) + \rho^{\text{bound},d}(T, \mu) + \rho^{\text{bound},\alpha}(T, \mu)$ such that α-particles determine the behavior of symmetric nuclear matter at densities below $\rho_\alpha^{\text{Mott}}$ and temperatures below the binding energy per nucleon of the α-particle. The formation of deuteron clusters alone gives an incorrect description because the deuteron binding energy is small, and, thus, the abundance of d-clusters is small compared with that of α-clusters. In the low density region of the phase diagram, α-matter emerges as an adequate model for describing the nuclear-matter equation of state.

With increasing density, the medium modifications—especially Pauli blocking—will lead to a deviation of the critical temperature $T_c(\rho)$ from that of an ideal Bose gas of α-particles (the analogous situation holds for deuteron clusters, i.e., in the isospin-singlet channel) [64].

Symmetric nuclear matter is characterized by the equality of the proton and neutron chemical potentials, i.e., $\mu_p = \mu_n = \mu$. Then an extended Thouless condition based on the relation for the four-body T-matrix (in principle equivalent to Eq. (5.55) at eigenvalue 4μ)

$$
T_4(1234, 1''2''3''4'', 4\mu) = \frac{1}{2} \sum_{1'2'3'4'} \left\{ \frac{\bar{V}(12, 1'2')[1 - f(1) - f(2)]}{4\mu - E_1 - E_2 - E_3 - E_4} \delta(3, 3')\delta(4, 4') \right.
$$

$$
\left. + \text{cycl.} \right\} T_4(1'2'3'4', 1''2''3''4'', 4\mu)
$$

(5.57)

serves to determine the onset of Bose condensation of α-like clusters, noting that the existence of a solution of this relation signals a divergence of the four-particle correlation function. An approximate solution has been obtained by a variational approach, in which the wave function is taken as Gaussian incorporating the correct solution for the two-particle problem [63].

On the other hand, Eq. (5.57), respectively Eq. (5.55) at eigenvalue 4μ, has also been solved numerically exactly by the Faddeev–Yakubovsky method employing the Malfliet–Tjon force [140, 141]. The results for the critical temperature of α-condensation is presented in Fig. 5.29 as a function of the chemical potential μ (see also Ref. [63]). The exact solution could only be obtained for negative μ, i.e. when there exists a bound cluster. It is, therefore, important to try yet another approximate solution of the in-medium four-body equation. Since the α-particle is strongly bound, we make a momentum projected mean field ansatz for the quartet wave function [142–144]

$$
\Psi_{1234} = (2\pi)^3 \delta^{(3)}(\mathbf{k}_1 + \mathbf{k}_2 + \mathbf{k}_3 + \mathbf{k}_4) \prod_{i=1}^{4} \varphi(\mathbf{k}_i)\chi^{ST},
$$

(5.58)

where χ^{ST} is the spin-isospin function which we suppose to be the one of a scalar ($S = T = 0$). We will not further mention it from now on. We work in momentum space and $\varphi(\mathbf{k})$ is the as-yet unknown single particle $0S$ wave function. In position space, this leads to the usual formula [24] $\Psi_{1234} \rightarrow \int d^3R \prod_{i=1}^{4} \tilde{\varphi}(\mathbf{r}_i - \mathbf{R})$ where $\tilde{\varphi}(\mathbf{r}_i)$ is the Fourier transform of $\varphi(\mathbf{k}_i)$. If we take for $\varphi(\mathbf{k}_i)$ a Gaussian shape, this gives: $\Psi_{1234} \rightarrow \exp[-c \sum_{1 \leq i < k \leq 4}(\mathbf{r}_i - \mathbf{r}_k)^2]$ which is the translationally invariant ansatz often used to describe α-clusters in nuclei. For instance, it is also employed in the α-particle condensate wave function of Tohsaki et al. (THSR) in Ref. [5].

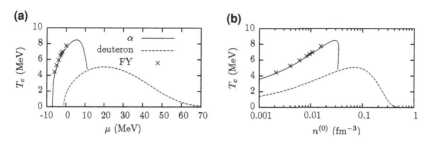

Fig. 5.29 Critical temperature of *alpha* and *deuteron* condensations as functions of **a** chemical potential and **b** density of free nucleons [65]. *Crosses* (×) correspond to the solution of Eq. (5.55) with the Malfliet–Tjon interaction (MT I–III) using the Faddeev–Yakubovski method

Inserting the ansatz (5.58) into (5.55) and integrating over superfluous variables, or minimizing the energy, we arrive at a Hartree–Fock type of equation for the single particle $0S$ wave function $\varphi(k) = \varphi(|\mathbf{k}|)$ which can be solved. However, for a general two body force $V_{\mathbf{k}_1\mathbf{k}_2,\mathbf{k}'_1\mathbf{k}'_2}$, the equation to be solved is still rather complicated. We, therefore, proceed to the last simplification and replace the two body force by a unique separable one, that is

$$V_{\mathbf{k}_1\mathbf{k}_2,\mathbf{k}'_1\mathbf{k}'_2} = \lambda e^{-k^2/k_0^2} e^{-k'^2/k_0^2} (2\pi)^3 \delta^{(3)}(\mathbf{K} - \mathbf{K}'), \qquad (5.59)$$

where $\mathbf{k} = (\mathbf{k}_1 - \mathbf{k}_2)/2$, $\mathbf{k}' = (\mathbf{k}'_1 - \mathbf{k}'_2)/2$, $\mathbf{K} = \mathbf{k}_1 + \mathbf{k}_2$, and $\mathbf{K}' = \mathbf{k}'_1 + \mathbf{k}'_2$. This means that we take a spin-isospin averaged two body interaction and disregard that in principle the force may be somewhat different in the $S, T = 0, 1$ or $1, 0$ channels. It is important to remark that for a mean field solution the interaction only can be an effective one, very different from a bare nucleon–nucleon force. This is contrary to the usual gap equation for pairs, to be considered below, where, at least in the nuclear context, a bare force can be used as a reasonable first approximation.

We are now ready to study the solution of Eq. (5.55) for the critical temperature T_c^α, defined by the point where the eigenvalue equals 4μ. For later comparison, the deuteron (pair) wave function at the critical temperature is also deduced from Eqs. (5.55) and (5.59) to be

$$\phi(k) = -\frac{1 - 2f(\varepsilon)}{k^2/m - 2\mu} \lambda e^{-k^2/k_0^2} \int \frac{d^3k'}{(2\pi)^3} e^{-k'^2/k_0^2} \phi(k'), \qquad (5.60)$$

where $\phi(k)$ is the relative wave function of two particles given by $\Psi_{12} \to \phi(|\frac{\mathbf{k}_1-\mathbf{k}_2}{2}|)$ $\delta^{(3)}(\mathbf{k}_1 + \mathbf{k}_2)$, and $\varepsilon = k^2/(2m)$. We also neglected the momentum dependence of the Hartree–Fock mean field shift in Eq. (5.60). It, therefore, can be incorporated into the chemical potential μ. With Eq. (5.60), the critical temperature of pair condensation is obtained from the following equation:

$$1 = -\lambda \int \frac{d^3k}{(2\pi)^3} \frac{1 - 2f(\varepsilon)}{k^2/m - 2\mu} e^{-2k^2/k_0^2}. \qquad (5.61)$$

In order to determine the critical temperature for α-particle condensation, we have to adjust the temperature so that the eigenvalue of (5.55) and (5.57) equals 4μ. The result is shown in Fig. 5.29a. In order to get an idea how this converts into a density dependence, we use for the moment the free gas relation between the density $n^{(0)}$ of uncorrelated nucleons and the chemical potential

$$n^{(0)} = 4 \int \frac{d^3k}{(2\pi)^3} f(\varepsilon). \tag{5.62}$$

We are well aware of the fact that this is a relatively gross simplification, for instance at the lowest densities, and we intend to generalize our theory in the future so that correlations are included into the density. This may be done along the work of Noziéres and Schmitt-Rink [145]. The two open constants λ and k_0 in Eq. (5.59) are determined so that binding energy (-28.3 MeV) and radius (1.71 fm) of the free ($f_i = 0$) α-particle come out right. The adjusted parameter values are: $\lambda = -992$ MeV fm^3, and $k_0 = 1.43$ fm^{-1}. The results of the calculation are shown in Fig. 5.29.

In Fig. 5.29, the maximum of critical temperature $T^\alpha_{c,\text{max}}$ is at $\mu = 5.5$ MeV, and the α-condensation can exist up to $\mu_\text{max} = 11$ MeV. It is very remarkable that the results obtained with (5.58) for T^α_c very well agree with the exact solution of (5.55) and (5.57) using the Malfliet–Tjon interaction (MT I-III) [140, 141] with the Faddeev–Yakubovski method also shown by crosses in Fig. 5.29 (the numerical solution only could be obtained for negative values of μ). This indicates that T^α_c is essentially determined by the Pauli blocking factors.

In Fig. 5.29 we also show the critical temperature for deuteron condensation derived from Eq. (5.61). In this case, the bare force is adjusted with $\lambda = -1305$ MeV fm^3 and $k_0 = 1.46$ fm^{-1} to get experimental energy (-2.2 MeV) and radius (1.95 fm) of the deuteron. It is seen that at higher densities deuteron condensation wins over the one of α-particle. The latter breaks down rather abruptly at a critical positive value of the chemical potential. Roughly speaking, this corresponds to the point where the α-particles start to overlap. This behavior stems from the fact that Fermi–Dirac distributions in the four body case, see Eq. (5.55), can never become step-like, as in the two body case, even not at zero temperature, since the pairs in an α-particle are always in motion. As a consequence, α-condensation generally only exists as a BEC phase and the weak coupling regime is absent, see also discussion in Sect. 5.5.4.

Figure 5.30 shows the normalized self-consistent solution of the wave function in momentum space derived from Eq. (5.55) with the mean field ansatz (5.58) and the wave function in position space defined by its Fourier transform $\tilde{\varphi}(r)$. Figures 5.30(a1) and (b1) represent the wave functions of the free α-particle. The wave function resembles a Gaussian and this shape is approximately maintained as long as μ is negative, see Fig. 5.30(a2). On the contrary, the wave function of Fig. 5.30(a3), where the chemical potential is positive, has a dip around $k = 0$ which is due to the Pauli blocking effect. For the even larger positive chemical potential of Fig. 5.30(a4) the wave function develops a node. The maximum of the wave function shifts to higher momenta and follows the increase of the Fermi momentum k_F, as indicated on Fig. 5.30.

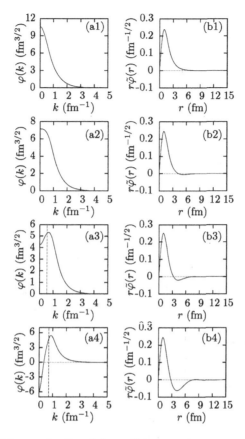

Fig. 5.30 Single particle wave functions (**a1** ∼ **a4**) in momentum space $\varphi(k)$ and (**b1** ∼ **b4**) in position space $r\tilde{\varphi}(r)$ at chemical potential (μ), critical temperature (T_c), and density (n), which are obtained by solving Eq. (5.55) with the mean field ansatz (5.58) [65]: for (**a1**) [(**b1**)] $\mu = -7.08$ MeV, $T_c = 0$ MeV, $n = 0$ fm^{-3}, for (**a2**) [(**b2**)] $\mu = -2.22$ MeV, $T_c = 6.61$ MeV, $n = 9.41 \times 10^{-3}$ fm^{-3}, for (**a3**) [(**b3**)] $\mu = 6.17$ MeV, $T_c = 8.45$ MeV, $n = 3.07 \times 10^{-2}$ fm^{-3}, and for (**a4**) [(**b4**)] $\mu = 10.6$ MeV, $T_c = 5.54$ MeV, $n = 3.34 \times 10^{-2}$ fm^{-3}. Figs. (**a1**) and (**b1**) correspond to the wave functions for free α-particle. The *vertical lines* in Figs. (**a3**) and (**a4**) are at the Fermi wave length $k_F = \sqrt{2m\mu}$

On the other hand, the wave functions in position space in Figs. 5.30(b2), (b3) and (b4) develop an oscillatory behavior, as the chemical potential increases. This is reminiscent to what happens in BCS theory for the pair wave function in position space [146].

An important consequence of this study is that at the lowest temperatures, Bose–Einstein condensation occurs for α particles rather than for deuterons. As the density increases within the low-temperature regime, the chemical potential μ first reaches -7 MeV, where the α's Bose-condense. By contrast, Bose condensation of deuterons would not occur until μ rises to -1.1 MeV.

The *"quartetting"* transition temperature sharply drops as the rising density approaches the critical Mott value at which the four-body bound states disappear. At that point, pair formation in the isospin-singlet deuteron-like channel comes into play, and a deuteron condensate will exist below the critical temperature for BCS pairing up to densities above the nuclear-matter saturation density ρ_0, as described in the previous Section. Of course, also isovector n–n and p–p pairing develops. The critical density at which the α condensate disappears is estimated to be $\rho_0/3$. There-fore, α-particle condensation primarily only exists in the Bose–Einstein-Condensed (BEC) phase and there does not seem to exist a phase where the quartets acquire a large extension as Cooper pairs do in the weak coupling regime. However, the vari-ational approaches of Ref. [63] and of Eq. (5.58) on which this conclusion is based represent only a first attempt at the description of the transition from quartetting to pairing. The detailed nature of this fascinating transition remains to be clari-fied. Many different questions arise in relation to the possible physical occurrence and experimental manifestations of quartetting: Can we observe the hypothetical "α condensate" in nature? What about thermodynamic stability? What happens with quartetting in asymmetric nuclear matter? Are more complex quantum condensates possible? What is their relevance for finite nuclei? As discussed, the special type of microscopic quantum correlations associated with quartetting may be important in nuclei, its role in these finite inhomogeneous systems being similar to that of pairing.

On the other hand, if at all, α-condensation in compact star occurs at strongly asymmetric matter. It is, therefore, important to generalize the above study for sym-metric nuclear matter to the asymmetric case. This can be done straight forwardly again using our momentum projected mean field ansatz (5.58) generalized to the asymmetric case. This implies to introduce two chemical potentials, one for neu-trons and for protons. We also have to distinguish two single particle wave functions in our product ansatz which now reads

$$
\begin{aligned}
\psi_{1234} \to{} & \varphi_p(\mathbf{k}_1)\varphi_p(\mathbf{k}_2)\varphi_n(\mathbf{k}_3)\varphi_n(\mathbf{k}_4)\chi_0 \\
& \times (2\pi)^3\delta(\mathbf{k}_1 + \mathbf{k}_2 + \mathbf{k}_3 + \mathbf{k}_4)
\end{aligned}
\tag{5.63}
$$

where $\varphi_\tau(\mathbf{k}_i) = \varphi_\tau(|\mathbf{k}_i|)$ is the s-wave single particle wave functions for protons ($\tau = p$) and neutrons ($\tau = n$), respectively. χ_0 is the spin-isospin singlet wave function. This now leads to two coupled equations of the Hartree–Fock type for φ_n and φ_p. For the force we use the same as in the symmetric case.

Figure 5.31a shows the critical temperature of α condensation as a function of the total chemical potential $\mu_{\text{total}} = \mu_p + \mu_n$. We see that T_c decreases as the asymmetry, given by the parameter

$$
\delta = \frac{n_n - n_p}{n_n + n_p},
\tag{5.64}
$$

increases. This is in analogy with the deuteron case (also shown) which already had been treated in Refs. [147, 148]. On the other hand, in Fig. 5.31b, it is also interesting to show T_c as a function of the free density which is

Fig. 5.31 Critical
temperature as a function of
the total chemical potential
$\mu_{\text{total}} = \mu_p + \mu_n$ (*top*) and
the total free density n_{total}
(*bottom*) [66]. *Thick* (*thin*)
lines are for α-particle
(deuteron). *Solid, dashed,*
and *dotted lines* are
respectively for
$\delta = 0.0$, $\delta = 0.5$, and
$\delta = 0.9$, where the density
ratio δ is as in Eq. (5.64)

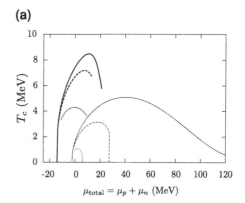

(a)

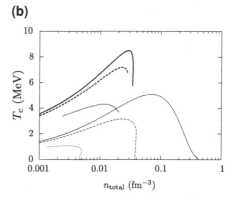

(b)

$$n_{\text{total}}^{(0)} = n_p^{(0)} + n_n^{(0)} \tag{5.65}$$

$$n_p^{(0)} = 2 \int \frac{d^3k}{(2\pi)^3} f_p(k) \tag{5.66}$$

$$n_n^{(0)} = 2 \int \frac{d^3k}{(2\pi)^3} f_n(k), \tag{5.67}$$

where the factor two in front of the integral comes from the spin degeneracy, and
$f_{p,n}(k) = [1 + \exp(\hbar^2 k^2/2m - \mu_{p,n})]^{-1}$. It should be emphasized, however, that
in the above relation between density and chemical potential, the free gas relation is
used and correlations in the density have been neglected. In this sense the dependence
of T_c on density only is indicative, more valid at the higher density side. The very
low density part where the correlations play a more important role shall be treated
in a future publication. It should, however, be stressed that the dependence of T_c on
the chemical potential as in Fig. 5.31a, stays unaltered.

The fact that for more asymmetric matter the transition temperature decreases,
is natural, since as the Fermi levels become more and more unequal, the proton–

neutron correlations will be suppressed. For small δ's, i.e., close to the symmetric case, α condensation (quartetting) breaks down at smaller density (smaller chemical potential) than deuteron condensation (pairing). This effect has already been discussed in our previous work for symmetric nuclear matter [63, 65]. For large δ's, i.e. strong asymmetries, the behavior is opposite, i.e., deuteron condensation breaks down at smaller densities than α condensation, because the small binding energy of the deuteron can not compensate the difference of the chemical potentials.

More precisely, for small δ's, the deuteron with zero center of mass momentum is only weakly influenced by the density or the total chemical potential as can be seen in Fig. 5.31. However, as δ increases, the different chemical potentials for protons and neutrons very much hinders the formation of proton–neutron Cooper pairs in the isoscalar channel for rather obvious reasons. The point to make here is that because of the much stronger binding per particle of the α-particle, the latter is much less influenced by the increasing difference of the chemical potentials. For the strong asymmetry $\delta = 0.9$ in Fig. 5.31 then finally α-particle condensation can exist up to $n_{\text{total}} = 0.02\,\text{fm}^{-3}$ ($\mu_{\text{total}} = 9.3\,\text{MeV}$), while the deuteron condensation exists only up to $n_{\text{total}} = 0.005\,\text{fm}^{-3}$ ($\mu_{\text{total}} = 6.0\,\text{MeV}$).

Overall, the behavior of T_c is more or less as expected. We should, however, remark that the critical temperature for α-particle condensation stays quite high, even for the strongest asymmetry considered here, namely $\delta = 0.9$. This may be of importance for the possibility of α-particle condensation in neutron stars and supernovae explosions [149, 150].

We also show the single particle wave functions of protons and neutrons, entering the quartet wave function (5.63), for various ratios of Fermi surface imbalance and chemical potentials in Fig. 5.32. In most cases of Fig. 5.32, the momentum-space wave functions with negative chemical potentials are monotonically decreasing whereas the ones with positive chemical potentials have a dip at $k = 0$. However, the momentum-space wave functions also develop a dip at $k = 0$ even at a negative chemical potential as the asymmetry takes on stronger values. This can be seen in Fig. 5.32(a3) and (c2). Furthermore, the neutron wave function in k-space with large positive chemical potential develops a node. This behavior is similar to the wave functions in Ref. [65]. As shown in Fig. 5.32, the dissymmetry of proton and neutron wave functions increases as δ increases. As a consequence, the critical temperature decreases, and the α condensation breaks down at a more dilute density, see Fig. 5.31. We also present in Fig. 5.32(a4), (b4) and (c4) the proton and neutron wave functions in real space. In spite of the sometimes strong dissymmetry in momentum space, the proton and neutron wave functions are relatively more similar to one another in r-space. The neutron wave function develops a node as the total chemical potential $\mu_{\text{total}} = \mu_p + \mu_n$ increases, but the negative values of the wave function remain rather moderate.

In conclusion the α-particle (quartet) condensation was investigated in homogeneous symmetric nuclear matter as well as in asymmetric nuclear matter. We found that the critical density at which the α-particle condensate appears is estimated to be around $\rho_0/3$ in the symmetric nuclear matter, and the α-particle condensation can occur only at low density. This result is consistent with the fact that the Hoyle

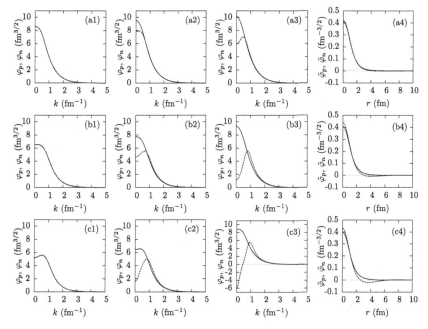

Fig. 5.32 Momentum-space single particle wave functions for proton φ_p (*solid line*), and for neutron φ_n (*dashed line*) for the critical temperature as a function of k for $\delta = 0.0$, 0.5, 0.9, and the real-space wave functions for proton $\tilde{\varphi}_p$ (*solid line*), for neutron $\tilde{\varphi}_n$ (*dashed line*) as a function of r for $\delta = 0.9$ derived from the Fourier transform of $\varphi_{p,n}(k)$ with $\tilde{\varphi}_{p,n}(r) = \int d^3k e^{i\mathbf{k}\cdot\mathbf{r}} \varphi_{p,n}(k)/(2\pi)^3$ [66]. The *top, middle* and *bottom* figures are for $\mu_{\text{total}} = \mu_p + \mu_n \sim -11\,\text{MeV}$, $\sim 0.0\,\text{MeV}$, and $\sim 9.0\,\text{MeV}$, respectively. The wave functions are normalized by $\int d^3k \varphi_{p,n}^2(k)/(2\pi)^3 = 1$. The details of data for respective figures are following: **a1** $\delta = 0.0$, $\mu_{\text{total}} = -11.1\,\text{MeV}$, $\mu_p = -5.53\,\text{MeV}$, $\mu_n = -5.53\,\text{MeV}$, $T_c = 4.52\,\text{MeV}$. **a2** $\delta = 0.5$, $\mu_{\text{total}} = -11.5\,\text{MeV}$, $\mu_p = -8.18\,\text{MeV}$, $\mu_n = -3.35\,\text{MeV}$, $T_c = 4.07\,\text{MeV}$. **a3, a4** $\delta = 0.9$, $\mu_{\text{total}} = -11.0\,\text{MeV}$, $\mu_p = -10.8\,\text{MeV}$, $\mu_n = -0.163\,\text{MeV}$, $T_c = 3.35\,\text{MeV}$. **b1** $\delta = 0.0$, $\mu_{\text{total}} = 0.028\,\text{MeV}$, $\mu_p = -0.014\,\text{MeV}$, $\mu_n = -0.014\,\text{MeV}$, $T_c = 7.46\,\text{MeV}$. **b2** $\delta = 0.5$, $\mu_{\text{total}} = 0.11\,\text{MeV}$, $\mu_p = -4.65\,\text{MeV}$, $\mu_n = 4.76\,\text{MeV}$, $T_c = 6.74\,\text{MeV}$. **b3, b4** $\delta = 0.9$, $\mu_{\text{total}} = -0.02\,\text{MeV}$, $\mu_p = -8.18\,\text{MeV}$, $\mu_n = 8.16\,\text{MeV}$, $T_c = 4.29\,\text{MeV}$. **c1** $\delta = 0.0$, $\mu_{\text{total}} = 8.80\,\text{MeV}$, $\mu_p = 4.40\,\text{MeV}$, $\mu_n = 4.40\,\text{MeV}$, $T_c = 8.44\,\text{MeV}$. **c2** $\delta = 0.5$, $\mu_{\text{total}} = 8.93\,\text{MeV}$, $\mu_p = -1.12\,\text{MeV}$, $\mu_n = 10.0\,\text{MeV}$, $T_c = 7.16\,\text{MeV}$. **c3, c4** $\delta = 0.9$, $\mu_{\text{total}} = 8.94\,\text{MeV}$, $\mu_p = -4.21\,\text{MeV}$, $\mu_n = 13.2\,\text{MeV}$, $T_c = 3.72\,\text{MeV}$

state (0_2^+) of ^{12}C also has a very low density $\rho \sim \rho_0/3$. On the other hand, in the asymmetric nuclear matter, the critical temperature T_c for the α-particle condensation was found to decrease with increasing asymmetry. However, T_c stays relatively high for very strong asymmetries, a fact of importance in the astrophysical context. The asymmetry affects deuteron pairing more strongly than α-particle condensation. Therefore, at high asymmetries, if at all, α-particle condensate seems to dominate over pairing at all possible densities.

5.5.3 Reduction of the α-Condensate with Increasing Density

The properties of α matter can be used to frame the discussion of the structure of $n\alpha$ nuclei. As described in the preceding section, computational studies of these nuclei based on THSR cluster states have demonstrated that an α condensate is established at low nucleon density. More specifically, states lying near the threshold for decomposition into α particles, notably the ground state of ^{8}Be, the Hoyle state (0_2^+) in ^{12}C, and corresponding states in ^{16}O and other $n\alpha$ nuclei are *dilute*, being of low mean density and unusually extended for their mass numbers. We have shown quantitatively within a variational approach that α-like clusters are well formed, with the pair correlation function of α-like clusters predicting relatively large mean distances. For example, in determining the sizes of the ^{12}C nucleus in its 0_1^+ (ground) state and in its 0_2^+ excited state, we obtained the r.m.s. radii of 2.44 and 3.83 fm, respectively. The corresponding mean nucleon densities estimated from $36/4\pi r_{rms}^3$ are close to the nuclear-matter saturation density $\rho_0 = 0.16$ nucleon/fm^3 in the former state and 0.03 nucleon/fm^3 in the latter. The expected low densities of putative alpha-condensate states are confirmed by experimental measurements of form factors [151].

All of our considerations indicate that quartetting is possible in the low-density regime of nucleonic matter, and that α condensates can survive until densities of about 0.03 nucleons/fm^3 are reached. Here, we are in the region where the concept of α matter can reasonably be applied [152–155]. It is then clearly of interest to use this model to gain further insights into the formation of the condensate, and especially the reduction or suppression of the condensate due to repulsive interactions [110]. We will show explicitly that in the model of α matter, as in our studies of finite nuclei, condensate formation is diminished with increasing density. Already within an α-matter model based on a simple $\alpha - \alpha$ interaction, we can demonstrate that the condensate fraction—the fraction of particles in the condensate—is significantly reduced from unity at a density of 0.03 nucleon/fm^3 and essentially disappears approaching nuclear matter-saturation density.

The quantum condensate formed by a homogeneous interacting boson system at zero temperature has been investigated in the classic 1956 paper of Penrose and Onsager [156] who characterize the phenomenon in terms of off-diagonal long-range order of the density matrix. Here we recall some of their results that are most relevant to our problem. Asymptotically, i.e., for $|\mathbf{r} - \mathbf{r}'| \sim \infty$, the nondiagonal density matrix in coordinate representation can be decomposed as

$$\rho(\mathbf{r}, \mathbf{r}') \sim \psi_0^*(\mathbf{r})\psi_0(\mathbf{r}') + \gamma(\mathbf{r} - \mathbf{r}'). \tag{5.68}$$

In the limit, the second contribution on the right vanishes, and the first approaches the condensate fraction, formally defined by

$$\rho_0 = \frac{\langle \Psi | a_0^\dagger a_0 | \Psi \rangle}{\langle \Psi | \Psi \rangle}. \tag{5.69}$$

Penrose and Onsager showed that in the case of a hard-core repulsion, the condensate fraction is determined by a filling factor describing the ratio of the volume occupied by the hard spheres. They applied the theory to liquid ^{4}He, and found that for a hard-sphere model of the atom–atom interaction yielding a filling factor of about 28%, the condensate fraction at zero temperature is reduced from unity (its value for the noninteracting system) to around 8%. (Remarkably, but to some extent fortuitously, this estimate is in rather good agreement with current experimental and theoretical values for the condensate fraction in liquid ^{4}He.)

To make a similar estimate of the condensate fraction for α matter, we follow Ref. [150] and assume an "excluded volume" for α particles of $20\,\mathrm{fm}^3$. At a nucleonic density of $\rho_0/3$, this corresponds to a filling factor of about 28%, the same as for liquid ^{4}He. Thus, a substantial reduction of the condensate fraction from unity (for a noninteracting α-particle gas at zero temperature) is also expected in low-density α matter.

Turning to a more systematic treatment, we proceed in much the same way as Clark and coworkers [152–154], referring especially to the most recent study with M. T. Johnson. Adopting the $\alpha - \alpha$ interaction potential

$$V_{\alpha\alpha}(r) = 475\, e^{-(0.7\,r/\mathrm{fm})^2}\,\mathrm{MeV} - 130\, e^{-(0.475r/\mathrm{fm})^2}\,\mathrm{MeV} \qquad (5.70)$$

introduced by Ali and Bodmer [118], we calculate the reduction of the condensate fraction as function of density within what is now a rather standard variational approach. Alpha matter is described as an extended, uniform Bose system of interacting α particles, *disregarding* any change of the internal structure of the α clusters with increasing density. In particular, the dissolution of bound states associated with Pauli blocking (Mott effect) is not taken into account in the present description.

The simplest form of trial wave function incorporating the strong spatial correlations implied by the interaction potential (5.70) is the familiar Jastrow choice,

$$\Psi(\mathbf{r}_1, \ldots, \mathbf{r}_A) = \prod_{i<j} f(|\mathbf{r}_i - \mathbf{r}_j|). \qquad (5.71)$$

The normalization condition

$$4\pi\rho_\alpha \int_0^\infty \left[f^2(r) - 1 \right] r^2 dr = -1, \qquad (5.72)$$

in which ρ_α is the number density of α-particles, is imposed as a constraint on the variational wave function, in order to promote the convergence of the cluster expansion used to calculate the energy expectation value [157]. In the low-density limit, the energy functional [binding energy per α cluster as a functional of the correlation factor $f(r)$] is given by

$$E[f] = 2\pi\rho_\alpha \int_0^\infty \left\{ \frac{\hbar^2}{m_\alpha} \left(\frac{\partial f(r)}{\partial r} \right)^2 + f^2(r) V_{\alpha\alpha}(r) \right\} r^2 dr, \qquad (5.73)$$

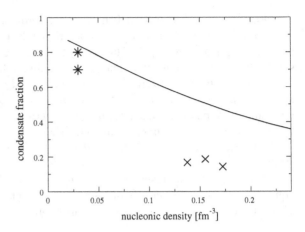

Fig. 5.33 Reduction of condensate fraction in α matter with increasing nucleon density. Exploratory calculations (*full line*) are compared with HNC calculations of Johnson and Clark [152–154] (*crosses*). For comparison, we show estimates of the condensate fraction in the 0_2^+ (Hoyle) state of ^{12}C, according to Refs. [25, 26] (*stars*)

where m_α is the α-particle mass, while the condensate fraction is given by

$$\rho_0 = \exp\left\{-4\pi\rho_\alpha \int_0^\infty [f(r) - 1]^2 r^2 dr\right\}. \tag{5.74}$$

The variational two-body correlation factor f was taken as one of the forms employed by Clark and coworkers [152–154], namely

$$f(r) = (1 - e^{-ar})(1 + be^{-ar} + ce^{-2ar}). \tag{5.75}$$

At given density ρ, the expression for the energy expectation value is minimized with respect to the parameters a, b, and c, subject to the constraint (5.72). It is important to note that these approximations, based on truncated cluster expansions, are reliable only at densities low enough so that the length scale associated with decay of $f^2(1) - 1$ is sufficiently small compared to the average particle separation, which is inversely proportional to the cubic root of the density [152–155, 157, 158].

To give an example, for the nucleon density $4\rho_\alpha = 0.06\,\text{fm}^{-3}$, a minimum of the energy expectation value (73) was found at $a = 0.616\,\text{fm}^{-1}$, $b = 1.221$, and $c = -5.306$, with a corresponding energy per α cluster of -9.763 MeV and a condensate fraction of 0.750. The dependence of the condensate fraction on the nucleon density $\rho = 4\rho_\alpha$ as determined in this exploratory calculation is displayed in Fig. 5.33.

The reduction of the condensate fraction of α matter to roughly 0.8 as given by our calculation at nucleonic density $0.03\,\text{fm}^{-3}$ agrees well with results of Suzuki [25] and Yamada [26] for ^{12}C in the Hoyle 0_2^+ state. Using many-particle approaches to the ground-state wave function and to the THSR (0_2^+) state of ^{12}C, the occupation of the inferred natural α orbitals is found to be quite different in the two cases. Roughly 1/3 shares (approaching equipartition) are found for the S, D, and G orbits in the ground (0_1^+) state, with α-cluster occupations of 1.07, 1.07, and 0.82, respectively (see Sect. 5.4.1.3). On the other hand, in the Hoyle (0_2^+) state, one sees enhanced

Fig. 5.34 Occupation of the S orbital as a function of density using the 3α OCM for ^{12}C [26]

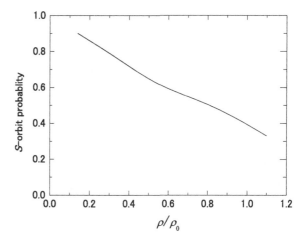

occupation (2.38) of the S orbit and reduced occupation (0.29, 0.16, respectively) of the D and G orbits (see also Sect. 5.4.1.3). This corresponds to an enhancement of about 70% compared with equipartition.

To get a more extended analysis, OCM calculations have been performed [26] for studying the density dependence of the S-orbit occupancy in the Hoyle state on the different densities $\rho/\rho_0 \sim (R(0_1^+)_{exp}/R)^3$, in which the rms radius (R) of ^{12}C is taken as a parameter and $R(0_1^+)_{exp} = 2.56$ fm. A Pauli-principle respecting OCM basis $\Psi_{0^+}^{OCM}(\nu)$ with a size parameter ν is used, in which the value of ν is chosen to reproduce a given rms radius R of ^{12}C, and the α density matrix $\rho(\mathbf{r}, \mathbf{r}')$ with respect to $\Psi_{0^+}^{OCM}(\nu)$ is diagonalized to obtain the S-orbit occupancy in the 0^+ wave function. The results are shown in Fig. 5.34. The S-orbit occupancy is $70 \sim 80\%$ around $\rho/\rho_0 \sim (R(0_1^+)_{exp}/R(0_2^+)_{THSR})^3 = 0.21$, while it decreases with increasing ρ/ρ_0 and amounts to about $30 \sim 40\%$ in the saturation density region. Figure 5.35 shows the radial behaviors of the S-orbit with given densities. A smooth transition of the S-orbit is observed, with decreasing ρ/ρ_0, from a two-node S-wave nature ($\rho/\rho_0 \sim 1.18$) in Fig. 5.35a to the zero-node S-wave one ($\rho/\rho_0 \simeq 0.15$) in Fig. 5.35d [26]. The feature of the decrease of the enhanced occupation of the S orbit is in striking correspondence with the density dependence of the condensate fraction calculated for nuclear matter (see Fig. 5.33).

A more accurate and reliable variational description of α matter can be realized within the hypernetted-chain (HNC) approach to evaluate correlation integrals; this approach [152–154, 157] largely overcomes the limitations of the cluster-expansion treatment, including the need for an explicit normalization constraint. Such an improved approach is certainly required near the saturation density of nuclear matter, where it predicts only a small condensate fraction [152–154]. Of course, at high densities the simple Ali-Bodmer interaction [118] ceases to be valid, and it becomes crucial to include the effects of Pauli blocking. Once again, this conclusion reinforces

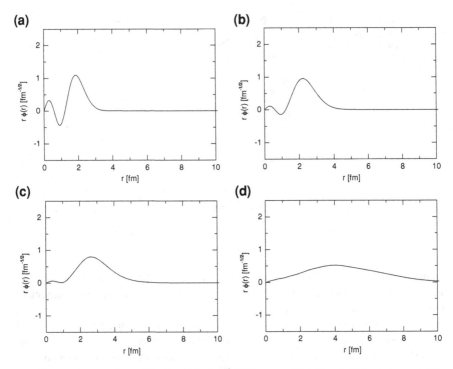

Fig. 5.35 Radial behaviors of the S orbit in the $^{12}C(0^+)$ state with **a** $R = 2.42$ fm ($\rho/\rho_0 \sim 1.18$), **b** $R = 2.70$ fm ($\rho/\rho_0 \sim 0.85$), **c** $R = 3.11$ fm ($\rho/\rho_0 \sim 0.56$), and **d** $R = 4.84$ fm ($\rho/\rho_0 \sim 0.15$), where R denotes the nuclear radius of the $^{12}C(0^+)$ state [26]

the point of view that we can expect signatures of an α condensate only for dilute nuclei near the threshold of $n\alpha$ decay.

5.5.4 'Gap' Equation for Quartet Order Parameter

For macroscopic α condensation it is, of course, not conceivable to work with a number projected α particle condensate wave function as we did when in finite nuclei only a couple of α particles were present. We rather have to develop an analogous procedure to BCS theory but generalized for quartets. In principle a wave function of the type $|\alpha\rangle = \exp[\sum_{1234} z_{1234} c_1^+ c_2^+ c_3^+ c_4^+]|\text{vac}\rangle$ would be the ideal generalization of the BCS wave function for the case of quartets. However, unfortunately, it is unknown so far (see, however, Ref. [159]) how to treat such a complicated many body wave function mathematically in a reasonable way. So, we rather attack the problem from the other end, that is with a Gorkov type of approach, well known from pairing but here extended to the quartet case. Since, naturally, the formalism is complicated, we only will outline the main ideas and refer for details to the literature.

Fig. 5.36 Graphic representation of the BCS mass operator in Eq. (5.78)

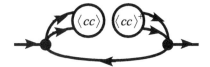

Actually one part of the problem is written down easily. Let us guide from a particular form of the gap equation in the case of pairing. We have at zero temperature [131, 132]

$$(\varepsilon_1 + \varepsilon_2)\kappa_{12} + (1 - n_1 - n_2)\frac{1}{2}\sum_{1'2'}\bar{V}_{121'2'}\kappa_{1'2'} = 2\mu\kappa_{12}, \qquad (5.76)$$

where $\kappa_{12} = \langle c_1 c_2 \rangle$ is the pairing tensor, $n_i = \langle c_i^+ c_i \rangle$ are the BCS occupation numbers, and $\bar{V}_{121'2'}$ denotes the antisymmetrized matrix element of the two-body interaction. The ε_i are the usual mean field energies. Equation (5.76) is equivalent to the usual gap equation in the case of zero total momentum and opposite spin, i.e. in short hand: $2 = \bar{1}$ where the bar stands for 'time reversed conjugate'. The extension of (5.76) to the quartet case is formally written down without problem

$$(\varepsilon_{1234} - 4\mu)\kappa_{1234} = (1 - n_1 - n_2)\frac{1}{2}\sum_{1'2'}\bar{V}_{121'2'}\kappa_{1'2'34}$$

$$+ (1 - n_1 - n_3)\frac{1}{2}\sum_{1'3'}\bar{V}_{131'3'}\kappa_{1'23'4} + \text{all permutations.} \qquad (5.77)$$

with $\kappa_{1234} = \langle c_1 c_2 c_3 c_4 \rangle$ the quartet order parameter. This is formally the same equation as in Eq. (5.55) with, however, the Fermi–Dirac occupation numbers replaced by the zero temperature quartet correlated single particle occupation numbers, similar to the BCS case. For the quartet case, the crux lies in the determination of those occupation numbers. Let us again be guided by BCS theory or rather by the equivalent Gorkov approach [160]. In the latter, there are two coupled equations, one for the normal single particle Green's function (GF) and the other for the anomalous GF. Eliminating the one for the anomalous GF in inserting it into the first equation leads to a Dyson equation with a single particle mass operator,

$$M_{1;1'}^{\mathrm{BCS}}(\omega) = \sum_2 \frac{\Delta_{12}\Delta_{1'2}^*}{\omega + \varepsilon_2} \quad \text{with} \quad \Delta_{12} = -\frac{1}{2}\sum_{34}\bar{V}_{12,34}\langle c_4 c_3 \rangle. \qquad (5.78)$$

This can be graphically represented in Fig. 5.36, where $\langle cc \rangle$ stands for the order parameter κ_{12} and the dot for the two body interaction.

The generalization to the quartet case is considerably more complicated but schematically the corresponding mass operator in the single particle Dyson equation can be represented graphically as in Fig. 5.37, with the quartet order parameter $\langle cccc \rangle$. Put aside the difficulty to derive a manageable expression for this 'quartet'

Fig. 5.37 Graphical
representation of the
approximate α-BEC mass
operator M^{quartet} of
Eq. (5.80)

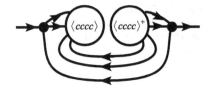

single-particle mass operator, what immediately strikes is that instead of only one
'backward going line' with $(-\mathbf{p}, -\sigma)$ as in the pairing case, we now have three back-
wards going lines. As a consequence, the three momenta $\mathbf{k}_1$, $\mathbf{k}_2$, $\mathbf{k}_3$ in these lines
are only constrained so that their sum be equal to $\mathbf{k}_1 + \mathbf{k}_2 + \mathbf{k}_3 = -\mathbf{p}$ and, thus, the
remaining freedom has to be summed over. This is in strong contrast to the pairing
case where the single backward going line is constrained by momentum conservation
to $-\mathbf{p}$. So, no internal summation occurs in the mass operator belonging to pairing.
The consequence of this additional momentum summation in the mass operator for
quartetting leads with respect to pairing to a completely different analytic structure
of the mass operator in case of quartetting. This is best studied with the so-called
three hole level density $g_{3h}(\omega)$ which is related to the imaginary part of the three
hole Green's function $G^{3h}(k_1, k_2, k_3; \omega) = (\bar{f}_1 \bar{f}_2 \bar{f}_3 + f_1 f_2 f_3)/(\omega + \varepsilon_{123})$ with
$\varepsilon_{123} = \varepsilon_1 + \varepsilon_2 + \varepsilon_3$ and $\bar{f} = 1 - f$ by

$$g_{3h}(\omega) = -\int \frac{d^3k_1}{(2\pi)^3} \frac{d^3k_2}{(2\pi)^3} \frac{d^3k_3}{(2\pi)^3} \text{Im} G^{(3h)}(k_1, k_2, k_3; \omega + i\eta)$$

$$= \int \frac{d^3k_1}{(2\pi)^3} \frac{d^3k_2}{(2\pi)^3} \frac{d^3k_3}{(2\pi)^3}$$
$$\times (\bar{f}_1 \bar{f}_2 \bar{f}_3 + f_1 f_2 f_3)\pi \delta(\omega + \varepsilon_1 + \varepsilon_2 + \varepsilon_3).) \tag{5.79}$$

In Fig. 5.38 we show the level density at zero temperature $(f(\omega) = \theta(-\omega))$,
where it is calculated with the proton mass $m = 938.27$ MeV (natural units) [67].
Two cases have to be considered, chemical potential μ positive or negative. In the
latter case we have binding of the quartet. Let us first discuss the case $\mu > 0$.
We remark that in this case, the $3h$ level density goes through zero at $\omega = 0$, i.e., since
we are measuring energies with respect to the chemical potential μ, just in the region
where the quartet correlations should appear. This is a strong difference with the
pairing case where the $1h$ level density, $g_{1h}(\omega) = \int d^3k/(2\pi\hbar)^3(\bar{f}_k + f_k)\delta(\omega + \varepsilon_k) =$
$\int d^3k/(2\pi\hbar)^3\delta(\omega + \varepsilon_k)$, does not feel any influence from the medium and, therefore,
the corresponding level density varies (neglecting the mean field for the sake of the
argument) like in free space with the square root of energy. In particular, this means
that the level density is *finite* at the Fermi level. This is a dramatic difference with
the quartet case and explains why Cooper pairs can strongly overlap whereas for
quartets this is impossible as we will see below. We also would like to point out that
the $3h$ level density is just the mirror to the $3p$ level density which has been discussed
in Ref. [161].

Fig. 5.38 $3h$ level densities defined in Eq. (5.79) for various values of the chemical potential μ at zero temperature [67]

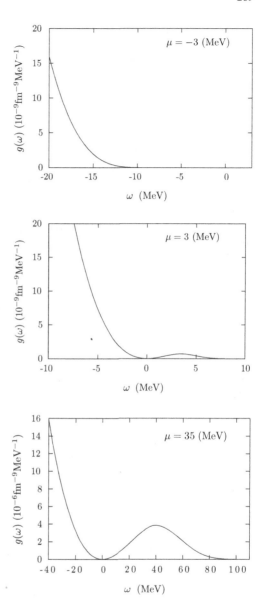

For the case where $\mu < 0$ there is nothing very special, besides the fact that it only is non-vanishing for negative values of ω and that the upper boundary is given by $\omega = 3\mu$. Therefore, the level density of Eq. (5.79) is zero for $\omega > 3\mu$.

With these preliminary but crucial considerations we now pass to the evaluation of the single-particle mass operator with quartet condensation. Its expression can be shown to be of the following form

$$M_{1;1}^{\text{quartet}}(\omega) = \sum_{234} \frac{\tilde{\Delta}_{1234}(\bar{f}_2\bar{f}_3\bar{f}_4 + f_2f_3f_4)\tilde{\Delta}_{1234}^*}{\omega + \varepsilon_{234}} \tag{5.80}$$

with

$$\tilde{\Delta}_{1234} = \frac{1}{2}\bar{V}_{12,1'2'}\delta_{33'}\delta_{44'}\langle c_{1'}c_{2'}c_{3'}c_{4'}\rangle. \tag{5.81}$$

Again, comparing the quartet single-particle mass operator (5.80) with the pairing one (5.78), we notice the presence of the phase space factors in the former case while in Eq. (5.78) they are absent. As already indicated above, this fact implies in the quartet case that only the Bose–Einstein condensation phase is born out whereas a 'BCS phase' (long coherence length) is absent. The complexity of the calculation in Eq. (5.80) is much reduced using for the order parameter $\langle cccc\rangle$ our mean field ansatz projected on zero total momentum, as it was already very successfully employed with Eq. (5.58),

$$\langle c_1c_2c_3c_4\rangle \rightarrow \phi_{\mathbf{k}_1\mathbf{k}_2,\mathbf{k}_3\mathbf{k}_4}\chi_0,$$
$$\phi_{\mathbf{k}_1\mathbf{k}_2,\mathbf{k}_3\mathbf{k}_4} = \varphi(|\mathbf{k}_1|)\varphi(|\mathbf{k}_2|)\varphi(|\mathbf{k}_3|)\varphi(|\mathbf{k}_4|)$$
$$\times (2\pi)^3\delta(\mathbf{k}_1 + \mathbf{k}_2 + \mathbf{k}_3 + \mathbf{k}_4), \tag{5.82}$$

where χ_0 is the spin-isospin singlet wave function. It should be pointed out that this product ansatz with four identical $0S$ single particle wave functions is typical for a ground state configuration of the α particle. Excited configurations with wave functions of higher nodal structures may eventually be envisaged for other physical situations. We also would like to mention that the momentum conserving δ function induces strong correlations among the four particles and (5.82) is, therefore, a rather non trivial variational wave function.

For the two-body interaction of $\bar{V}_{12,1'2'}$ in Eq. (5.81), we employ the same separable form (5.59) as done already for the quartet critical temperature.

At first let us mention that in this pilot application of our selfconsistent quartet theory, we only will consider the zero temperature case. As a definite physical example, we will treat the case of nuclear physics with the particularly strongly bound quartet, the α particle. It should be pointed out, however, that if scaled appropriately all energies and lengths can be transformed to other physical systems. For the nuclear case it is convenient to measure energies in Fermi energies $\varepsilon_F = 35$ MeV and lengths in inverse Fermi momentum $k_F^{-1} = 1.35^{-1}$ fm.

The single particle wave functions and occupation numbers obtained from the above cycle are shown in Fig. 5.39. We also insert the Gaussian wave function with same r.m.s. momentum as the single particle wave function in the left figures in Fig. 5.39. As shown in Fig. 5.39, the single particle wave function is sharper than a Gaussian.

We could not obtain a convergent solution for $\mu > 0.55$ MeV. This difficulty is of the same origin as in the case of our calculation of the critical temperature for α particle condensation. In the r.h.s. panels of Fig. 5.39 we also show the corresponding

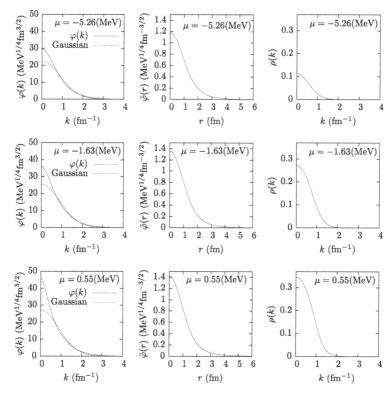

Fig. 5.39 Single particle wave function $\varphi(k)$ in k-space (*left*), for r-space $\tilde{\varphi}(r)$ (*middle*), and occupation numbers (*right*) at $\mu = -5.26$ (*top*), -1.63 (*middle*), and 0.55 (*bottom*), with zero temperature. The r-space wave function $\tilde{\varphi}(r)$ is derived from the Fourier transform of $\varphi(k)$ by $\tilde{\varphi}(r) = \int d^3k e^{i\mathbf{k}\cdot\mathbf{r}} \varphi(k)/(2\pi)^3$. The *dashed line* in the *left* panels correspond to the Gaussian with same norm and r.m.s. momentum as $\varphi(k)$ [67]

occupation numbers. We see that they are very small. However, they increase for increasing values of the chemical potential. For $\mu = 0.55$ MeV the maximum of the occupation still only attains 0.35 what is far away from the saturation value of one. What really happens for larger values of the chemical potential, is unclear. Surely, as discussed in Sect. 5.5.2 the situation for the quartet case is completely different from the standard pairing case. This is due to the fact, as already mentioned, that the 3h level density goes through zero at $\omega = 0$, i.e. just at the place where the quartet correlation should build up for positive values of μ. Due to this fact, the inhibition to go into the positive μ regime is here even stronger than in the case of the critical temperature [65].

The situation in the quartet case is also in so far much different, as the 3h Green's function produces a considerable imaginary part of the mass operator. Figure 5.40 shows the imaginary part of the approximate quartet mass operator of Eq. (5.80) for $\mu < 0$ and $\mu > 0$. These large values of the damping rate imply a strong violation

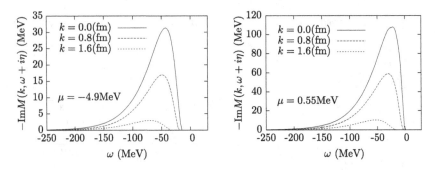

Fig. 5.40 $-\mathrm{Im}M^{\mathrm{quartet}}(k_1, \omega + i\eta)$ in Eq. (5.80) as a function of ω for $\mu = -4.9\,\mathrm{MeV}$ (*left*) and for $\mu = 0.55\,\mathrm{MeV}$ (*right*) at zero temperature [67]

of the quasiparticle picture. In Fig. 5.41 we show the spectral function of the single particle GF. Contrary to the pairing case with its sharp quasiparticle pole, we here only find a very broad distribution, implying that the quasiparticle picture is completely destroyed. How to formulate a theory which goes continuously from the quartet case into the pairing case, is an open question. One solution could be to start right from the beginning with an in medium four body equation which contains a superfluid phase. When the quartet phase disappears, the superfluid phase may remain. Such investigations shall be done in the future.

5.6 Summary and Conclusions

We discussed α condensation in nuclear systems. One remarkable manifestation is the Hoyle state (0_2^+) in $^{12}\mathrm{C}$ at 7.65 MeV with a gas-like structure of three α-particles, trapped by a shallow self-consistent mean field of wide extension, in which the c.m. motion of the α particles occurs dominantly in the lowest $0S$ orbit. We found that a simple wave function of the α-condensate type, called the THSR wave function, describes very nicely the structure of the Hoyle state and reproduces the inelastic form factor of $^{12}\mathrm{C}(e, e')$ and others quantities. The condensate feature of the Hoyle state was confirmed by the calculation of the bosonic occupation numbers in diagonalizing the bosonic density matrix. It was shown that the occupation of the $0S$ state of the α-particles is over 70% for the Hoyle state, and the remaining component (30%) comes from residual correlations, mostly of the Pauli type, among the α particles. In spite of the very different number of particles and other important differences, the situation has some analogy with the case of cold bosonic atoms.

We conjectured that the α-particle condensates also exist in heavier self-conjugate nuclei. Theoretical calculations of the OCM type indicate that the 0_6^+ state at 15.1 MeV in $^{16}\mathrm{O}$ is a strong candidate. So far we do not dispose of sufficient experimental data to confirm its nature. Experiments are under way and being analyzed. This analogue of the Hoyle state in $^{16}\mathrm{O}$ has many similarities with the original one

Fig. 5.41 Spectral function of the single particle GF, $-2\text{Im}G(k, \omega + i\eta)$, as function of ω for $\mu = -4.9\,\text{MeV}$ (*top*) and for $\mu = 0.55\,\text{MeV}$ (*bottom*) at a zero temperature [67]

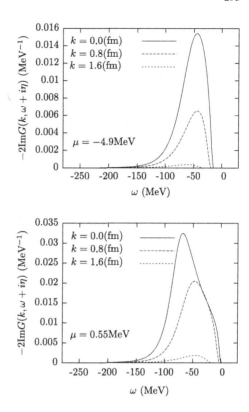

in ^{12}C: it lies a couple of hundred keV above the 4α disintegration threshold. It is quite strongly excited by (e, e'). Its width is 160 keV. This is much larger than for the Hoyle state in ^{12}C but with respect to its energy it is still unusually small. The large width stems from a position higher up in Coulomb barrier and also the Coulomb barrier itself has become slightly lower. The situation in ^{16}O with respect to alpha clustering is considerably more complicated than in ^{12}C. Results from the 4α OCM calculations showed that 2nd up to 5-th 0^+ states in ^{16}O have $\alpha + ^{12}$C structures. Only the 6-th 0^+ state is the analogue to Hoyle state. We also discussed the results of the THSR wave function for ^{16}O.

As for the heavier α-particle condensates, we found first that they are predicted to be slightly above their $n\alpha$ threshold in the $A = 4n$ nuclei but below the Coulomb barrier, and second the phenomenon will terminate at about eight to ten α's as the confining Coulomb barrier fades away. However, the concept of α condensation in nuclei can be generalized to non self-conjugated nuclei ($A \neq 4n$). Since the nuclear α-particle condensation is described dominantly as a product state of α particles occupying the lowest $0S$ orbit, the counterpart in $A \neq 4n$ nuclei should still be presented as a product state of the constituent clusters in the $0S$ state. For instance, we can conjecture product states composed of α's, a few neutrons and/or s-wave clusters $(d, t, ^3\text{He})$ such as $(0S)^2_\alpha (0S)_t$ in ^{11}B and $(0S)^3_\alpha (0S)_n$ in ^{13}C etc. Indeed, our OCM calculations indicate that they appear slightly above their three- and four-cluster disintegrated thresholds, $2\alpha + t$ for ^{11}B and $3\alpha + n$ for ^{13}C, as positive-parity states with $J^\pi = 1/2^+$. These results encourage us to conjecture that cluster-gas-like states described by antisymmetrized product wave functions of constituent clusters, all in the $0S$ level, can exist in general in excited states of low density in light nuclei.

We dwell on the fact that concepts developed for infinite nuclear matter are of value also to interpret properties in finite nuclei and to construct useful approximations. As examples, we refer to pairing, two and more body correlations, and one body occupation numbers. Pairing definitely also is a useful concept for many finite nuclei, in spite of the fact that nuclei are by far not macroscopic objects. For example, the strong reduction of measured moments of inertia of such nuclei compared with the classical values are explained as a consequence of superfluidity [24, 162, 163]. In this sense, we discussed nuclear α-particle condensation as the analogue to the number-projected BCS wave function, replacing Cooper pairs by α particles. A real macroscopic phase of condensed α's may be formed during the cooling process of compact stars [164, 165], where one predicts the presence of α-particle condensates [150]. On the other hand, a possibility of quartetting with cold atoms in which fermions are trapped in four different magnetic substates also have been discussed [142, 144, 166, 167]. Theoretical and experimental works in this direction will also be useful and helpful to investigate the low-density bosonic α-particle gas states in nuclei.

In conclusion, the idea of α-particle condensation in nuclei is novel. A completely new nuclear phase in which α particles move like in a gas as quasi-elementary constituents is surely intriguing. In order to bring deeper insights into the role of clustering and quantum condensates in the systems of strongly interacting fermions, it is hoped that more α-particle states in nuclei and/or many α's around a nuclear core, including cluster-gas-like states composed of α's, t's and n's etc., will be observed in the near future.

Acknowledgments The authors would like to thanks to B. Borderie, M. Freer, Y. Hatanaka, K. Ikeda, M. Itoh, T. Kawabata, K. Katō, W. von Oertzen, M. F. Rivet, T. Sogo, and T. Wakasa for useful discussions and comments.

References

1. Wildermuth, K., Tang, Y.C.: A Unified Theory of the Nucleus. Vieweg, Braunschweig (1977)
2. Ikeda, K., Horiuchi, H., Saito, S., Fujiwara, Y., Kamimura, M., Katō, K.K., Suzuki, Y., Uegaki, E., Furutani, H., Kanada, H., Kaneko, T., Nagata, S., Nishioka, H., Okabe, S., Sakuda, T., Seya, M., Abe, Y., Kondō, Y., Matsuse, T., Tohsaki-Suzuki, A.: Prog. Theor. Phys. Suppl. **68** (1980)
3. von Oertzen, W., Freer, M., Kanada-Enyo, Y.: Phys. Rep. **432**, 43 (2006)
4. Hiyama, E., Yamada, T.: Prog. Part. Nucl. Phys. **63**, 339 (2009)
5. Tohsaki, A., Horiuchi, H., Schuck, P., Röpke, G.: Phys. Rev. Lett. **87**, 192501 (2001)
6. Hiura, J., Tamagaki, R.: Prog. Theor. Phys. Supple. **52**, 25 (1972)
7. Wiringa, R.B., Pieper, S.C., Carlson, J., Pandharipande, V.R.: Phys. Rev. C **62**, 014001 (2000)

8. Uegaki, E., Okabe, S., Abe, Y., Tanaka, H.: Prog. Theor. Phys. **57**, 1262 (1977)
9. Uegaki, E., Abe, Y., Okabe, S., Tanaka, H.: Prog. Theor. Phys. **59**, 1031 (1978)
10. Uegaki, E., Abe, Y., Okabe, S., Tanaka, H.: Prog. Theor. Phys. **62**, 1621 (1979)
11. Fukushima, Y., Kamimura, M.: In: Tokyo T. Marumori (ed.) Proceedings of International Conference on Nuclear Structure (1977)
12. Fukushima, Y., Kamimura, M.: J. Phys. Soc. Jpn. Suppl. **44**, 225 (1978)
13. Kamimura, M.: Nucl. Phys. A **351**, 456 (1981)
14. Funaki, Y., Tohsaki, A., Horiuchi, H., Schuck, P., Röpke, G.: Phys. Rev. C **67**, 051306(R) (2003)
15. Chernykh, M., Feldmeier, H., Neff, T., von Neumann-Cosel, P., Richter, A.: Phys. Rev. Lett. **98**, 032501 (2007)
16. Neff, T.: Talk at YIPQS International Molecule Workshop on Alpha- and Dineutron-Correlation in Nuclear Many-Body Systems Program, 6–24 Oct, Kyoto (2008)
17. Hoyle, F.: Astrophys. J. Suppl. **1**, 121 (1954)
18. Cook, C.W., Fowler, W.A., Lauritsen, C.C., Lauritsen, T.B.: Phys. Rev. **107**, 508 (1957)
19. Ajzenberg-Selove, F.: Nucl. Phys. A **506**, 1 (1990)
20. Schuck, P., Funaki, Y., Horiuchi, H., Röpke, G., Thosaki, A., Yamada, T.: Prog. Part. Nucl. Phys. **59**, 285 (2007)
21. Funaki, Y., Horiuchi, H., Röpke, G., Schuck, P., Tohsaki, A., Yamada, T.: Nucl. Phys. News **17**(04), 11 (2007)
22. Funaki, Y., Horiuchi, H., von Oertzen, W., Ropke, G., Schuck, P., Tohsaki, A., Yamada, T.: Phys. Rev. C **80**, 064326 (2009)
23. Funaki, Y., Yamada, T., Horiuchi, H., Röpke, G., Schuck, P., Tohsaki, A.: In: Brenner, M. (ed.) Cluster Structure of Atomic Nuclei (Research Signpost/Transworld Research Network, ISBN: 978-81-308-0403-3), vol. 1 (2010)
24. Ring, P., Schuck, P.: The Nuclear Many-Body Problem. Springer, Berlin (1980)
25. Matsumura, H., Suzuki, Y.: Nucl. Phys. A **739**, 238 (2004)
26. Yamada, T., Schuck, P.: Eur. Phys. J. A. **26**, 185 (2005)
27. Funaki, Y., Yamada, T., Tohsaki, A., Horiuchi, H., Röpke, G., Schuck, P.: Phys. Rev. C **82**, 024312 (2010)
28. Funaki, Y., Tohsaki, A., Horiuchi, H., Schuck, P., Röpke, G.: Eur. Phys. J. A. **24**, 321 (2005)
29. Funaki, Y., Horiuchi, H., Tohsaki, A.: Prog. Theor. Phys. **115**, 115 (2006)
30. Kurokawa, C., Katō, K.: Phys. Rev. C **71**, 021301 (2005)
31. Kurokawa, C., Katō, K.: Nucl. Phys. A **792**, 87 (2007)
32. Navrátil, P., Vary, J.P., Barrett, B.R.: Phys. Rev. Lett. **84**, 5728 (2000)
33. Navrátil, P., Vary, J.P., Barrett, B.R.: Phys. Rev. C **62**, 054311 (2000)
34. Barrett, B.R., Mihaila, B., Pieper, S.C., Wiringa, R.B.: Nucl. Phys. News **13**, 17 (2003)
35. Navrátil, P., Quaglioni, S., Stetcu, I., Barrett, B.R.: J. Phys. G **36**, 083101 (2009)
36. Itoh, M. et al.: Nucl. Phys. A **738**, 268 (2004)
37. Freer, M. et al.: Phys. Rev. C **71**, 047305 (2005)
38. Freer, M. et al.: Phys. Rev. C **76**, 034320 (2007)
39. Freer, M. et al.: Phys. Rev. C **80**, 041303 (R) (2009)
40. Kokalova, T.z. et al.: Eur. Phys. J. A **23**, 19 (2005)
41. Kokalova, T.z. et al.: Phys. Rev. Lett. **96**, 192502 (2006)
42. Ohkubo, S., Hirabayashi, Y.: Phys. Rev. C **70**, 041602 (R) (2004)
43. Ohkubo, S., Hirabayashi, Y.: Phys. Rev. C **75**, 044609 (2007)
44. Ohkubo, S., Hirabayashi, Y.: Phys. Lett. B **684**, 127 (2010)
45. Takashina, M., Sakuragi, Y.: Phys. Rev. C **74**, 054606 (2006)
46. Takashina, M.: Phys. Rev. C **78**, 014602 (2008)
47. Kanada-Enyo, Y.: Prog. Theor. Phys. **117**, 655 (2007)
48. Wakasa, T. et al.: Phys. Lett. B **653**, 173 (2007)
49. Ad Raduta, R. et al.: Talk at 2nd Workshop on State of the Art in Nuclear Cluster Physics, Universite Libre de Bruxelles, pp. 25–28 (2010) May

50. Khoa, D.T., Cuonga, D.C., Kanada-En'yo, Y.: Phys. Lett. B **695**, 469 (2011)
51. Funaki, Y., Yamada, T., Horiuchi, H., Röpke, G., Schuck, P., Tohsaki, A.: Phys. Rev. Lett. **101**, 082502 (2008)
52. Saito, S.: Prog. Theor. Phys. **40**, 893 (1968)
53. Saito, S.: Prog. Theor. Phys. **41**, 705 (1969)
54. Saito, S.: Prog. Theor. Phys. Supple. **62**, 11 (1977)
55. Tohsaki, A., Horiuchi, H., Schuck, P., Röpke, G.: Nucl. Phys. A **738**, 259 (2004)
56. Yamada, T., Schuck, P.: Phys. Rev. C **69**, 024309 (2004)
57. Kawabata, T. et al.: Phys. Lett. B **646**, 6 (2007)
58. Kanada-En'yo, Y.: Phys. Rev. C **75**, 024302 (2007)
59. Yamada, T., Funaki, Y.: Phys. Rev. C **82**, 064315 (2010)
60. Kawabata, T. et al.: Int. J. Mod. Phys. E **17**, 2071 (2008)
61. Yamada, T., Funaki, Y.: Int. J. Mod. Phys. E **17**, 2101 (2008)
62. Yoshida, T., Itagaki, N., Otsuka, T.: Phys. Rev. C **79**, 034308 (2009)
63. Röpke, G., Schnell, A., Schuck, P., Nozières, P.: Phys. Rev. Lett. **80**, 3177 (1998)
64. Beyer, M., Sofianos, S.A., Kurths, C., Röpke, G., Schuck, P.: Phys. Lett. B **488**, 247 (2000)
65. Sogo, T., Lazauskas, R., Röpke, G., Schuck, P.: Phys. Rev. C **79**, 051301(R) (2009)
66. Sogo, T., Röpke, G., Schuck, P.: Phys. Rev. C **82**, 034322 (2010)
67. Sogo, T., Röpke, G., Schuck, P.: Phys. Rev. C **81**, 064310 (2010)
68. Doucot, B., Vidal, J.: Phys. Rev. Lett. **88**, 227005 (2002)
69. Capponi, S., Roux, G., Lecheminant, P., Azaria, P., Boulat, E., White, S.R.: Phys. Rev. A **77**, 013624 (2008)
70. Lecheminant, P., Boulat, E., Azaria, P.: Phys. Rev. Lett. **95**, 240402 (2005)
71. Ottenstein, T.B., Lompe, T., Kohnen, M., Wenz, A.N., Jochim, S.: Phys. Rev. Lett. **101**, 203202 (2008)
72. Huckans, J.H., Williams, J.R., Hazlett, E.L., Stites, R.W., O'Hara, K.M.: Phys. Rev. Lett. **102**, 165302 (2009)
73. Wheeler, J.A.: Phys. Rev. **52**, 1083, 1107 (1937)
74. Ikeda, K., Tamagaki, R., Saito, S., Horiuchi, H., Tohsaki-Suzuki, A., Kamimura, M.: Prog. Theor. Phys. Supple. **62** (1977)
75. Brink, D.M.: In: Proceedings of International School of Physics Enrico Fermi, Course 36, p. 247. Academic Press, New York (1966)
76. Margenau, H.: Phys. Rev. **59**, 37 (1941)
77. Bayman, B.F., Bohr, A.: Nucl. Phys. **9**, 596 (1958/59)
78. Yamada, T., Funaki, Y., Horiuchi, H., Ikeda, K., Tohsaki, A.: Prog. Theor. Phys. **120**, 1139 (2008)
79. Hill, D.L., Wheeler, J.A.: Phys. Rev. **89**, 1102 (1953)
80. Griffin, J.J., Wheeler, J.A.: Phys. Rev. **108**, 311 (1957)
81. Horiuchi, H.: Prog. Theor. Phys. **51**, 1266 (1974)
82. Horiuchi, H.: Prog. Theor. Phys. **53**, 447 (1975)
83. Suzuki, Y.: Prog. Theor. Phys. **55**, 1751 (1976)
84. Suzuki, Y.: Prog. Theor. Phys. **56**, 111 (1976)
85. Fukatsu, K., Katō, K., Tanaka, H.: Prog. Theor. Phys. **81**, 738 (1989)
86. Fukatsu, K., Katō, K.: Prog. Theor. Phys. **87**, 151 (1992)
87. Horiuchi, H.: Prog. Theor. Phys. **58**, 204 (1977)
88. Horiuchi, H.: Prog. Theor. Phys. Supple. **62**, 90 (1977)
89. Kamimura, M.: Phys. Rev. A **38**, 621 (1988)
90. Hiyama, E., Kino, Y., Kamimura, M.: Prog. Part. Nucl. Phys. **51**, 223 (2003)
91. Hiyama, E., Kamimura, M., Motoba, T., Yamada, T., Yamamoto, Y.: Prog. Theor. Phys. **97**, 881 (1997)
92. Kukulin, V.I., Krasnopol'sky, V.M., Voronchev, V.T., Sazonov, P.B.: Nucl. Phys. A **417**, 128 (1984)
93. Suzuki, Y., Takahashi, M.: Phys. Rev. C **65**, 064318 (2002)

94. Suzuki, Y., Horiuchi, W., Orabi, M., Arai, K.: Few-Body Syst. **42**, 33 (2008)
95. Yamada, T., Funaki, Y., Horiuchi, H., Röpke, G., Schuck, P., Tohsaki, A.: Phys. Rev. A **78**, 035603 (2008)
96. Yamada, T., Funaki, Y., Horiuchi, H., Röpke, G., Schuck, P., Tohsaki, A.: Phys. Rev. C **79**, 054314 (2009)
97. Pethick, C.J., Pitaevskii, L.P.: Phys. Rev. A **62**, 033609 (2000)
98. Yamaya, T., Katori, K., Fujiwara, M., Kato, S., Ohkubo, S.: Prog. Theor. Phys. Suppl. **132**, 73 (1998)
99. Michel, F., Ohkubo, S., Reidemeister, G.: Prog. Theor. Phys. Suppl. **132**, 7 (1998)
100. Wada, T., Horiuchi, H.: Phys. Rev. C **38**, 2063 (1988)
101. Itagaki, N., Otsuka, T., Ikeda, K., Okabe, S.: Phys. Rev. Lett. **92**, 142501 (2004)
102. Funaki, Y., Horiuchi, H., Tohsaki, A., Schuck, P., Röpke, G.: Prog. Theor. Phys. **108**, 297 (2002)
103. Volkov, A.B.: Nucl. Phys. A **74**, 33 (1965)
104. Dalfovo, F., Giorgini, S., Pitaevskii, L.P., Stringari, S.: Rev. Mod. Phys. **71**, 463 (1999)
105. Zinner, N.T., Jensen, A.S.: Phys. Rev. C **78**, 041306(R) (2008)
106. Libert-Heinemann, M., Baye, D., Heenen, P.-H.: Nucl. Phys. A **339**, 429 (1980)
107. Horiuchi, H., Ikeda, K.: Prog. Theor. Phys. **40**, 277 (1968)
108. Hasegawa, A., Nagata, S.: Prog. Theor. Phys. **45**, 1786 (1971)
109. Tanabe, F., Tohsaki, A., Tamagaki, R.: Prog. Theor. Phys. **53**, 677 (1975)
110. Funaki, Y., Horiuchi, H., Röpke, G., Schuck, P., Tohsaki, A., Yamada, T.: Phys. Rev. C **77**, 064312 (2008)
111. Lane, A.M., Thomas, R.G.: Rev. Mod. Phys. **30**, 257 (1958)
112. Tohsaki, A.: Phys. Rev. C **49**, 1814 (1994)
113. Pitaevskii, L.P.: Zh. Eksp. Teor. Fiz. **40**, 646 (1961)
114. Pitaevskii, L.P.: Sov. Phys. JETP **13**, 451 (1961)
115. Gross, E.P.: Nuovo Cimento **20**, 454 (1961)
116. Pitaevskii, L.P.: J. Math. Phys. **4**, 195 (1963)
117. Gogny, D.: In: Ripka, G., Porneuf, M. (eds.) Proceedings of the International Conference on Nuclear Selfconsistent Fields, Trieste, 1975, Noth Holland, Amsterdam (1975)
118. Ali, S., Bodmer, A.R.: Nucl. Phys. A **80**, 99 (1966)
119. von Oertzen, W. Cluster in nuclei. In: Beck, C. (ed.) Nuclei Lecture Notes in Physics vol. 818. p. 109. Springer, Heidelberg (2010)
120. Ajzenberg-Selove, F.: Nucl. Phys. A **490**, 1 (1988)
121. Aguilar, J., Combes, J.M.: Commun. Math. Phys. **22**, 269 (1971)
122. Balslev, E., Combes, J.M.: Commun. Math. Phys. **22**, 280 (1971)
123. Simon, B.: Math. Phys. **27**, 1 (1972)
124. Kruppa, A.T., Lovas, R.G., Gyarmati, B.: Phys. Rev. C **37**, 383 (1988)
125. Kruppa, A.T., Katō, K.: Prog. Theor. Phys. **84**, 1145 (1990)
126. Aoyama, S., Myo, T., Katō, K., Ikeda, K.: Prog. Theor. Phys. **116**, 1 (2006)
127. Soić, N. et al.: Nucl. Phys. A **742**, 271 (2004)
128. Curtis, N. et al.: Phys. Rev. C **72**, 044320 (2005)
129. Charity, R.J. et al.: Phys. Rev. C **78**, 054307 (2008)
130. Kraeft, W.D., Kremp, D., Ebeling, W., Röpke, G.: Quantum Statistics of Charged Particle Systems. Akademie-Verlag, Berlin (1986)
131. Röpke, G., Münchow, L., Schulz, H.: Nucl. Phys. A **379**, 536 (1982)
132. Röpke, G., Schmidt, M., Münchow, L., Schulz, H.: Nucl. Phys. A **399**, 587 (1983)
133. Akaishi, Y., Bandō, H.: Prog. Theor. Phys. **41**, 1594 (1969)
134. Brink, D.N., Castro, J.J.: Nucl. Phys. A **216**, 109 (1973)
135. Tohaski, A.: Prog. Theor. Phys. **81**, 370 (1989)
136. Tohaski, A.: Prog. Theor. Phys. **88**, 1119 (1992)
137. Tohaski, A.: Prog. Theor. Phys. **90**, 871 (1993)
138. Tohaski, A.: Phys. Rev. Lett. **76**, 3518 (1996)

139. Takemoto, H., Fukushima, M., Chiba, S., Horiuchi, H., Akaishi, Y., Tohsaki, A.: Phys. Rev. C **69**, 035802 (2004)
140. Malfliet, R.A., Tjon, J.A.: Nucl. Phys. A **127**, 161 (1969)
141. Payne, G.L., Friar, J.L., Gibson, B.F.: Phys. Rev. C **26**, 1385 (1982)
142. Kamei, H., Miyake, K.: J. Phys. Soc. Jpn. **74**, 1911 (2005)
143. Schuck, P.: State of the Art in Nuclear Cluster Physics. Strasbourg, May 2008. Int. J. Mod. Phys. E **17**, 2136 (2008)
144. Stepanenko, A.S., Gunn, J.M.F.: arXiv: cond-mat/9901317
145. Noziéres, P., Schmitt-Rink, S.: J. Low Temp. Phys. **59**, 195 (1985)
146. Matsuo, M.: Phys. Rev. C **73**, 044309 (2006)
147. Alm, T., Friman, B.L., Röpke, G., Schulz, H.: Nucl. Phys. A **551**, 45 (1993)
148. Lombardo, U., Noziéres, P., Schuck, P., Schulze, H.-J., Sedrakian, A.: Phys. Rev. C **64**, 064314 (2001)
149. Lattimer, J.M., Swesty, F.D.: Nucl. Phys. A **535**, 331 (1991)
150. Shen, M., Toki, H., Oyamatsu, K., Sumiyoshi, K.: Prog. Theor. Phys. **100**, 1013 (1998)
151. Funaki, Y., Tohsaki, A., Horiuchi, H., Schuck, P., Roepke, G.: Eur. Phys. J. A **28**, 259 (2006)
152. Johnson, M.T., Clark, J.W. Kinam **2**, 3 (1980) (PDF available at Faculty web page of Clark, J.W. at http://wuphys.wustl.edu)
153. Clark, J.W., Wang, T.P.: Ann. Phys. (N.Y.) **40**, 127 (1966)
154. Mueller, G.P., Clark, J.W.: Nucl. Phys. A **155**, 561 (1970)
155. Sedrakian, A., Müther, H., Schuck, P.: Nucl. Phys. A **766**, 97 (2006)
156. Penrose, O., Onsager, L.: Phys. Rev. **140**, 576 (1956)
157. Clark, J.W.: Prog. Nucl. Part. Phys. **2**, 89 (1979)
158. Pentförder, R., Lindenau, T., Ristig, M.L.: J. Low Temp. Phys. **108**, 245 (1997)
159. Jemai M., Schuck P.: Phys. Atom. Nucl. **74**, 1139 (2011)
160. Fetter, A.L., Walecka, J.D.: Quantum Theory of Many-Particle Systems. Dover, New York (2003)
161. Blin, A.H., Hasse, R.W., Hiller, B., Schuck, P., Yannouleas, C.: Nucl. Phys. A **456**, 109 (1986)
162. Bohr, A., Mottelson, B.R.: Nuclear Structure Vol. 2. Benjamin, New York (1975)
163. Blaizot, J.-P., Ripka, G.: Quantum Theory of Finite Systems. MIT, Cambridge, MA (1986)
164. Shapiro, S.L., Teukolsky, S.A.: Black holes, white Dwarfs and Neutron Stars: The Physics of Compact Objects. Wiley, N.Y (1983)
165. Pines, D., Tamagaki, R., Tsuruta, S. (eds): Neutron Stars. Addison-Wesley, N.Y (1992)
166. Doucot, B., Vidal, J.: Phys. Rev. Lett. **88**, 227005 (2002)
167. Capponi, S., Roux, G., Lecheminant, P., Azaria, P., Boulat, E., White, S.R.: Phys. Rev. A **77**, 013624 (2008)

Chapter 6
Cluster in Nuclei: Experimental Perspectives

P. Papka and C. Beck

Abstract This lecture notes treat some experimental aspects of nuclear cluster states studies, ranging from traditional techniques to some of the most recent developments and emerging methods. Experimental investigations, in the field of nuclear clusters are discussed in terms of detection techniques and associated electronics. Recent developments in accelerator technology and targetry are also presented in the scope of new opportunities in cluster studies. The nature of cluster states makes exclusive measurements crucial. It requires the simultaneous detection of nucleons, light, intermediate-mass and heavy fragments, and possibly γ-rays together with timing information. Precise measurements of angular correlations and energy distributions between emitted particles are needed for kinematic reconstruction in order to achieve a detailed study of the decay modes and the underlying dynamics. Within this scope, highly segmented and high-efficiency detection systems are depicted. Developments in digital signal processing have made possible major advances in experimental nuclear physics. The combination of large numbers of channels with fast data acquisition systems is one of the key aspects of this modern technology. Nuclear reactions play a key role in the study of the structure of nuclear clusters. Therefore, aspects of acceleration, including high-intensity, low-energy stable and radioactive beams are presented. Targetry has received a renewed interest with the advent of active targets (ACTAR). The combination of radioactive beams and active targets for the study of nuclear clustering is certainly opening new horizons in this field of physics.

P. Papka (✉)
Department of Physics, University of Stellenbosch,
Private Bag X1, Merensky Building, Merriman Avenue,
Stellenbosch 7600, South Africa
e-mail: papka@tlabs.ac.za

C. Beck
Département de Recherches Subatomiques,
Institut Pluridisciplinaire Hubert Curien IN2P3/CNRS and Université de Strasbourg,
23 rue du Loess, BP.28, 67037 Strasbourg Cedex 2, France
e-mail: christian.beck@iphc.cnrs.fr

C. Beck (ed.), *Clusters in Nuclei, Vol.2*, Lecture Notes in Physics 848,
DOI: 10.1007/978-3-642-24707-1_6, © Springer-Verlag Berlin Heidelberg 2012

A number of current experimental setups and computer codes are cited to illustrate some of the techniques described but this list is by no means exhaustive.

6.1 Introduction

The clearest evidence for the occurrence of cluster states in nuclei is observed through their decaying modes above the particle threshold via neutrons, protons, α particles or heavier cluster emissions. Abnormal form factor measurements or the effects on reaction mechanisms are some of the other manifestations of cluster states in nuclei. Some of the striking examples are the characteristic breakup of ^{6}He and the $2n$ transfer reaction rate to target nuclei [1]; halo nuclei such as ^{9,11}Li [2]; or ^{12}C [3] for its famous Hoyle state just above the three α particle threshold [4].

We begin this chapter with an overview of the different aspects of cluster states as investigated from natural or exotic radioactivity to various nuclear reactions induced either by stable or radioactive beams. The shape evolution of cluster configurations and internal structure of cluster states have also been investigated by means of the electron probe with (e, e') reactions.

The experimental techniques in cluster studies do not differ strongly from standard nuclear physics methods where neutrons, γ-rays or charged particles need to be detected. However, the dedicated experimental setups often use sophisticated charged-particle arrays where particle identification, detection efficiency and energy resolution are some of the key aspects.

6.2 Population of Cluster States

6.2.1 Radioactive Decay

6.2.1.1 Heavy Cluster Radioactive Decays

Alpha radioactivity can be considered as the first known manifestation of cluster emission. Initially discovered in the Uranium decay series [5–7], α radioactivity requires the pre-formation of an ^{4}He nucleus emitted through the Coulomb barrier. Decay modes involving much heavier clusters were only discovered in 1984 by Rose and Jones [8] from Oxford University.

More exotic emissions, such as ^{14}C-cluster radioactivity of ^{223}Ra [9, 10], were also measured almost simultaneously at Orsay by using the superconducting spectrometer SOLENO [11]. SOLENO—see Fig. 6.1—has been employed to detect and identify ^{14}C clusters spontaneously emitted from ^{222,223,224,226}Ra parent nuclei. Thanks to the excellent energy resolution of the magnetic spectrometer a structure in the kinetic energy spectrum of ^{14}C emitted by ^{223}Ra was discovered [12]. Even heavier clusters

Fig. 6.1 Layout of the SOLENO magnetic spectrometer: 1 iron shield, 2 solenoidal coil, 3 vacuum chamber, 4 source, 5 iris, 6 shutters, 7 detector. The arrows indicate the trajectory of $^{14}C^{6+}$ ions [11]

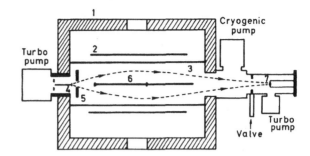

[5–7, 13–16] were discovered later on but their full identification is more complicated for essentially two reasons. The branching ratios for such decay processes decrease rapidly with the mass of the cluster, namely about 10^6 to 10^{15} times less intensely than for α particles. Due to the larger atomic numbers of the heavy-clusters, the absorption owing to large energy loss in the foil which contains the radioactive nuclei is much higher. Only fragments with sufficient energy escape the foil and deposit some residual energy in a detector.

Only a few types of experiments were undertaken to study exotic cluster decays [9, 13–15]. Experimental arrangements were developed to deal with the large ratio between the dominant emission of α particles, or spontaneous fission, and exotic heavy-cluster decay in the late 1980s, early 1990s [6, 7].

Two important aspects for these experiments to be successful were carefully investigated: i.e. the preparation of the radioactive source and the sensitivity of the related experimental setup to be optimised for the detection of heavy clusters rather than for α-particles or fission fragments. The radioisotope of interest can either be found naturally, or produced through a beam-induced nuclear reaction, and chemically separated before preparing the source. Two methods for source preparation are identified: via implantation in a substrate directly after production using beam-induced nuclear reaction or via chemical separation, a method applicable to both naturally occurring or synthesised radioisotopes. A relatively large number of radioactive nuclei are necessary to overcome low heavy-cluster decay probabilities; meanwhile the radioactive material must be spread over a large area to allow the heavy clusters to escape from the substrate.

Magnetic Spectrometers

The use of semiconductor detectors for energy measurement is limited by the total incoming flux of charged particles. This is not only because of count rate limitations from the data acquisition system but mostly due to radiation damage. This applies especially in detecting heavy fission fragments as the radiation damage increases dramatically with Z [17].

The separation of unwanted particles can be achieved using a magnetic spectrometer. The cluster emitted from the radioactive source is stripped from most of its electrons when ejected from the foil and the magnetic field settings of the spectrometer are tuned in order to transport the particles of interest with a defined rigidity to a focal plane. Particles with a different rigidity are dumped in a stopper. The SOLENO setup shown in Fig. 6.1 [11] probably was the unique spectrometer dedicated to this type of measurement. The selectivity, in terms of a heavy cluster to α separation, had a limit of 10^{-8} [10, 13].

Glass Detectors

The plate irradiation technique is relatively ancient. The tracks of the particles are physically recorded in a substrate during irradiation and they are optically analysed after a given exposure time. This method has a long history and is still used in a number of applications, for example neutron dosimetry and high-flux neutron measurements, or in some neutrino experiments. The latest techniques make use of automated optical scanning with sophisticated algorithms to identify the tracks of the particles. Regarding heavy-cluster studies, nuclear track detectors were essentially made of glass phosphate. Heavy clusters impinging on the glass with high incident energy create defects in the crystal due to the displacement of atoms, changing the local properties of the glass. The track left on the glass is revealed through the etching method; a dip is formed and the features of the track relate to the energy loss of the particle through the material. Information on Z and E of the decay products can be deduced from the depth and the angle of the revealed cone.

Radioactive sources are prepared in the form of large-area thin foils placed against the glass detector for an exposure time in the order of several hundreds of days. These measurements were performed in a low background-radiation environment, especially because high-energy cosmic rays can produce tracks resembling heavy-cluster events. Several experiments undertaken by Bonetti et al. [15, 16] were performed at the Gran Sasso Underground Laboratory making use of the mountain as a very thick shield against muon cosmic rays. This technique allows high sensitivity to heavy clusters to α-particles with a ratio ranging from 10^{-8} to 10^{-16} [15]. However, for extremely large flux irradiation, or long exposure, not only cosmic rays but also α-particles produce defects in the glass that mimic heavy-cluster events.

Accelerator Mass Spectrometry Measurement

Accelerator Mass Spectrometry (AMS) [18] was developed together with the advent of electrostatic accelerators. AMS is nowadays used for a large range of applications concerned with very low concentrations (down to one in 10^{18}) of elements or isotopes. AMS techniques make use of the charge-to-mass ratio to separate the species of interest. In addition, the nuclei are accelerated with sufficient energy to provide direct charge and mass identifications by using $\Delta E - E$ and time-of-flight measurements, respectively. Owing to the extreme sensitivity, only a limited number of detected ions are sufficient to characterise a sample. In the field of heavy-cluster radioactivity, investigations have been carried out on Uranium samples where the heavy clusters are contained after radioactive decay. Fragments emitted from ^{14}C radioactivity [19] could be trapped in the sample and the idea to count these atoms using AMS measurement emerged [15]. Direct measurement of the isotope of interest is possible if its half-life is long enough to allow for concentration evaluation at the time of the preparation of the sample. The decay rate is extracted from the concentration using the secular equilibrium formula, taking into account the half-lives of the parent and daughter nuclei.

6.2.1.2 Ternary and Quaternary Fission

Ternary fission was suggested shortly after the discovery of binary fission, mostly because it is more energetically favourable. Binary fission can be observed relatively easily with cheap solar cells placed in opposing directions with respect to a spontaneous fission source but ternary and quaternary fissions are experimentally much more challenging. It took a great deal of effort to observe these events, which are less frequent than binary fission, at a ratio of $\approx 10^{-3}$ for ternary and $\approx 10^{-7}$ for quaternary fission of the well studied ^{252}Cf spontaneous fission source. Note that ^{252}Cf, with a half-life of 2.645 years, does not occur naturally but is produced in high-flux neutron reactors. Following spontaneous ternary fission, the fragments were initially expected to be emitted in three distinct directions, separated with $\approx 120°$, and with the energy shared between the fragments. It was discovered later that, at the scission, the fragments adopt a collinear configuration with the formation of two or three necks. In ternary fission the fragments at the tips of the collinear configuration are emitted approximately in opposite directions. The middle fragment carries then relatively low energy and is emitted nearly at rest in the laboratory frame. A number of setups dedicated to this specific decay mode were devised, for example NESSI [20, 21] or FOBOS [22]. The NEutron Scintillator tank and SIlicon ball detector (NESSI) consists of two 4π detectors for neutrons, the Berlin Neutron Ball, and the Berlin Silicon Ball for charged particles. Neutron multiplicity is measured with high efficiency in a liquid Gadolinium-loaded scintillator detector and the fission fragments detected with the silicon detector array. The FOBOS setup (see Fig. 6.2) is composed of modular detectors placed opposite one another with respect to a spontaneous fission source, ^{252}Cf, or a target bombarded with thermal neutrons, ^{235}U(n_{th}, f) or α particles ^{238}U(α, f). Owing to the difficulties in detecting the third fragment, the general approach is to determine the missing momentum via precise measurement of the velocity and energy of the two detected fragments to reconstruct the missing mass. The mass resolution on the mass reconstruction is crucial. The mass and charge of the fragments are deduced using a sophisticated method some aspects of which are detailed in [23, 24]. In the FOBOS setup the time of flight is recorded with fast detectors, noted as 1 and 3 in Fig. 6.2. The charge and mass are deduced by combining the Time-of-Flight (ToF) measurement and pulse shape analysis of the energy deposition within the Large Ionization Detectors 2. Large energy loss of the fragments in the entrance windows and mass loss due to neutron emission must be taken into account and deduced from a multi-iteration procedure. Further developments in ternary fission measurements with semi-conductor charged particle detectors and efficient neutron detection with ^{3}He gas-filled detectors are implemented in the COMETA setup under the FOBOS collaboration.

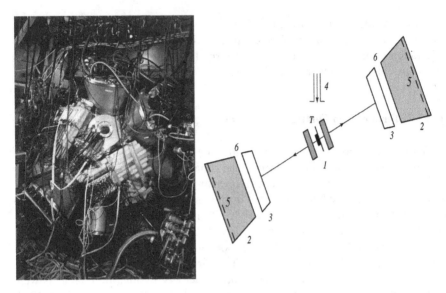

Fig. 6.2 Left panel: FOBOS (see Refs. [23, 24]) in a 12 module arrangement setup. Right panel: schematics of a pair of FOBOS modules with, *T* target, 1 start avalanche counters, 2 Large Ionization Chambers (LIC), 3 stop position-sensitive counters, 4 neutron collimator, 5 Frisch grid of the LIC, 6 cathode combined with the LIC entrance window

6.2.1.3 Study of Light Cluster Nuclei via β-Decay

A number of studies explored the β-decay [25, 26] to populate unbound resonances in light cluster nuclei as, for instance, in Ref. [27]. The radioactive nuclei are populated through primary nuclear reactions and implanted in a substrate where excited states of the daughter nuclei are populated through radioactive decay. The excited states are populated through the β-decay selection rules and the highest excitation energy will be dictated by the Q-value of the decay.

Charged-particle detectors surround the substrate where the radioactive ions are implanted. Each implanted ion leads to an event of interest, which is useful in dealing with extremely low beam intensities in the order of a few thousand ions per second. This technique is similar to a measurement using a radioactive source, but in this case the lifetime of the radioactive isotopes can be lower than a millisecond. Contaminating reactions are considerably reduced and scattered particles from ion beam interactions are negligible. However, isobaric nuclei can be transported to the implantation site, which is one of the main concern when using Rare Ion Beams as detailed in a further section. Cluster states can be studied nearly at rest in the laboratory frame with little kinematic distortion. The daughter nucleus populated in the primary β-decay has a very low recoil velocity owing to its mass of a few thousand times larger than the emitted lepton–antilepton pair. Large coverage solid angle and large efficiency are also made possible because of non-kinematic narrowing. The kinetic energy of the break-up particles originates from the difference between the excitation energy of the level state populated in the β decay and the Q-value of the break-up channel. For this reason, the study of low-lying states is not suitable when the radioactive ions are implanted in a substrate. The available energy shared between the break-up particles must allow sufficient energy to overcome the energy loss through the various layers before energy deposition within the active part of the detector. In one hand, only a limited number of states are populated, due to the selection rules of β-decay, but very selective measurements can be performed due to a simplified excitation energy spectrum.

In Ref. [28] the excited states of ^{9}Be were populated via β-decay of ^{9}Li with two α-particles emitted and one neutron emitted in the subsequent break-up. The identification of the ^{9}Be states relies on the reconstruction of the missing momentum to deduce the position and energy of the neutron. As a consequence, the states at $E_x \leq 2.7$ MeV could not be identified properly, leading to low recoil energy. However, the higher excited states could be clearly studied through angular distribution measurements.

This technique is restricted to a limited number of nuclei, such as ^{8,9}Be, and an extensive study of ^{12}C states was undertaken by means of ^{12}N and ^{12}B beam decay studies [27]. To overcome the detection thresholds, direct irradiation of a highly segmented detector was used [29, 30]. Secondary ^{12}N and ^{12}B ion beams were produced at the Kernfysisch Versneller Instituut (KVI), Groningen, and separated with the magnetic separator TRIμP. The beam was defocused and homogeneously implanted in a finely segmented 48×48 silicon strip detector of a 16×16 mm^2 total area and a thickness of 78 μm (2304 pixels, of 300×300 μm^2 size) through a degrader and primary DSSSD detector ("detector 2" of Fig. 6.3). With a low implantation rate and owing to the fine segmentation of the detector, the occurrence of pile-up events (two radioactive decays in the same pixel) is very restricted. In this experiment the states of interest in ^{12}C are located in an excitation energy region $E_x \approx 10$ MeV. The available energy for the α-particles in ^{12}C$^* \rightarrow 3\alpha$ can be, at most, of $2(E_x - Q)/3 \leq 3$ MeV considering a two-step decay with ^{8}Be emitted in its ground state. With such low energy, the α particles are stopped within less than 40 μm of silicon. The incident beam energy is chosen in order to implant the ions half-way through the detector. The three α-particles are emitted from inside the detector and their energy is not deteriorated as no dead layer is encountered by the particles. The energy is dissipated in a single pixel and the total energy can be measured precisely with virtually no detection threshold.

More recently, thanks to its well-established theory [25, 26], β-decay has been found to be a useful tool for studying peculiar features of the halo structure of nuclei [31].

6.2.1.4 Two-Proton Radioactivity

The diproton radioactivity was predicted by Goldansky [32] in the 1960s as the "true three-body decay". Such two-proton radioactivity [33, 34] occurs for a number of very neutron-deficient nuclei on the neutron drip line or beyond. Some nuclei, as for

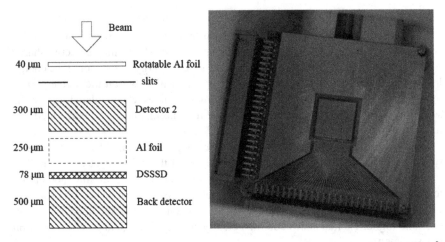

Fig. 6.3 Right panel: finely segmented DSSSD detector (48×48 silicon strip detector, 16×16 mm^2 area, $78\,\mu$m thickness). Left panel: detector and degrader arrangement for homogeneous implantation of radioactive ions close to the centre of the $78\,\mu$m-thick DSSSD detector. Excited states in ^{12}C are populated through β-decay of the ^{12}N and ^{12}B radioactive ions [30, 31]

example ^{45}Fe [35, 36], have a forbidden one-proton emission for Q-value reason but can possibly emit two protons via the direct emission of an unstable diproton. Though the prediction was postulated about five decades ago the experimental techniques only now allow the identification of such an exotic cluster decay. Many candidates, such as ^{16}Ne and ^{19}Mg [37] for instance, were identified in long-lived to short-lived nuclei or from excited states populated via β-decay.

Typical Rare Ion Beams for two-proton emission involve primary beam fragmentation followed by ion separation. The difference between two-proton emission and sequential two-proton emission is obtained from the typical correlation between the two protons from in-flight decay methods or knock-out reactions [37–39]. More details on two-proton radioactivity can be found in the excellent reviews of Blank [33, 34].

6.2.2 In-Beam Induced Reactions

In-beam induced nuclear reactions are mostly used for synthetising unstable nuclei populated in the excited states of interest. Nuclei close to the valley of stability and on the β^+ unstable side are traditionnaly populated with stable ion beams. Cluster studies, away from the valley of stability with large values of isospin, are populated with the Rare Ion Beams (RIB) [40]. The reaction mechanism depends on some of the basic characteristics of the system: projectile and target species and beam energy. The adequate reaction is chosen for the maximum production cross section and specific population of excited states. The experimental constraints considering a

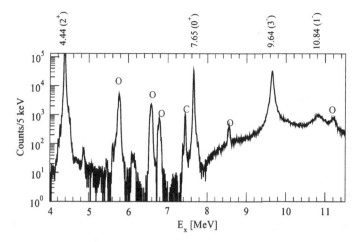

Fig. 6.4 Excitation energy landscape in ^{12}C with ^{13}C and ^{16}O contamination indicated with C and O labels. Excitation energy, spin and parity of the excited states in ^{12}C are reported on the spectrum. This figure has been adapted from Ref. [43]

particular reaction must also be taken into account. For example, low-lying break-up states are preferably investigated by using complete kinematic measurements [41]. In inverse kinematics, when the projectile is heavier than the target nucleus, the break-up particles are emitted in the centre-of-mass frame, moving with a velocity close to the beam velocity. Limitations may arise when particles need to be measured at angles approaching the beam axis at 0° [42] and detection efficiency is not favoured.

6.2.2.1 Inelastic Scattering Reactions

Inelastic scattering converts kinetic energy into excitation energy within the projectile and/or target nuclei. For example the α-unbound channels in ^{12}C were studied with great care using hadronic interaction with (p, p') or (α, α') reactions with high energy-resolution spectrometers [43]. The technique relies on a precise momentum measurement of the scattered particles, assuming the missing momentum has been transferred to the recoil nucleus and the excitation energy is deduced from energy conservation. The energy spectrum obtained at a given scattering angle is therefore used to extract the excitation energy of the target nucleus. Figure 6.4 shows a ^{12}C spectrum at finite angle with a natural carbon target being bombarded with a 66 MeV incident energy proton beam [43]. The line shape of the excited state can be characterised with precise determination of the width and interference between excited states. The angular distribution of the scattered particles informs on the spin and parity of the states. Weakly bound projectiles such as ^{6}Li, ^{7}Li or ^{9}Be have been investigated with such reactions to study some exotic states in the light-mass region. Typical cases such as ^{6}He [44] or ^{11}Li [45] show very pronounced halo structures deduced from elastic and inelastic cross section measurements.

6.2.2.2 Electron Scattering

Electron scattering [46, 47] has long been known as one of the best probes for charge distribution measurement within the atomic nucleus because of the point-like structure of the charge of the projectile. The search for exotic cluster states in light nuclei is reported in numerous papers [48–50] including the study of the properties of the Hoyle state of ^{12}C [51]. Electron scattering cross sections are often plotted as the ratio of the measured to the Mott cross section, the equivalent of the Rutherford cross section for ion scattering. The Hoyle state being predicted to have a well-developed three α structure, should show a large form factor if one could measure the elastic scattering on the Hoyle state itself. This is not possible as the ^{12}C target nuclei are found naturally in their ground state. The population of this state can be produced via inelastic collision with an electron. The scattering cross sections on the 7.65 MeV state of ^{12}C from Ref. [51] are well interpreted in the framework of the Fermionic Molecular Model (FMD) and α cluster model. The large form factor calculated in the bottom panel of Fig. 6.5 indicates a diffuse state of nuclear matter supported by the very good agreement obtained for the 0^+ and 0^+ to 2^+ transition data.

6.2.2.3 Transfer Reactions

In single-nucleon transfer reactions, neutron and/or proton removal allows the population of nuclei slightly away from the valley of stability. New information for light nuclei (^{7,8}Be, ^{10}C) was obtained by using this type of reaction. The reaction mechanism informs on the structure of a particular nucleus. Recent studies have shown the strong correlation of the two neutrons in ^{6}He through an enhanced $2n$ transfer channel in the ^{6}He $+^{65}$Cu system at 22.6 MeV incident beam energy [1]. Cross sections for transfer reactions are relatively high, from a few MeV/u up to several tens of MeV/u incident beam energy.

Multi-nucleon transfer reactions offer interesting ways of populating specific states in residual nuclei. The transfer of α particles is interesting in populating α-like nuclei where strong resonances are observed. The case of two nucleon transfer reaction using (^{3}He, n) or (p, t) reactions can be used selectively to populate target-like nuclei by adding two correlated nucleons on specific orbitals. Some more exotic reactions (^{4}He, ^{8}He) [52] and (^{3}He, ^{8}He) [53] were successfully performed but rarely used due to low characteritic event rates.

6.2.2.4 Charge–Exchange Reaction

Heavy-ion charge–exchange reaction, through Gamow–Teller or Fermi transitions, is a powerful tool for spectroscopic studies in exotic nuclei, and may be used to investigate the isovector response of near drip line nuclei. In charge-exchange reactions, and due to the nature of the interaction, isobaric analogue states are favourably populated which can be of interest for the purpose of selectivity (isoscalar, Fermi–

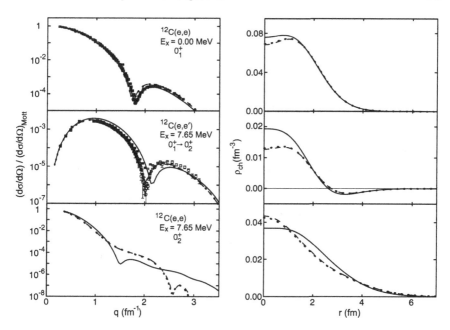

Fig. 6.5 Right column: Charge density in ^{12}C calculated with the Fermionic Molecular Dynamics model (*solid lines*), α cluster model (dashed lines) and Bose-Einstein Condensation (*dotted lines*). Left Column: Ratio of the experimental cross sections to the Mott cross sections (*open square*) compared to DWBA calculations performed with the potentials from the right column. Comparisons are shown for the 0^+ state (top panels) and 0^+ to 2^+ transition (middle panels) and the theoretical charge distributions and cross sections are shown for the 2^+ state in the bottom panels. This figure has been adapted from Ref. [51]

Gamow and analogue states features). Nuclei such as ^{6,7}Be, ^{10}C and ^{16}F have been investigated using (p, n), $(^3$He, ^{3} H) reactions on various targets (see for instance [54–56]). Charge exchange reaction is a route to study exotic ions and to populate even more exotic nuclei towards both neutron-rich and neutron-deficient regions in inverse kinematic reactions using secondary beams.

6.2.2.5 Knock-Out Reactions and Fragmentation Reactions

Knock-out reactions occur at incident beam energies from 100 MeV/u and higher. Using light targets, the main contribution in knock-out reactions corresponds to those events where single nucleons from the projectile interact with the target nucleus in surface-grazing collisions through inelastic break-up, or stripping [57]. Ground state and excited states are populated in the projectile but minimum re-arangement within the projectile/target nuclei is expected. The spread in momentum of the projectile after the knock-out reaction not only depends on the multiple scattering through the target, but also carries information related to the bound state wave function of the ejected nucleon. For example the knock-out of a loosely bound nucleon causes less spread than a tightly bound nucleon. High value of orbital angular momentum of the ejected nucleon also increases the momentum spread of the quasi-projectile. Therefore, a measurement of the momentum spread characterizes the wave function of a single particle. The longitudinal spread is a more sensitive measure of the momentum dispersion than transversal, which is affected by scattering mechanisms. With heavy targets and higher energy beams, 1 GeV/u, Coulomb knock-out plays an important role [58].

Fragmentation occurs between a fast moving projectile and a target, in inverse kinematics reaction. Beam energies are well above 100 MeV/u as for example the 1 GeV/u Ar beam at the FRS/GSI facility. Further away from the valley of stability nuclei are efficiently produced via fragmentation of the primary beams. This technique implies that the fragmented nuclei are separated and transported in a beam for secondary collisions where the knock-out reactions previously described present relatively high cross sections ($\approx$10 mb). Recent studies have employed single neutron knock-out reaction to populate ^{16}Ne and ^{19}Mg from ^{17}Ne and ^{20}Mg fragmented beams respectively [37].

6.2.2.6 Ion Beams

Most of the stable isotopes, if not all, were accelerated within stable ion beam facilities in the last nearly 80 years by using oscillating or electrostatic accelerating techniques. Electrostatic machines are mostly making use of negative ion beams that allow pre-acceleration with a positive high voltage followed by efficient stripping of the ions for a second acceleration with a higher charge state. Such machines are called tandem van de Graff accelerators. Radio Frequency resonators in LINAC, synchrotron or cyclotron make use of positive ions produced in ECR sources. Some of the stable beam facilities have explored the acceleration of pre-synthesised isotopes separated in offline chemical separation. Some very interesting projectiles such as ^{7}Be, ^{10}Be can be produced by using primary irradiation of some stable elements or isolated from naturally occurring radioactive isotopes such as ^{14}C. Beryllium-7 was prepared at Louvain-la-Neuve using the ^{7}Li$(p, n)^{7}$Be reaction, separated using offline chemical separation techniques, and introduced in an ECR source [59]. A number of other isotopes can be employed for ion beam generation or for radioactive targets, for example Americium or Californium.

Recent RIB facilities produce rare ion beams with down to ms half-life for secondary reactions. Such short-lived isotopes are found far from stability and are crucial to populate nuclei in un-explored regions in the both neutron-deficient region, especially for exotic $N = Z$ nuclei, and the neutron-rich region with high N/Z ratio. In terms of cluster nuclei studies, a wide range of interesting projectiles are available and still being developed in a number of facilities worldwide, for example, radioactive Helium isotopes produced at JINR and GANIL with intensities up to $\approx$10^{7} ^{6}He and 10^{5} pps ^{8}He. Facilities around the world use two main principles for rare ion beam

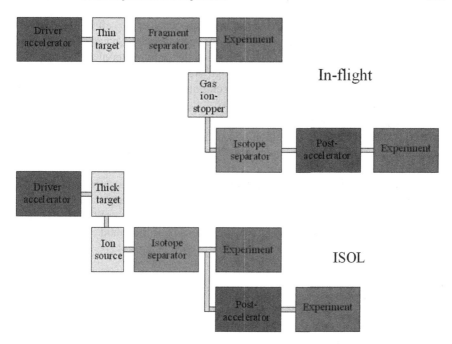

Fig. 6.6 General layout of in-flight fragment separation and ISOL Rare Ion Beam facilities. Experiments can be performed with a low-energy rare ion beam for β-decay studies and material research

production: in-flight fragment separation and Isotope Separation On-Line beams as schematically depicted in Fig. 6.6.

In the first case high-energy projectiles are stripped from a number of nucleons through fragmentation reaction within a relatively thin target and are separated from a cocktail of particles with a sophisticated mass separator to produce a clean beam. The main challenge is to eliminate isobaric nuclei, which is achieved at the cost of a multi stage separator. Due to the spread in energy and multiple scattering of the ions in the target, plus possible degraders in the mass separator, the beam quality is relatively poor. Alternately, the ions can be stopped just after the production target in a gas catcher. The ions are slowed down but kept in a low charge state for extraction and post acceleration. In this way the beam quality is greatly improved. The first in-flight separation was performed in the 1970s, at the Lawrence Berkeley Laboratory, and some of the facilities that routinely employ such techniques include LISE/SISSI/GANIL (France), FAIR/FRS/GSI (Germany), Notre Dame and NSCL/MSU (USA), ACCULINNA/JINR (Russia) and ETNA/LNS (Italy). Neutron-rich and neutron-deficient nuclei can be produced with this technique.

ISOL techniques were implemented for the first time in the late 1950s, at the Niels Bohr Institute in Copenhagen. The technique makes use of intense primary beams from a driver accelerator or reactor impinging on a thick target for the production of radioactive nuclei, typically through fission or spallation reactions. Note that the primary beam can be used to generate fast neutrons which causes efficient fissioning thus the production of fission products that are neutron-rich by nature. Diffusion of the newly formed isotopes is obtained by heating the target material and feeding the unstable isotopes to an ion source for charge breeding and post acceleration. The first implementation of the modern ISOL technique at ISOLDE CERN made use of a 600 MeV proton beam with 4 μA intensity. Large amounts of unstable and stable nuclei are produced and it is essential to select the isotope of interest. Ionization through two- or three- step excitation using adequately tuned laser allows selective ionization of virtually only one species then extracted with a voltage potential. High resolution mass spectrometers are also employed to select the isotope of interest from the inevitable isobars. Such techniques require special developments to attain maximum extraction efficiency and transmission through the accelerator and separators. A number of facilities such as ISAC/TRIUMF (Canada) or SPIRAL/GANIL (France) for instance, deliver radioactive ion beams by this method.

6.3 Targetry

Irrespective of the effort invested in delivering either stable or Radioactive Ion Beams (RIB), a proper choice of the target is essential to produce clean nuclear interactions with incident projectiles. Indeed, the traditional high-energy physics colliders have not been used often for applications in the low-energy nuclear physics domain and nuclear interaction studies have been limited to fixed-target setups. Targetry, in the jargon, is a subject of intense developments at the interface of material research and nuclear physics. Chemistry, laser physics, material science and cryogenics, together with mechanical engineering, are involved in a discipline sometimes called an art by the target makers. Regarding the abundance of the related literature [60], an exhaustive review in targetry would require a dedicated series of text books. This chapter will be restricted to some developments in targets that concern cluster and resonant states studies.

Targets, used in conjunction with charged particle spectroscopy, are traditionally kept relatively thin in order to reduce multiple scattering of both incoming and outgoing particles, typically from just under $10\,\mu g/cm^2$ to $10\,mg/cm^2$ ($\approx 10^{18}$ to $\approx 10^{21}$ atm/cm^2) depending of the features of the incident beam and the outgoing particles to be detected, and according to the cross section of the reaction of interest. The targets preferably are self-supported to avoid contamination from a substrate, as well as to reduce the amount of material to be penetrated. The field of targetry is subject to new experimental challenges owing to the rapid increase of Rare Ion Beam facilities. This, depending on the technology adopted, is mostly in relation to the features of secondary beams possibly characterised by large emittance and energy spread, but mostly with limited intensities. New developments in RIB require both high power dissipation targets for secondary beam production and very specific targets for secondary reactions. They are used in conjunction with sophisticated detection devices if not as part of the detector itself.

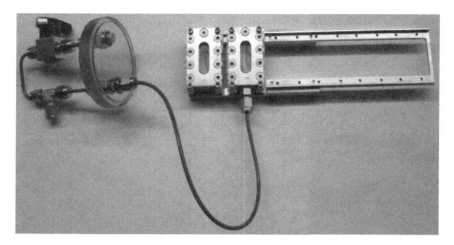

Fig. 6.7 Gas target with $1 \times 3 \, cm^2$ Aramid $7.0 \, \mu m$ thick entrance and exit windows. The gas target is 1 cm thick and pressure up to 1 atm is sustained [Neveling R., et al., private communication]

6.3.1 Gas Targets

Gas targets are not necessarily compatible with the high-vacuum requirement for beam transport. Straightforward gas targets consist of a cell enclosed between two thin windows to retain the gaseous target material. The windows of a gas cell introduce unwanted contaminants but, on the other hand, any type of gas can be bombarded. Polymer materials, like Kapton ($C_{22}H_{10}N_2O_5$) Aramid ($C_{12}H_{10}N_2O_2$) or Mylar($C_8H_{10}O_4$), are suitable for making gas target windows owing to their high strength at relatively thin thicknesses down to $1.5 \, \mu m$. A gas target for enriched isotopes is displayed in Fig. 6.7, which shows a volume of $3 \, cm^3$ of material contained between $1 \times 3 \, cm^2$ Aramid $7.0 \, \mu m$ windows placed 1 cm apart. A pressure of 1 bar with a low leakage rate is sustained, insuring good vacuum in the target chamber and beam pipe and a relatively thick target $\approx 1 \, mg/cm^2$.

The number of atoms/cm^2 for such a gas target is approximately 3×10^{19} atoms/cm^2 at a pressure of 1 bar at room temperature. HAVAR (Co $\approx 45\%$, Fe $\approx 20\%$, Cr $\approx 20\%$, Ni $\approx 15\%$) is an alternative to polymers and the strength of this material is even better. This type of material is used to avoid light H, C, O, N contaminants.

Gas cells are not necessarily suitable for hydrogen or Helium gasses due to the small size of these atoms/molecules. Hydrogen atoms migrate relatively easily through thin film material. One way to reduce the leakage and to increase the density of atoms is to cool down the gas cell. While reducing temperature, the optimal pressure is kept at a desired value by increasing the number of moles in the cell according to the ideal gas law. The thermal motion being slowed down, both the H atoms migrate less efficiently through materials, and materials offer better permeability.

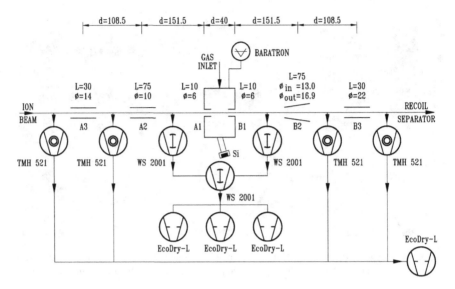

Fig. 6.8 Window less gas target pumping system of the European Recoil Separator ERNA [60]

If the windows of the gas cells induce background events, even using the thinnest material, the way forward is to remove them but to keep the vacuum at a reasonable level sounds somewhat unnatural. Windowless gas cell targets consist of an extended volume where a low pressure over a relatively long distance enables target thicknesses to be made in the order of 10^{17}–10^{18} atoms/cm^2 (5 cm at 4 mbar). Differential pressure by means of a multi-chamber connected with small apertures placed to increase the vacuum impedance enable a pressure gradient from 1 to 10^{-6} mbar within short distances. With such pressure the target thickness is typically in the order of a few µg/cm^2 or less. In the case of astrophysical nuclear reactions, the energy loss through the target is a very important parameter. Self-supporting targets require a minimum thickness in order to keep their integrity and mechanical strength. Carbon foils are amongst the thinnest self-supported targets with commercially available 1 µg/cm^2 foils (a layer of nearly 25 atoms). Windowless gas targets offer the possibility of producing extremely thin targets, in principle down to residual gas pressure if needed. The window less gas target of the ERNA mass separator illustrated in Fig. 6.8 shows the complexity of the system between pumping stages and pump arrangement [61]. The labels TMH indicate turbomolecular pumps, and WS the high pumping speed pumps. Dry pumps, also called oil-free pumps, are backing the TMH and WS pumps. The differential pressure is maintained by means of the apertures denoted by L.

Jet-gas targets are of interest for increased thicknesses together with a better determination of the interaction point. This is a very important requirement for both γ-ray and magnetic spectrometer experiments. The gas target is injected at supersonic velocity in a differential pressure system obtained through evacuated chambers that are vigorously pumped in order to keep the best vacuum possible in the beam lines. A high-pressure nozzle blows the target material pretty much in the same way as a high-pressure water cleaner. The jet is directed into a funnel connected to a high pumping speed system to evacuate the bulk amount of gas. Owing to the momentum of the particles within the jet, the main part of the gas material remains within the jet axis and is evacuated through the dedicated pumping system. Residual gas lost from the divergence of the jet is evacuated through differential pumping in an equivalent system to that detailed for the gas cell target.

Gas targets are non-destructible because they are continuously regenerated and allow the use of relatively high-intensity beams. However, a large deposition of energy in the gas induces rapid dilation and, consequently, a loss of efficiency [62]. For both gas cell and jet-gas targets, the gaseous material has to be constantly fed in order to maintain constant thickness against pumping. Expensive material must be recovered and recirculated to keep the cost of the measurement within a reasonable budget. Helium-3 material with a natural abundance just above 1 ppm, for example, presents large variations in cost depending on the availability of the parent ^{3}H together with the demand for ^{3}He-based neutron detectors or other large-scale applications. In a gas target, optimised recovery together with the purification system allows an economical use of isotopically enriched gas. Extensive nuclear astrophysics measurements were performed with the windowless gas target facility RHINOCEROS [62, 63]. Isotopically enriched targets such as ^{15}N [64] were used with purification of the recycled gas. Another well-known example is the $\alpha(^{12}C, ^{16}O)$ radiative capture reaction which has been extensively studied, in particular with the use of the ERNA mass separator with different types of windowless and gas jet targets [61]. The S-factor is measured at lowest possible centre-of-mass energy. The use of inverse kinematics is of particular interest as it allows higher beam energy, E_{lab}, compared to direct kinematics for identical centre-of-mass energy, E_{cm}, as indicated in Eq. 6.1 as a function of the mass of the projectile M_p and target M_t. The recoils produced with more recoiling energy are easier detected.

$$E_{cm} = E_{lab} \frac{M_t}{M_t + M_p} \tag{6.1}$$

6.3.2 Solid Hydrogen Targets

Proton targets are of very high importance for reactions involving RIBs [41]. Radioactive nuclei with sub-millisecond lifetime can be delivered in a form of RIB but certainly not in the form of a fixed solid target. As a consequence, a number of nuclear reactions require hydrogen targets to investigate cluster states far from stability in inverse kinematics reactions. Proton beams were traditionally used for (p, p'), (p, n), (p, d), (p, t), $(p, 2p)$ reactions, and deuteron beams for (d, p), (d, n) reactions. With the advent of RIB facilities there is a renewed interest in using this type of reactions but in inverse kinematics, and this is especially the case for the study of light neutron-deficient nuclei.

In this respect, plastic targets certainly are the most affordable and easiest hydrogen targets to manufacture in the form of CH_2 and CD_2 polymers in any thickness and dimension. However, beam intensities have to be limited due to the easy disintegration of plastic-type materials but this is not really a concern with RIBs. In terms of purity, the areal thickness of hydrogen accounts for approximately 15% of H in CH_2, or 33% of D in CD_2, which is quite ineffective in terms of contaminants. Therefore, cryogenic solid targets are preferred for experiments that are highly sensitive to background contamination.

Early developments have proposed solid hydrogen targets in the form of nearly self-supporting cryogenic hydrogen material [65]. With the advent of RIB facilities, cryogenic hydrogen targets [66, 67] have been refined in order to provide relatively thinner thicknesses with a limited amount of contaminating materials.

Hydrogen freezing in itself presents an interesting thermodynamical problem. Indeed, the freezing of such material must be adequately performed in order to produce a homogeneous layer without structures, pockets of liquid material or bubbles. The hydrogen should preferably be grown in a crystalline form far away from the triple point where sudden changes in density are known to induce undesirable large defects in the layer. The gas is generally fed at rather low temperature ($\approx$80 K), and special care is required to avoid clogging of the inlet pipes while freezing the material.

Windowless solid–gas targets have also been the subject of interesting developments. A substrate is required to freeze the material, in other words, to grow the crystal. Material with a relatively high boiling point, together with rather low vapour pressure is suitable for use in high-vacuum chambers, otherwise a gradual reduction of the target thickness is observed through sublimation. Concerning H windowless gas targets, a gold substrate with the sufficient thickness of 60 mg/cm^2, cooled down to a temperature < 3K with a liquid Helium cryostat, can hold a homogeneous layer of hydrogen ice on its surface [67]. This kind of setup is not necessarily ideal with regard to cluster studies, as the gold backing would induce a large energy loss from the incoming or outgoing particles depending on the orientation of the target.

In recent experiments at the RIKEN RIB facility [68], H targets [69] were produced between two substrates that were removed after freezing, leaving a solid windowless H solid layer as shown in Fig. 6.9. The target holder was made of copper and the windows, used in the specific study of Ref. [69], were made of stainless steel plates coated with Teflon. The cell was sealed with a ductile metal, Indium, compressed between the target holder and the stainless steel plate. The 5–10 mm thick H cell is cooled down at He liquid temperature while the H crystal is grown. Once the freezing process is completed, the windows are moved away from the beam axis with no damage to the target; as a result a very homogeneous 40 mg/cm^2 thick target can be sustained under high vacuum at a temperature of 3 K with a very slow sublimation rate.

Thinner solid H and D targets are detailed in Ref. [66]. The princjple resides in the supply of gaseous H in a liquid He temperature cell with an appropriate design for the pressure and temperature gradient to freeze the H material gradually from the bottom to the top of the cell. The target, depicted in Fig. 6.10 is composed of

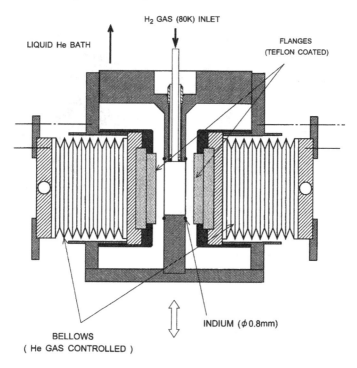

H₂ GAS (80K) INLET

LIQUID He BATH

FLANGES
(TEFLON COATED)

BELLOWS
(He GAS CONTROLLED)

INDIUM (φ 0.8mm)

Fig. 6.9 Schematics of the solid H target from [69]. The two removable bellows are cooled with liquid He during feeding of H to be gradually frozen until complete coverage of the hole

three cells, the middle one of which accommodates the H material. The He coolant is circulated in the two adjacent outside cells contained between Mylar windows. In this type of target, the Mylar windows are necessary to contain the coolant during the hydrogen phase transition. After freezing, the temperature is kept below the freezing point of H and the Helium coolant is not needed anymore. The H target is then sustained through the cooling of the cryostat, whereas the Mylar windows avoid the sublimation of the material, keeping the pressure within the solid target cell under the triple point. Although such a solution requires a containing windows, the thickness of the target can be kept below 10 mg/cm^2. The purity of the target is largely improved, as compared to a solid plastic target, but the Mylar windows, in contrast to self-supported targets, induce contaminating reactions. Further developments to remove the two Mylar windows from the He cooling with the solution described in Ref. [69] can be considered in order to reduce the contaminants by a factor of two and increase the thickness homogeneity.

Solid hydrogen can be made to large thicknesses compared to gas targets. The main advantage is the very good spatial resolution (1–10 mm) which is a crucial parameter with regard to kinematical reconstruction of many-body decay events, or using a magnetic spectrometer, for which the interaction point must be known as accurately as possible.

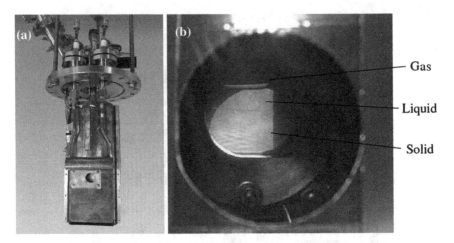

Fig. 6.10 Picture of the solid H target with the He cooling system a) and detail of the cryogenic cell b) [66], [Roussel-Chomaz, P., et al., Private communication]

6.3.3 Active Targets

Statistics governs the quality of a measurement. The event rate is calculated with the general expression:

$$N_{evt} = N_p N_t \sigma \eta_{det} \qquad (6.2)$$

where N_{evt}, the number of recorded events per second, is the product of the efficiency of the detector η_{det}, the number of target nuclei per cm^2, N_t, the number of projectiles per second N_p and the cross section σ of the reaction of interest. The beam intensity, especially when dealing with Rare Ion Beams, is in the hands of the accelerator engineers and will increase with technology and developments over the next decades. The cross section is a physical parameter and dealing with low beam intensity reactions with relatively large cross sections are preferred. The efficiency of the detector, η_{det}, must be pushed to the highest value as close as 100%. The last number to consider is N_t, which can easily be increased by adding more target nuclei, in other words increasing the target thickness. But this, in principle, is against good energy and position resolution. The first implementation of an active target probably emerged from the need of a proton target where low-energy detection thresholds were required in high-energy physics [65]. In cluster physics studies, recent investigations made use of a thick target to investigate the properties of resonant states. A beam impinged on a thick gas target was followed by two annular silicon strip detectors. The two detectors allowed some degree of tracking, namely a straight line intersecting two points. The interaction point was deduced by assuming that the projectile kept to a linear trajectory in the gas target along the beam axis. This information was sufficient to correct the energy of the detected particle before

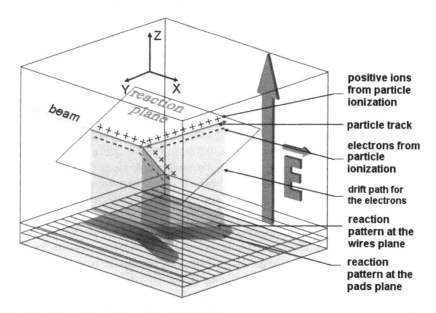

positive ions from particle ionization

particle track

electrons from particle ionization

drift path for the electrons

reaction pattern at the wires plane

reaction pattern at the pads plane

Fig. 6.11 Schematic rendition of the MAYA active-target. A beam projectile enters the detector volume where it reacts with a nucleus in the gas. The particles involved in the reaction may produce enough ionization to induce a pattern in the segmented cathode, after traversing a Frisch grid and a plane of amplification wires. A set of ancillary detectors is used in the exit side of the detector

it penetrated the two detectors and the gas target. The energy of the incident particle must also be calculated using the beam energy before penetrating the gas target and the thickness of gas material before the interaction point.

The archetype of active targets, in the domain of secondary beams is the IKAR detector [70] used at GSI (Germany) to study elastic scattering of exotic beams at relativistic energies. Other examples, such as the MSTPC detector [71] designed at RIKEN (Japan) to study fusion and astrophysical nuclear reactions in low-energy regions, or the early active target MAYA (see Fig. 6.11), paved a new avenue in combining the detector and the target in the same apparatus.

Active Targets (ACTAR) [72] work on the basic principle that the gas of a Time-charge Projection Chamber (TPC) [65] is also the target for nuclear reactions and the beam is impinging inside the chamber. The gas target can be pure or composed of a mixture of standard gas detection C_4H_{10} for instance or H, D and ^{3}He. Such an apparatus offers both very high efficiency and low detection thresholds; particle identification and complete reconstruction of events can be performed with deduction of the energy of the incident particle at the interaction point. This has the advantage of making full use of a low-intensity ion beam and a single-beam energy is sufficient to perform a complete excitation function measurement.

Early developments were based on square-shaped chambers such as the MAYA detector at GANIL shown in Fig. 6.11 [73] or the Bordeaux TPC setup [74]. It is interesting to mention that the Bordeaux TPC setup has been very successful with the first identification of the two-proton radioactivity [33, 34]. A number of the present developments are based on barrel-shape geometry of the chamber to comply with cylindrical coordinates of nuclear reactions occurring along the beam axis. It is mechanically more suitable to handle various pressures in the target ranging from a few mb to several bars with a minimum amount of material. This configuration presents an advantage if the target is combined with γ-ray detectors; the walls of the target can be kept relatively thin to reduce the absorption of the γ-rays. The diameter of the barrel-shaped ACTAR typically is 0.5 m and the length around 1 m. This geometry is ideal for applying a longitudinal magnetic field along the beam axis for particle identification. This would preferably be achieved by means of conventional (or even superconducting) magnets to sufficiently bend particles such as intermediate energy protons.

In the cylindrical geometry, as for example TACTIC at TRIUMF (see Fig. 6.12), the electric field of the TPC can be longitudinal or radial. Charged particles with enough kinetic energy, ionize the gas along their path. The electrons are drifting, due to the HV potential, and are detected with highly segmented detector systems placed on the end cap of the barrel or on the side walls. The electrons drift in the chambers with the velocity characteristic of the gas mixture and High Voltage (HV). Typical CO_2/Ar gas detection has been proven to be very suitable for use in drift chambers, especially, due to the high HV break down allowing large gains. The gain is limited by HV discharges but recent developments allow higher gain together with higher granularity with the Gas Electron Multiplier (GEM) technology developed at CERN by Sauli [76].

Drift electron velocity typically is in the order of 1–10 m/μs. Incident particles with relatively low energy at nuclear scale, i.e. 1 MeV/A, travel much faster than the drifting electrons, namely with velocity > 0.5%c, and the ionization can be considered to occur instantly compared to the drift time. The time difference, together with a 2D projection on the cap or side of the barrel, allows the reconstruction of 3D tracks. The curvature of the track, together with the total energy of the particle and path length, allow particle identification and the measurement of momentum projected on the three axes. Reconstruction of complete events requires the identification of the track of every particle. The readout chamber defines the count rate capability of the system, granularity, position resolution, drift times and energy.

Amongst the ACTAR setups currently operational or under development [72], the Active Target Time Projection Chamber (AT-TPC) at MSU/NSCL is designed to run in two different modes, namely as ACTAR or as a conventional detector [77]. The AT-TPC is a dual-functionality device containing both traditional active-target and time-projection chamber capabilities. The detector consists of a large gas-filled chamber installed in an external magnetic field.

Finally, it can be noted that in an ancillary detector mode, an exit window allows the ejectiles to exit the target chamber to combine the TPC for track identification and solid state detectors to stop energetic particles, as for example the SAMURAI TPC from Riken [68].

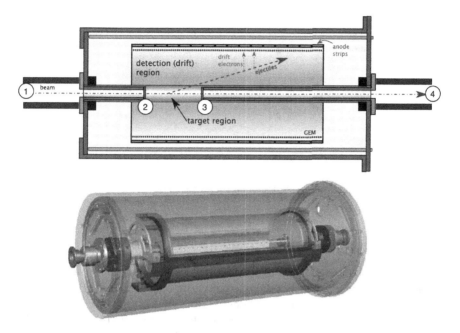

Fig. 6.12 Side view (top) and 3D representation (bottom) of the TACTIC ACtive TARget at TRI-UMF. The beam enters and exits the chamber at positions labelled 1 and 4. The entrance and exit windows to keep the target/detection gas within the chamber are labelled 2 and 3 [75]

6.4 Detection Techniques

Radiation detectors have been, and currently are, the subject of extensive research. Details concerning traditional radiation detectors may be found in two exhaustive review documents [78, 79]. The following section reports on some of the latest developments in highly segmented detection systems and advances in radiation detectors relevant in cluster studies.

6.4.1 Gamma-Ray Spectroscopy

Gamma-ray spectroscopists are confronted with a great dilemma between efficiency and resolution. High resolution is obtained at a cost of lower efficiency detectors and higher efficiency materials offer poorer energy resolution. In this section, some of the recent advances in the field of Germanium and $LaBr_3$ technologies with typical resolution of $\approx 2\,keV$ and $25\,keV$ at $1.332\,MeV$, respectively, are discussed. Ultimately, high resolution γ-ray spectrometers are assembled in large-coverage angle setups to overcome the intrinsic low efficiency of the detecting material.

Fig. 6.13 Comparison of
LaBr$_3$: Ce, NaI(Tl) and
BaF$_2$ energy resolutions with
a ^{60}Co source [79]

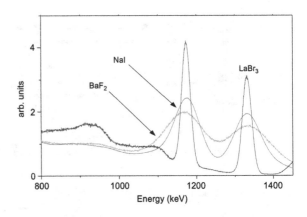

State-of-the-Art Scintillation Detectors

A recent break-through in scintillation technology has significantly improved the energy resolution in γ-ray spectroscopy, compared to traditional NaI(Tl) scintillators as shown in Fig. 6.13 where the resolutions of LaBr$_3$, NaI(Tl) and BaF$_2$ detectors are compared [80]. With a high yield of $\approx$60,000 photons per MeV, LaBr$_3$: Ce(0.5%Ce) scintillating material offers increased energy resolution for $E_\gamma > 100$ keV as well as high efficiency. The resolution at 662 keV is just less than 3% FWHM and improves with higher energy. The material shows excellent timing resolution in the order of 250 ps (decay time $\approx$30 ns) [81]. However, intrinsic background radiation in LaBr$_3$ originates from naturally occurring EC and β^- emitter ^{138}La(0.09%), the α emitter ^{227}Ac introduced while growing the crystal, and its daughter β^- and α emitters.

PARIS [82] is a $4\pi\gamma$-ray calorimeter project based on combined scintillating material. This detector array is under study to perform high energy γ-ray spectroscopy. Such an instrument will offer high selectivity capabilities and high detection efficiency with a resolution of near 1% at $E_\gamma \approx 10$ MeV. The still costly LaBr$_3$ scintillators are expected to be combined in sandwiched detectors, called phoswich, by which a CsI(Tl) with a long decay constant and a fast LaBr$_3$ are optically coupled to a unique light sensor. Pulse shape analysis, based on the very different decay constants, enables determination of the location of the interaction point in one or both crystals. A large number of physics cases are listed, amongst them, radiative capture through molecular state resonances and γ-transitions between molecular states.

Segmented Germanium Detectors

The search for γ-ray transitions between molecular states has been the subject of extensive investigations, but up to now there still is a paucity of concluding results. Modern γ-ray spectrometers will allow increased sensitivity of high resolution germanium detectors. Two major 4π segmented germanium detector arrays GRETINA/GRETA [83] and AGATA [84, 85] are being implemented. The AGATA demonstrator (see Fig. 6.15) and GRETINA, at a quarter of the GRETA array, present

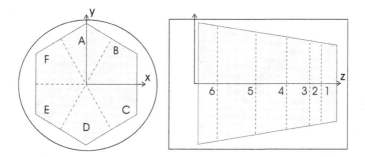

Fig. 6.14 Electric segmentation of a GRETA single Germanium crystal. The high resolution signal is collected from the inner core contact, not shown in the figure, and the outer junction is segmented into 36 elements [82]

the first steps in 4π sr germanium detector systems and, at this point in time, in-beam measurements have already been performed with the two instruments. The TIGRESS array composed of 16 Compton suppressed segmented clover detectors, with potential full γ-tracking, is already in use at TRIUMF [86]. The 4π γ-spectrometers will allow the measurement of rare events with unprecedented efficiency and signal-to-noise ratio owing to an extremely low Compton escape background. The γ tracking technology is based on the electric segmentation of a single Germanium crystal, as shown in Fig. 6.14. The high energy resolution signal is obtained from the core contact at which the electrons are collected. The holes drifting in the opposite direction induce a signal on several outer contacts. The point of interaction can be deduced from the line shape and amplitude of the signals by using sophisticated algorithms.

In theory, the 3D position of the interaction point can be deduced to within a few cubic millimetre-size resolution. In this way the accurate determination of the emission angle of the γ-ray allows precise Doppler correction. This is of interest when dealing with a fast beam where the broadening of γ-rays depends strongly on the opening angle of the non-segmented detectors. Full γ-tracking consists in the reconstruction of multiple hit events originating from Compton scattering or pair production. This is a very challenging task mainly due to the large number of possible combinations increasing very rapidly with the number of interaction points and number of incident γ-rays. AGATA and GRETA not only are based on the highest resolution detecting material currently available, but those instruments will be about 40% efficient at 1.332 MeV, with an angular resolution of $\approx 1°$. In calorimeter mode, events can be selected on the basis of multiplicity or total energy considerations.

The reference energy line at 1.332 MeV, together with the 1.173 MeV transition, originate from the ^{60}Co radioactive source feeding the excited states in ^{60}Ni. Traditional sources for calibration are ^{137}Cs for its single 661.7 keV transition; ^{152}Eu for its X-rays and numerous transitions between 121.8 and 1408.0 keV. This source is also useful for efficiency calibration as the strength of the various transitions is known very precisely. Low energy transition at 14 keV from ^{57}Co, or higher energies from ^{56}Co with high energy transitions up to 3.611 MeV, cover a

Fig. 6.15 Five clusters of the AGATA demonstrator at LNL, Legnaro, Italy

wide range of energy with relatively long-lived isotopes. Gallium-66 ($t_{1/2} = 9.49\,\mathrm{h}$) offers 18 strong transitions up to 4.806 MeV with well defined intensities [87]. For calibration points larger than 5 MeV, the first excited state of ^{16}O at 6.19 MeV can be populated through inelastic scattering. Transitions in the 10–15 MeV range are obtained using radiative capture (p, γ) or (n, γ) reactions, such as ^{11}B$(p, \gamma)^{12}$C with E_γ up to 13.92 MeV.

6.4.2 Charged Particle Detectors

Charged particle detection certainly is the main experimental probe in cluster studies. This is in relation to the decay mode of the states involved as well as the nature of the interaction of charged particles with matter. Semi-conductor detectors can be manufactured in thin layers, from 15 μm to 2 mm in thickness for silicon detectors. The technology has been developed extensively and detectors are made available in a large variety of shapes, sizes and segmentation. These detectors are suitable for the Particle Identification (PID) method based on the measurement of energy loss, where the thinner detector absorbs only a fraction of the incident energy, thus allowing very low detection thresholds. Germanium detectors for charged particles are less practical owing to the low operating temperature requiring a cryostat or a Helium cooler to be attached to the diode usually surrounded by a capsule. They are used in some applications where high-energy particles need to be measured with high-energy resolution.

The energy resolution of semi-conductor detectors is excellent, being in the order of 25 keV at 8.784 MeV for silicon. The position resolution is also crucial when reconstructing the kinematics of a cluster decay. In this respect, the fine electric segmentation on single-crystal silicon detectors is very useful. However, some imperfections due to the pulse height defect function of charge Z of the incident particle or the inevitable dead layer, required for charge collection purposes, induces some non-linear effects in the energy determination. Using the segmented detectors, interstrip energy deposition induces charges in two adjacent channels and possibly mimics two distinct particles. Accumulated radiation dose induces dark current in semi conductor detectors which in turns deteriorates gradually the energy resolution [17].

Some of the traditional calibration points are obtained from α emitters, usually parent radioactive isotopes and the daughters in the α chain. Open sources such as ^{241}Am ($E_\alpha = 5.486$ MeV), or the shorter-lived ^{228}Th (1.912y) with one of the highest α energy lines available at $E_\alpha = 8.784$ MeV provide precise calibration points under 10 MeV. Those energies are useful for the lower energy calibration points and beam-induced particles are often required for higher energy calibration points. The latter involves kinematics and stopping power considerations. Elastic scattering of light projectiles on natural ^{197}Au thin targets offer high-energy calibration points from forward to backward angles. Nuclear reactions such as ^{12}C(^{16}O, α)^{24}Mg* or ^{12}C(^{12}C, α)^{20}Ne* between $E_{lab} = 30$–60 MeV incident energy offer discrete α spectra according to the discrete levels in the residual nuclei. Due to the kinematics, the energy depends on the detection angle, which must be calculated accordingly. Fission sources with total kinetic energy up to $\approx$200 MeV shared between the two fragments can be used in some cases, but the continuous character of the spectrum and the various types of particles emitted make the calibration somewhat complicated. Numerous methods for calibration that use a fission source are reported in the literature [87].

Position-Sensitive Silicon Detector

Position-sensitive Silicon-Strip Detectors (PSSSD) are composed of individual electrically isolated adjacent strips on a semi-conductor equipped with a resistive contact. Considering a given energy deposited in the strip, the signal read on both edges decreases proportionally with the distance between the hit and the edge of the diode. The total energy of the strip is deduced from the combination of the two signals when calibrated adequately. When the total energy is known, the position of the hit is deduced from the relative amplitude. Figure 6.16 shows typical α loci obtained from the discrete states in ^{20}Ne populated in the ^{12}C(^{12}C, α)^{20}Ne* reaction at $E_{lab} = 32$ MeV. Due to the kinematics, the α-particle energies vary with the

Fig. 6.16 Two-dimensional scatter plot e_h versus e_l measured with one strip of a 16×16 strip ($50 \times 50\,\text{cm}^2$) Position-sensitive Detector using the $^{12}\text{C}(^{12}\text{C}, \alpha)^{20}\text{Ne}^*$ reaction at $E_{lab} = 32\,\text{MeV}$

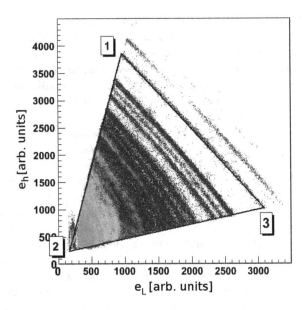

emission angle and show the typical curvatures with decreasing energy for increasing emission angle. The excited states in ^{20}Ne are well separated, the line defined by points 1 and 3 corresponds to the first excited state of this nucleus. The protons from elastic scattering, $^{1}\text{H}(^{12}\text{C},^{1}\text{H})$, should also be pointed out in the lower part of the spectrum. Hydrogen is contained from inevitable water contamination of the targets.

In principle each event falls within a triangle, determined by points 1, 2 and 3 shown in Fig. 6.16, which energy is defined by the equation:

$$E = L e_l + H e_h \tag{6.3}$$

where $e_{l,h}$ represent the raw energies, in channels, and the calibration factors are L and H. The intercept of the two lines (point 2) defined by the particles with various energies detected along the edges of the detector, represents the zero energy, thus defining the offset of the two channels. By using a mono energetic α source or beam-induced particles using elastic scattering of a very light particle on a gold target, one obtains the relationship between the calibration factors H and L as:

$$\frac{L}{H} = \frac{e_{h_1} - e_{h_2}}{e_{l_2} - e_{l_1}}. \tag{6.4}$$

L and H reduce to only one calibration factor, L as a function of H or vice versa, which allows plotting of an uncalibrated total energy spectrum. The remaining factor is deduced from the calibration points by matching the centroids of the peaks to the known energies.

The position information, x, is extracted from the two calibrated amplitudes $E_{l,h}$ as:

$$x = X \left(a + \frac{E_l - E_h}{E_l + E_h} \right) \tag{6.5}$$

where X is the length of the strip and a a calibration parameter ideally equal to 0.5. The position resolution along the strip depends on the energy resolution of the detector. A typical performance of 0.1 mm position resolution is achieved. Longitudinal resolution is much better than the position resolution across the strip, which depends directly on its physical width.

The detection threshold is a function of the position along the strip, which must be considered when measuring low-energy particles. The lowest threshold is at the centre of the strip where the signal experiences an equal amount of resistive material from the hit to both edges of the strip. The threshold is at the maximum for a hit close to the edges when the amplitude is most decreased through the complete strip. Two-dimensional position sensitive detectors are fitted with four contacts and offer good position resolution on two axes [89, 90]. Position and energy can be measured using relatively large detectors with a limited number of electronic channels.

Double-Sided Silicon Strip Detector

The P and N junctions of a semiconductor are segmented in parallel strips on both sides but perpendicularly to one another. A hit is then recorded in two individual strips and the particle position is identified in a pixel defined by the crossing point of the two strips. Great advances were performed with the development of the inner barrels of high-energy physics detectors such as ALICE at CERN, or the high-precision cosmic-ray detector AMS-02 of the International Space Station. The size of the pixel is defined by the pitch of the strip varying from the typical size of 50–3000 μm [91, 92] to the total detector size of up to 100×100 mm^2. The design of MUST/ MUST2 [93, 94] shown in Fig. 6.17 is based on 300 μm thick DSSSD, 128×128 strips with a total area of 100×100 mm^2.

Recent studies in 2-proton radioactivity have made use of multi layers of highly segmented detectors with a pitch of 100 μm [95, 96]. Tracking techniques developed for this instrument, part of the RB3 setup, allow the reconstruction of the vertex of the primary decay with high accuracy and determine the correlation between fast-moving particles. Life-time measurement of the state can be performed down to a few picoseconds when using position reconstruction. After interaction, the location of the decay with respect to the target position is obtained from the determination of the vertex with a precision of some tens of μm. The life time is determined according to the velocity of the break-up nucleus and the position distribution of the decay within the target.

Position assignment in single-particle detection is not ambiguous. Multi-hit events within one double sided detector have to be sorted adequately due to multiple combination of crossing strips as shown in Fig. 6.18. This is possible when particles are detected within a relatively large energy range. The pair of X–Y strips

Fig. 6.17 Highly segmented silicon-strip detector of the MUST2 setup [93]

of the same hit read an approximately equal energy. The true crossing points are deduced from the comparison of the energies. Identical particle energy events remain ambiguous and should be discarded. For multiple-hit events, the number of X and Y strips must be identical, otherwise more than one particle is detected in a single strip, and the energy of the particles must be well separated. The number of combinations increases rapidly as it is proportional to $N!$ with N the number of hits.

Annular Segmented Detectors

Annular Segmented detectors offer an ideal geometry for in-beam measurements. The technology is identical to the DSSSD, differing only in the geometry. Cylindrical coordinates are better adapted to in-beam measurement and the Silicon detector disc covers a very large portion of the azimuthal angles. This is of particular interest when detecting γ-rays in coincidence with charged particles for Doppler correction. The centre of the detector is placed in the beam axis passing through the hole and covers a large solid angle downstream or upstream of the target. The minimum angle is defined by the distance between the detector and the target. Annular detectors are found in a single detector unit, also called a CD on account of its similar size and aspect, or can be assembled from separated sectors. The sectors can be arranged in a flat or "lampshade" geometry as shown in Fig. 6.19 in single or double-sided segmentation.

Precise energy measurement must take into account the effective thickness of the detector. In large-acceptance setups the detectors can be placed relatively close to the target and the effective thickness observed by the particle must be taken into account

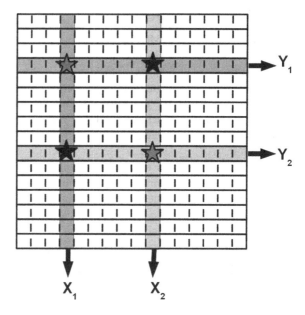

Fig. 6.18 Two-hit event in a 16 × 16 Double-sided Silicon-Strip Detector. Red stars indicate true pixels, black stars are mis-assigned events

Fig. 6.19 Separated sectors arranged in two different annular configurations

as a function of the incident angle with respect to the detector plane. This is very important, especially when correcting for energy loss in the junctions of the detector, the so called dead layer.

Micro-Channel Plates (MCP)

Micro-Channel Plates (MCP) are large area detectors pierced with a large number of small holes. A high voltage is placed across the front and back of the detector. When a particle strikes the front of the detector, secondary electrons are emitted and multiplied through the channels until collected on the back of the plate. Multiplication factors in the order of 10^6 are achieved with adequate X–Y readout for position measurement with excellent timing with a rise time of typically 2–3 ns. MCPs are used traditionally to multiply electrons emitted by an incident particle when impinging on a thin foil.

Charge identification is not sufficient to fully characterise a particle and mass identification is often required. Mass determination is accessed by using Time-of-Flight (ToF) and energy measurement techniques. MCP detectors are very useful for this purpose when a thin foil, Mylar or carbon foil, for instance, is placed in the path of the particles for electron production. Those electrons directed at an MCP, possibly with a magnetic field to bend the trajectory and enhance the collection efficiency, generate precise timing signals. The reference time signal is derived from a second MCP or by using the beam time structure, if available. When using very thin ^{12}C foils down to $4 \, \mu g/cm^2$ the path and energy of the particles experience very little alteration. The energy is measured by using a solid-state detector or a magnetic spectrometer placed downstream from the MCP detectors. This technique is employed from low-energy particles in Elastic Recoil Detection Analysis (ERDA) or RIB facilities from incident energies of some hundreds of keV/u to a few hundreds of MeV/u.

The large-area micro-channel plate entrance detector DANTE [97] of the magnetic spectrometer PRISMA installed at LNL (Legnaro, Italy) is used to determine the direction of the recoil nucleus by using the X–Y determination capability of MCP detectors.

Gaseous Detectors

A large variety of detectors are based on the ionisation of a detection gas enclosed in a chamber under high voltage bias in order to collect the positive and negative charges before recombination occurs. Geiger Müller counters, Parallel Plate Avalanche Counters (PPAC), Multi Wire Proportional Chambers (MWPC), Drift Chambers and Time Projection Chambers (TPC) are some examples of gaseous detectors based on ionisation chambers with various electric fields, electrode arrangements and granularity. Large density of strips allows high granularity of the detection system and increased position resolution. Micro-Strip Gas Chambers [98] (MSGCs) followed the multi-wire chambers where the anodes and cathodes are thin wires stretched between mechanical supports. The strips are preferably printed on an isolating substrate [98] allowing higher density and an easier manufacturing process. As a result, the density of wires has been increased largely together with gains and robustness. The life time and reliability of the detectors were greatly improved with MSGC technology.

Amplification in gaseous detectors mostly is a function of the electric field and pressure. Ionisation electrons, being emitted during interaction with an energetic particle, are accelerated under an electric field. Wire chambers are made of relatively thin wires, in the order of 50 μm in diameter, under potential differences of a few kV and a pressure of the ionisation gas of between a few mbar up to atmospheric pressure. Multiplication is possible when the maximum velocity of the electrons, which is the function of the acceleration and the mean free path, is enough to overcome the ionisation energy and create new electrons through collisions with the atoms and molecules of the gas. The multiplication mostly occurs when electrons approach the wires, as the electric field, inversely proportional to the distance, increases dramatically. Proportionality is obtained if the electric field around the wires is not affected by the density of electrons produced in the avalanche, otherwise saturation effects are experienced. Multiple processes that take place while multiplication occurs includes the production of photons able to ionise further atoms or molecules through photoelectric effect. Very large amplification factors, in other words large electric fields, imply possible break-down or sparking within the detector.

Gas Electron Multipliers (GEM) Detectors

Constant improvements of gaseous detectors has led to new generations of microstrip detectors, for instance the MICROMEGAS [99, 100] developed at Saclay or Secondary Electron Detectors (SED) based on the detection of electrons emitted from a Mylar foil used to track rare ion beam projectiles at a relatively high rate [101]. Gas Electron Multipliers (GEM) [76, 102] developed at CERN are based on MSGC technology with X–Y reading equipped with a preamplifying device to multiply the number of electrons before collection [98].

GEM detectors are based on multipliers essentially composed of a thin ($\approx$50 μm thick) isolating plastic foil, Kapton or Mylar for example, coated on both sides with an even thinner layer of conducting metal. A chemical process is used to pierce the material with micro holes, located $\approx$50 μm from each other and arranged in a regular pattern as shown in Fig. 6.20a. The cross section of the multiplier with electric field equipotential lines is shown in Fig. 6.20b.

The voltage applied between the opposite layers is typically in the order of 1000 V, offering a large electric field between the two layers of metal, greater than 10^7 V/m. The incident electrons, ejected by ionising particles in the detection gas, generate avalanches of ions and electrons through the holes. The amplification is operated at much lower HV than standard MSGCs but higher gradient owing to the small spacing between the two metal layers. Standard MSGCs are placed at ground potential, making the detection system relatively safe against HV discharges.

The third generation of GEM consists of a number of multiplying layers gradually incremented from 1 to 3. The gain is subsequently increased through multiple stages. Since the early developments in the late 1990s, GEM detectors have found a wide range of applications primarily in high-energy physics but also in medical scanning devices or nuclear physics within the newly developed ACtive TARgets, and have

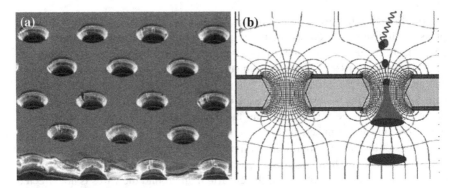

Fig. 6.20 Regular arrangement of micro holes through a foil of Kapton coated with Copper on both sides a), and cross section with electric field represented through the thin layer b)

become particularly interesting in nuclear cluster physics and nuclear astrophysics studies.

Segmented Scintillators

Scintillation detectors convert some of the energy of an incident particle into photons, typically on the UV side of the visible range. The secondary photons are converted into electrons and further processed through electronic systems. The energy resolution obtained from scintillation material is not competitive, compared to solid state detectors, mostly because of the lower number of electrons emitted per unit of energy deposited. Moreover, the response of the detectors depends on the type of incident particle. The light output is a function of the mass, charge and energy of the particles and is not linear. However, those detectors, especially the plastic scintillators, can be shaped at ease and to relatively large sizes. Inorganic scintillators, such as NaI(Tl), LaBr$_3$ or CsI(Tl) for example, are slightly to highly hygroscopic and must be housed accordingly. Large arrays of scintillation detectors were developed to register high-multiplicity events at intermediate beam energies. INDRA and CHIMERA [103] arrays are made of a large number of silicon and scintillator detectors assembled in rings. Such detectors have been used to search for α gas condensate in $N = Z$ nuclei.

Large position-sensitive photomultiplier tubes, of $50 \times 50\,mm^2$, were recently made available on the market. They offer new opportunities for position-sensitive scintillator detectors. An arrangement of needle scintillators, closely packed and optically coupled with the PM tube but optically isolated from each other, allows the measurement of the energy of the particles with a good determination of position, typically within the size of a needle. The signal of the position-sensitive photomultiplier tubes is obtained by means of only four channels derived from a network of resistors connected in a sophisticated manner. The sum of the four signals is proportional to the energy deposited and the relative amplitudes are used to determine the

2D position. This technique offers high granularity with a relatively low number of electronics channels.

The maximum thickness of segmented silicon detectors is in the order of 1 mm. This is sufficiently thick to stop ions with a few tens to few hundreds of MeV/u but light ions need substantially thicker detectors for complete energy deposition. Multi-layered detectors are not only necessary for $\Delta E - E$ discrimination, as discussed in the following section, but also to stop higher energy particles. A 1 mm silicon layer stops only 12 MeV protons, and 48 MeV α particles and higher energy particles are traditionally stopped in multi-stage silicon/scintillator telescopes.

6.4.3 Neutron Detection

Detection of fast neutrons requires large volume detectors due to their long mean-free path within matter. The type of interaction between neutron and matter also pre-empts high-energy resolution measurement using other means than ToF techniques. For these reasons, large neutron detector arrays are currently not capable of determining the position and energy of particles with a comparable degree of precision to charged particles with silicon detectors or γ-rays with germanium tracking detectors.

If thermal neutrons are traditionally detected with ^{3}He-, ^{6}Li- or ^{10}B-based detectors, fast neutrons are preferably detected with scintillator detectors or fission chambers. Part of the neutron momentum is converted into a moving charged particle using, for example, the (n,p) elastic scattering reaction in plastic scintillators. Those detectors are preferred in nuclear physics studies as the neutrons are rarely emitted with less than a few hundreds of keV. Liquid organic scintillators, such as NE213 material (Nuclear Enterprise Ltd), are employed in a number of arrays, for instance DEMON [104]. This neutron detector array has been used to investigate a hypothetical tetraneutron resonant state [105]. Thus far, such observation has not been confirmed, partly owing to the difficulty of firmly assigning such high-multiplicity neutron events. The neutron wall of EUROBALL made use of the BC501A (Bicron Radiation Measurement Products) liquid scintillator [106] that uses pulse shape discrimination to distinguish photons and neutrons. ToF measurement is used as another way of discriminating γ and neutrons but also to determine the neutron energy.

Plastic scintillators are very fast detectors but do not display particle pulse shape dependence. As a consequence, the γ-neutron discrimination is performed using only time-of-flight techniques. However, large volume detectors can be fitted with multiple light sensors. Both the time difference method and relative amplitude between two PM tubes fitted at both end of a scintillator bar can be employed to locate the interaction point, which allows a degree of position determination. The light collected by a PM tube is a rather complex function of the distance between the interaction point and the PM tube. The loss of light from the interaction point to the photocathode depends on the light absorption through the scintillator and reflecting material. A relatively good approximation consists in assuming a linear decrease of light intensity as a function of the distance between the PM tube and the interaction

Fig. 6.21 Compact geometry configuration of the MoNA neutron detector at MSU/NSCL [105] with 70% efficiency for E_n 50–250 MeV and position resolution of $\approx$12.5 mrad (7 cm within the scintillator bar)

point, hence, the position can be reconstructed using the same method detailed for the Position-sensitive Silison Detectors (PSD). Typical position resolution of about 5 cm is achieved.

The MoNA/LISA setup is designed for the detection of fast neutrons emitted from neutron-rich nuclei populated with the rare ion beams at the MSU/NSCL facility. The compact version of the MoNA detector is shown in Fig. 6.21. The first version of this modular neutron detector array that is based on time difference measurement for position determination, is composed of 144 scintillator detectors of $10 \times 10 \times 200 \, cm^3$ size [107]. The current upgrade of this system, naturally called LISA, is being implemented to provide a larger covering angle and efficiency.

A new neutron detection array at the RIB facility of the FLNR/JINR of Dubna, Russia, is based on stilbene crystals by which γ/n discrimination is made possible to even lower energy thresholds compared to liquid organic scintillators. However, the response of the crystal depends on the emission angle of the charged particle with respect to the crystal orientation. The efficiency of such detector material is superior to liquid scintillators, and more compact detectors with more granularity can be made.

6.4.4 Mass Spectrometers, Mass Separators and Combined Setup

Application of mass spectrometers to cluster state physics was mentioned earlier with some of the early measurements in heavy cluster emission with SOLENO. Magnetic mass spectrometers are very powerful instruments in terms of selectivity and energy resolution. A charged particle in a magnetic field is deflected according to its rigidity, given in units of Tm, which is the product of the magnetic field B times the radius of curvature ρ.

$$B\rho = \frac{p}{Q} \tag{6.6}$$

where for a given magnetic field the radius ρ is proportional to p, the momentum, and inversely proportional to Q, the charge of the particle in units of proton charge. The rigidity squared becomes a function of energy E in MeV, mass M in a.m.u. and charge Q of the particle. The K value characterises the maximum particle energy that a spectrometer or cyclotron can deflect within the boundaries of its magnetic field, following the expression:

$$K = \frac{EM}{Q^2}. \tag{6.7}$$

Zero degree measurements are very attractive in terms of population selectivity of states from the relationship between angular momentum transfer and incident beam energy [108, 109]. Typically low spin states are favourably populated at beam energies under $50\,\text{MeV/u}$ in (p, t) reaction measured at $0°$. Those measurements are much more challenging because the outgoing particles must be separated out of the ion beam. Rejection factors of 10^{10} are obtained in (p, p') reactions with a high resolution measurement and of 10^{13} or better in (p, t) reactions. Currently, only a few facilities perform routinely this type of measurements; the Grand Raiden spectrometer at RCNP, Japan [110], and the K600 spectrometer at iThemba LABS, South Africa [111].

Mass separators not only make use of the magnetic field, but also of electrostatic high voltage. Velocity filters, called Wien filters, are a combination of a magnetic field and an electric field called perpendicular to one another. A succession of magnetic dipoles (M), magnetic quadrupoles (Q), magnetic sextupoles (S) and electrostatic dipoles (E) of the mass separator DRAGON, TRIUMF [112], is depicted in Fig. 6.22.

In both mass spectrometers and mass separators, the particles deflected away from the focal plane must be dumped in a stopper. Unexpected background arises from the scattering of unwanted particles such as multiple-scattering particles within the target, beam halo, reflections on the walls of the vessel or from the stopper itself. The selectivity of the apparatus lies in its capability to reject unwanted particles.

The particles are transported through the magnetic and electric fields to a focal plane where they are detected. In a magnetic spectrometer, the energy is deduced from the position at the focal plane, calibrated from known reaction products obtained by means of Multi Wire Chambers or solid-state detectors with a high resolution position. Unlike total absorption energy measurements with solid state detectors, the energy resolution is independent of excitation energy and remains fairly constant throughout the whole spectrum. Energy resolutions of under $20\,\text{keV}$ are obtained in careful (p, p') measurements. Particle identification can also be performed with

Fig. 6.22 Layout of the DRAGON mass separator at TRIUMF with magnified ion trajectories [112]. The mass separator is a succession of magnetic dipoles (M), magnetic quadrupoles (Q), magnetic sextupoles (S) and electrostatic dipoles (E). Particles are detected at the focal plane with a segmented silicon detector

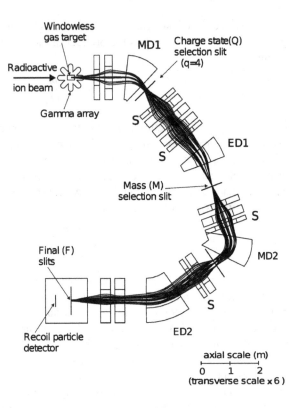

multi-layer detectors placed after the position-sensitive detectors. This allows some degree of discrimination when particles with identical rigidity are produced.

Magnetic spectrometers and mass separators are used in conjunction with ancillary detectors and are often part of complex experimental setups. The PRISMA large-acceptance magnetic spectrometer (≈ 80 msr, $\Delta p/p \approx 20\%$) at Legnaro was initially fitted with the CLARA array, 25 Clover Germanium detectors with total photopeak efficiency at 3% [113], then with the AGATA demonstrator. Internal PPAC detectors are placed at the focal plane and the DANTE detector is used for ToF measurements and to determine the direction of the ion entering the spectrometer, which is a very important information for Doppler correction. Precise determination of A and Z of the recoil nuclei is obtained from the trajectory of the particle, together with the range energy in the PPAC. The energy is then deduced from the mass and ToF measurements.

The population of the excited states in the two ^{24}Mg fragments originating from molecular resonant states in ^{48}Cr were studied through γ-ray spectroscopy with the PRISMA/CLARA setup [114]. Recent measurement at the DRAGON facility, with the γ-ray spectrometer based on high-efficiency BGO detectors, has shown the high radiative capture cross section in ^{12}C $+ ^{12}$C reaction with evidence of doorway states in ^{24}Mg [115].

MUST2/TIARA/VAMOS/EXOGAM/BTD is a sophisticated detector arrange-ment which acombines state-of-the-art detection techniques in a very complete setup. The incident rare ion beam particles are tracked with the Beam Tracking Detector, low pressure multi-wire proportional chamber for secondary beam tracking (CATS) [116], or with Secondary Electron Detectors [117]. The energy and position measurement of the light-charged particles is performed with TIARA/MUST2 [94], γ-ray spectroscopy with EXOGAM, and the detection of the recoils with the large-acceptance magnetic spectrometer VAMOS [118].

6.4.5 Particle Identification

$\Delta E - E$ and Time-of-Flight Methods

Traditional methods are based on time-of-flight discrimination by using energy versus time correlation. Absolute time reference is useful as, for example, the RF signal from cyclotron accelerators. Two MCPs can be employed when dealing with continuous ion beams of electrostatic machines when a pulsed beam is not necessarily available. The time difference between particles detected in coincidence can also be employed.

The energy loss of energetic particles depends on the mass A, charge Z, energy E and on the interacting material. The famous Bethe and Block formula describing the energy loss of particles can, in its simplest version, be reduced to:

$$\frac{dE}{dx} \propto \frac{AZ^2}{E}. \tag{6.8}$$

For particles with identical incident energy, enough to punch through a sufficiently thin detector, the energy loss is proportional to the mass and the square of the charge of the particle. This technique is widely used with Silicon telescopes or ionisation chambers backed with solid detectors for lower detection thresholds. Using silicon detectors the technique is limited by the lack of mechanical strength of thin ΔE diodes, typically not thinner than $15\,\mu$m, setting a limit on the detection thresholds. Some efforts were devoted to build monolithic detectors where the ΔE layer, as thin as $1\,\mu$m, is built using ion implantation on a $400\,\mu$m E detector [119].

Pulse Shape Discrimination

Pulse shape discrimination also has a long history, especially in using CsI detec-tors [120], organic liquid scintillators and even silicon detectors. Particles deposit their energy as a function of their interaction mode, with the rate of energy loss depending on the mass, the charge and incident energy. Differences in the rise time or decaying part of the pulses are caused by the increasing fraction of excitation with longer lifetimes of the molecules of some specific scintillating material with increas-ing stopping power. Pulse Shape Discrimination is crucial in neutron spectroscopy for identifying neutrons and γ-rays and was first suggested in the late 1950s by

Fig. 6.23 Particle identification with a single silicon detector in a rise time versus energy scatter plot [120]

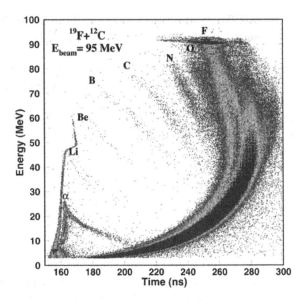

F. D. Brooks [121]. Some work on Silicon detectors by Ammerlaan in the early 1960s shows particle dependence of the signal. The basic principle of particle identification from a single signal consists in measuring the ratio of the charge accumulated under selected areas. A two-dimensional plot of the ratio, within the regions of interest, versus the total energy allows particle identification.

In Silicon detectors, the dynamics of the charge collection depends on a number of phenomena. The holes and electrons are produced by the ionising radiation and migrate towards the electrodes. The collection time for electrons is faster than for the holes and the shape of the signal, due to the interplay between negative and positive charges, depends on the penetration depth of the incident particle. Moreover, the linear energy transfer of a particle in matter depends strongly on its charge Z. Finally, the plasma created along the particle track partly shields the electric field felt by the electron-hole pairs and results in delayed charge collection time known as plasma erosion time. Therefore, the leading edge of the pulse is expected to carry some information on the particle type. A number of electronics manufacturers for Nuclear Physics applications provide multiple CFD output discriminators, namely two CFDs at 30 and 80% rise time that can be used in conjunction with TDCs to record the time information. The time difference between the two CFDs plotted against the amplitude, itself coded using a standard ADC, produces a particle identification spectrum. This technique has proven to be efficient with silicon detectors, which are relatively slow, where the leading edge experiences a difference in shape depending on the nature of the particle. Figure 6.23 shows such a PID matrix, namely rise time versus energy, recorded using the $^{19}F + {}^{12}C$ system at 95 MeV incident beam energy [122].

Pulse shape analysis has long relied on dedicated electronics designed for certain types of detectors, therefore, though the method is very powerful, the technique remained limited. With the advent of Digital Signal Processing, standard electronics has become much more versatile. At the time this document is written standard digital signal processing units digitise with a sampling rate in the order of 100 MHz to a few GHz. This contrasts with analog electronics where the analog-to-digital conversion is performed at the end of the electronic chain to capture the amplitude of a signal at a rate not exceeding tens of kHz. Note also that slow conversion induces long periods of time, called the dead time, during which the electronics is awaiting the processing of an event before being ready for a new event. With high sampling rate digital electronics, details on the pulses are obtained and pulse shape analysis is performed by means of an algorithm. As a consequence, Pulse Shape Analysis is much more flexible with digital electronics compared to analog electronics. Software QCDs are set to measure the integrated charge in various numbers of selected areas of an electronic pulse. The areas of interest are selected in relation to the detector in use or particles to be detected. Calculated CFDs, rise time, or analysis of the leading edge of a signal can be performed within the FPGA of the module to deduce useful information. Recent work has shown the discrimination between low-energy charged particles by plotting the ratio of the integrated charge to the total charge versus the energy of the particle [123]. Figure 6.24 displays such particle identification obtained for particles stopped in the silicon detector using E versus the Rise Time correlation. The right panel shows the traditional $\Delta E - E$ PID spectrum for those particles punching through the 300 μm silicon detector. The left panel displays the result of Pulse Shape Discrimination techniques for the particles stopped in the silicon detector. This technique has the enormous advantage of reducing the detection threshold, especially regarding high Z particles.

Particle identification in Bragg Ionization Chambers (BIC) is based on pulse shape analysis. The energy deposition profile of a charged particle shows a large part of the energy loss at the end of the path through the medium while the energy deposited per unit of distance along the path is fairly homogeneous. This is understood from the expression of the energy loss dE/dx of a particle, which is inversely proportional to its energy. The ratio of the energy deposited along the path to the energy deposited on the Bragg peak informs on the atomic number of the incident particle. This method is applicable to particles with incident energies $E \geq 0.5$ MeV/u. Those detectors allow sufficiently long paths along which ionisation electrons are collected radially. Such particle identification spectrum shown in Fig. 6.25 is obtained with the Binary Recoil Filter in the ^{24}Mg + ^{12}C system at $E_{lab} = 130$ MeV [124]. The identification thresholds are relatively low and the turning points on the higher energy side of the Z bands correspond to the ions punching through the ionisation chamber. The pressure within the chamber is optimised to stop the ions of interest indicated here with the locus on $Z = 12$. BIC detectors are used in the FOBOS [22] setup where the energy of the heavy ions is too low to efficiently use the method as the ions penetrate the chamber when already within the Bragg peak regime. However, using the Bohr-Willer empirical equation, the range of a particle R within the ionisation chamber is found to be:

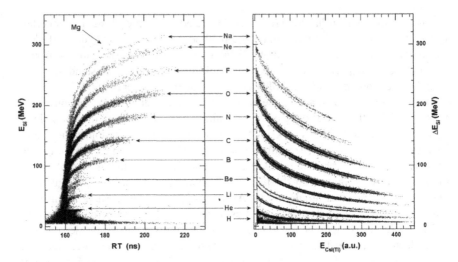

Fig. 6.24 Particle Identification obtained with a 420 MeV ^{20}Ne beam bombarding a ^{12}C target. Left Panel: PID using Energy versus Rise Time techniques from Digital Signal Processing electronics. Right panel: PID using $\Delta E - E$ techniques with Silicon and CsI(Tl) telescope [123]

$$R \propto \frac{\sqrt{EM}}{Z^{2/3}}. \qquad (6.9)$$

Complete Kinematics Measurements

Complete kinematics events measured with sufficiently high energy resolution can provide a degree of particle identification even if only the energies are recorded. Particle identification is made possible on the basis of momentum considerations. If the final state is known, through Total final state Kinetic Energy (TKE) selection, the calculated sum of the momentum of the individual particles must equal the momentum of the projectile. The momentum on the z-axis namely must equal the momentum of the beam and the sum of the momenta on both the x- and y-axes must equal zero. This implies calculating the momentum for all possible combinations until the conditions are fulfilled within the resolution of the experimental setup. A number of $N!$ combinations exist for N different particles in the exit channel. This is fairly straightforward for binary kinematics. In such a specific case, the mass of the two particles can even be deduced by means of Eq. 6.10 from the energy of the two individual fragments

$$M_2 = M_T \frac{E_1 sin^2(\theta_1)}{E_1 sin^2(\theta_1) + E_2 sin^2(\theta_2)} \qquad (6.10)$$

where M_T is the sum of the mass of the two fragments $M_{1,2}$ involved in the decay, $E_{1,2}$ and $\theta_{1,2}$ their energies and longitudinal angles in the laboratory frame. This is

Fig. 6.25 Particle identification in Bragg Ionization Chamber of the Binary Recoil Filter using the $^{24}Mg +^{12}C$ system at 130 MeV [122]. Pressure in the BIC is adjusted to stop the ^{24}Mg recoils as shown with the 2D gate

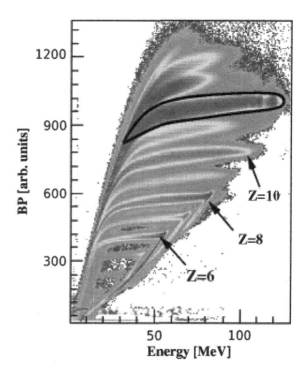

typically used to determine the mass of fission fragments produced in fusion fission reaction approximated as a binary decay omitting neutron emission.

Magnetic Spectrometers and Gas-Filled Spectrometers

Magnetic spectrometers make use of the magnetic rigidity to deflect a particle with a given charge, mass and energy to the focal plane. Full particle identification makes use of ToF and other techniques detailed in this section. As stated at the beginning of this section, the measurement of the energy loss of particles through a given thickness of material possibly is one of the most widely used methods for discriminating particles. A variation of that method consists of letting the ions of interest pass through a chamber with a partial vacuum in the mbar region. An average charge state distribution is populated as a function of the pressure and the energy of the ions. Such a chamber is placed in a magnetic field in order to deflect the ions according to their average mass over charge state ratio. This method has been implemented in a number of setups such as the BGS (Berkeley Gas-filled Spectrometer) [125] or the RITU Gas-filled spectrometer [126, 127]. Recently, some similar measurements with VAMOS, [128] and references therein, were performed in order to separate the Evaporation Residues from the projectiles at $0°$.

Magnetic Field

As already detailed in the section dedicated to Active Targets, most of the PID methods described in this section can be employed to determine the nature of particles. In high-energy physics, particles are bent in large toroidal magnets for particle identification and energy measurement purposes. Such techniques are relevant in ACTAR, in applying an axial magnetic field to deviate the particles not necessarily emitted perpendicularly to the magnetic field. The trajectory of the emitted particle is coiling, with the bending radius being a function of the mass, charge state and incident energy itself decreasing through the gas cell. The AT-TPC of MSU is designed to accommodate a 2T magnetic coil for PID purposes. Regarding the already large amount of information available, the insertion of magnets or coils is not entirely necessary, especially if some ancillary detectors, for example as γ-ray detectors, are to be placed around the target.

6.4.6 Electronics and Data AcQuisition (DAQ) Systems

Some of the large experimental setups presented in the previous sections require a few thousand electronics channels and standard modular electronics is not really suitable. For reasons of granularity, the pitch of the silicon-strip detectors has been reduced dramatically on increased size detectors and space constraint becomes a limiting factor. Every strip requires an electronic channel with preamplification, amplification and possibly digitisation capabilities. Integrated electronics allows the implementation of all these components very close to the detector.

Front-End Electronics

The signal processing of detection systems is operated by two main components, front-end and back-end electronics. The front-end electronics processes electric signals with characteristic rise time, decay time and amplitude delivered through the characteristic impedance depending on the type of detector. The preamplification stage must be of high quality and large bandwidth for high resolution and linearity. Careful shielding and a minimum length cable are prerequisites to minimise the electronic noise picked up from cables acting as aerials before feeding the preamplifiers. A linear response of the preamplifiers and amplifiers for the spectroscopy signal is crucial. Fast-timing is necessary when generating logic signals and for efficient pulse shape discrimination if such a method is employed. Single channel modules in NIM format tend to be ineffective in terms of space and rapidly increase the complexity of a system when a large number of channels are involved. Relatively compact electronics in NIM/VME/CAMAC format offer high-density modules, traditionally in a multiple number of 16 channels, 16/32/64/128, for linear amplifiers, ADC, TDC, QDC or other various functions. In analog electronics systems, the building and recording of an event is performed with one or several front-end computers communicating with the converters (ADC/TDC/QDC).

The communication is often managed by a trigger module requesting the coding, initiated by a trigger signal built according to the requirements of the experiment. While the transfer of the data is operated, the trigger module vetoes any new acquisition until the system has processed the event.

Alternately integrated electronics developed for highly segmented detectors, perform some of the signal processing, such as two stage amplification, timing output and, possibly, fast digitisation in electronic modules placed very close to the detectors. Such integrated systems are developed for dedicated setups where compactness and the characteristics of the detectors are unique. High-density electronics, however, is challenging in terms of parasitive induced signals between neighbouring components; reduced cross talk is one of the key features of such electronics. Customs application can be implemented using Application-Specific Integrated Circuits (ASIC). Unlike a dedicated integrated circuit, this kind of standard modern electronics makes use of adequately connected existing blocks. This technology makes the designing easier and more cost-effective compared to developing a new chip from scratch. Possible digitisation, for Digital Signal Processing, and some triggering can also be implemented and each module is connected to the back-end electronics for event building and communication between the modules. Digital electronics require a great deal of computing resources, especially when high-resolution digitising, 14-bit sampling, is combined with a high sampling rate creating large amount of data. Some systems are even run triggerless, which has the disadvantage of generating extremely large amounts of data but using time stamping, long time separation events can be retrieved off-line.

Front-end electronics is equipped with its own computing unit and operating system, for example VMS/VAX, UNIX/Linux, Windows or VXWorks. For a medium-sized and large experimental setup, more than one front-end computer can be used to operate each a section of the setup. Front-end computing units communicate with the back-end electronics via, for example, fast Ethernet connection.

Back-End Electronics/Data Acquisition Systems

The communication with the front-end electronics is performed through the back-end electronics. Online and offline analysis, monitoring of the detectors, slow control, configuration of the front-end electronics, and data storage are the typical operations performed by the back-end electronics. Numerous systems are available or often built in-house for dedicated applications. MIDAS, developed at the Paul Sherrer Institute, is a general Data Acquisition software tool making use of the ability of ROOT for data handling and can be run on multiple platforms [129]. The parameters of the DAQ are stored in an Online Data Base. Control is performed via a web interface which makes it accessible from any authorised machine. Online analysis allows ungated or gated histogramming by means of selections set in the data base. Communication with the front-end electronics is operated through the drivers of the front-end electronic modules; CAMAC, VME, Fastbus, GPIB and RS232 are part of the MIDAS distribution set.

6.5 Kinematics

In this section, the basic principles of momentum and total energy conservation are applied to a few cases useful for the analysis of high-energy and position-resolution data for kinematic reconstruction in cluster physics.

6.5.1 Complete Kinematics

Characterisation of cluster-states often involves the detection of fragments and break-up particles. Ideally, all outgoing particles are detected and the correlation between those particles within the centre-of-mass frame of the break-up nucleus can be deduced. In the rest frame, observables such as the angular distribution between the break-up particles and their relative energy give information about the decaying states. The velocity vector of a nucleus before particle decay is deduced from momentum conservation. If the nucleus decays to n particles, which are identified and detected at longitudinal and azimuthal angles, θ_n and ϕ_n, with energy E_n, its momentum p is obtained as the sum of the momenta of the n particles:

$$p = mV = \sum p_n \tag{6.11}$$

where m, the mass of the break-up nucleus, is the product of the invariant mass m_0 times the Lorentz factor γ

$$m = m_0 \frac{1}{\sqrt{1 - (V/c)^2}} = m_0 \gamma. \tag{6.12}$$

The same expression applies to the n break-up particles with mass m_n and velocity V_n. The norm of the velocity, V_n, is determined from energy measurement using the expression:

$$V_n = c \frac{\sqrt{1 + 2m_{0n}c^2/E_n}}{1 + m_{0n}c^2/E_n} \tag{6.13}$$

and the x, y, z projections in Cartesian coordinates obtained from the θ_n and ϕ_n angles:

$$V_{nx} = V_n sin\theta_n cos\phi_n, \quad V_{ny} = V_n sin\theta_n sin\phi_n, \quad V_{nz} = V_n cos\theta_n \tag{6.14}$$

The final operation is to determine the relative energy between the particles in the centre-of-mass frame of the nucleus. From the velocity vectors, the relative velocity between two particles n_1 and n_2 is deduced as follows:

$$\vec{v}_{rel} = \vec{V}_{n1} - \vec{V}_{n2} \tag{6.15}$$

The typical excitation energy of the break-up nucleus, converted into kinetic energy, does not justify the use of relativistic expressions within the c-o-m frame. The relative energy between two particles, denoted n_1 and n_2, is thus given by:

$$E_{rel} = \frac{1}{2}\left(\frac{m_{n1} \cdot m_{n2}}{m_{n1} + m_{n2}}\right) v_{rel}^2 = \frac{1}{2}\mu v_{rel}^2 \tag{6.16}$$

where μ is called the reduced mass. The excitation energy plus Q-value of the break-up state is obtained by adding the total relative energy of all the break-up particles.

The excitation energy resolution evolves with the square root of the excitation energy. This resolution is inherent to the position and energy resolution of the charged particle detector. Hence, charged particle spectroscopy shows limitations when reconstructing highly excited states as neighbouring states begin to overlap due to the resolving power which decreases with increasing excitation energy.

Intermediate decay steps can be deduced from known energy resonance identified in the relative energy spectra. This, for example, is the case of the very narrow ground state of ^{8}Be with only 92 keV relative energy in the c-o-m frame. Information can also be deduced from the angular correlation in multi-step decay [130]. The angle between two velocity vectors, i.e. between the directions of two-step decay, is expressed in terms of the ratio of the scalar product and the product of the vectors:

$$cos\beta = \frac{\vec{v}_{n1} \cdot \vec{v}_{n2}}{\vec{v}_{n1} \times \vec{v}_{n2}} \tag{6.17}$$

Figure 6.26 displays the break-up of unbound ^{6}Be decaying in two protons and one α particle as depicted in the diagram through the two-body decay channels namely ^{6}Be $\rightarrow$^{5}Li $+ p$ and ^{6}Be $\rightarrow$^{4}He $+$^{2}He [131]. The direction taken by the particles in the subsequent break-up of the ^{5}Li and ^{2}He resonances with respect to the primary decay are sensitive to the spin and the parity of the states involved in parent and daughter nuclei.

From complete kinematics measurements, a number of selections can be applied to isolate the events of interest, for example the total kinetic energy distribution in the final state. By considering the total momentum projected on the X- and Y-axes, fortuitous events can be discarded if the sum of the momenta is greater than an acceptable deviation from the zero value. Reconstruction of the azimuthal angle of the particles involved in the primary binary reaction allows the discarding of those events not contained within the reaction plane. The in-plane events are selected within a narrow angle around $\Delta\phi = 180°$ to discard those events originating mostly from contaminating reactions.

6.5.2 Particle Reconstruction

The derivation described above is still valid when considering nearly complete kinematics measurements, i.e. when one particle is missing in the event. Reconstruction

Fig. 6.26 Decay diagram of
^{6}Be following two two-body
decay paths namely
^{6}Be $\rightarrow ^5$Li $+ p$ and
^{6}Be $\rightarrow ^4$He $+^2$ He followed
by subsequent break-up of
the ^{5}Li and ^{2}He resonances
respectively [128]

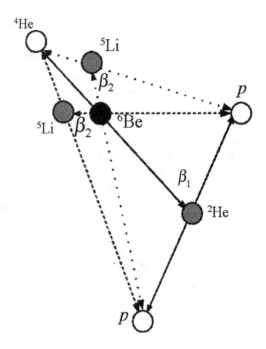

of the missing particle is achieved if N-1 particles are identified, with well defined
energy and position, the momentum vector of the missing particle is deduced as
follows:

$$P_{N_x} = 0 - \sum_{n=1}^{N-1} P_{n_x}, \quad P_{N_y} = 0 - \sum_{n=1}^{N-1} P_{n_y}, \quad P_{N_z} = P_{beam} - \sum_{n=1}^{N-1} P_{n_z}. \quad (6.18)$$

This method is powerful for neutral particle identification, as the granularity and
efficiency of neutron detectors are usually not as good as for charged particle detectors. The reconstruction of a missing neutron from charged particle information can
be performed with relatively good resolution. The draw-back of this method resides
in the cumulation of uncertainties from detected particles on the undetected particle;
the heavier the detected fragments compared to the reconstructed particle, the larger
the uncertainty of the reconstructed position and energy.

6.5.3 Total Final State Kinetic Energy (TKE)

From energy conservation principle, the total kinetic energy after a nuclear reaction
is equal to the beam energy plus the Q-value of the reaction. The Q-value relates the
amount of kinetic energy converted to mass (negative Q-value) or the amount of mass
converted into energy (positive Q-value). Total Kinetic Energy (TKE) conservation

Fig. 6.27 Q-value spectrum of the $^{12}C(^{12}C, 3\alpha)^{12}C$ reaction [132]. The peak indicated with a marker corresponds to the three α break-up of one ^{12}C nucleus and the ground state of the missing ^{12}C nucleus. The two peaks at lower energy correspond to the same exit channel, but the ^{12}C recoils are emitted in the 4.44 and 9.6 MeV excited states

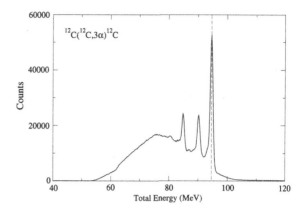

is written in Eq. 6.19 for N break-up particles considering possible dissipation of energy through undetected γ-ray emission (E_γ):

$$E_{TKE} = E_{beam} + Q - E_\gamma = \sum_{n=1}^{n=N} E_n - E_\gamma \qquad (6.19)$$

The Total Kinetic Energy distribution of the final state is also called a Q-value spectrum and is very useful for selecting the channels of interest, especially when mutual excitation plays an important role. Figure 6.27 shows the TKE spectrum in the $^{12}C + {}^{12}C$ reaction at $E_{lab} = 101.5$ MeV [132]. One of the carbon nuclei is identified from the three α break-up and the missing ^{12}C recoil is reconstructed by using momentum conservation. The prominent peak at 94 MeV corresponds to the ground state of the missing ^{12}C and the two peaks at lower energy correspond to the 4.44 and 9.6 MeV excited states respectively.

6.5.4 Dalitz Plots

In terms of theoretical calculations, three-body decay studies are much more challenging compared to two-body break-up channels. The Dalitz plot [133] is a powerful representation of the energy correlation between three break-up particles. The energy of the three particles is represented by means of a point in an equilateral triangle connected perpendicularly to the three sides of the triangle. As in Figure 6.28, the summed length of the three vectors, OD + OE + OF, remains constant wherever the point O is located within the triangle. The height of the triangle equals the total energy; the length of each segment connecting O to the sides of the triangle, OD, OE and OF, is equal to the energy of the respective particle [133]. Following geometrical manipulations, a 2D plot can be constructed with x- and y-values given by:

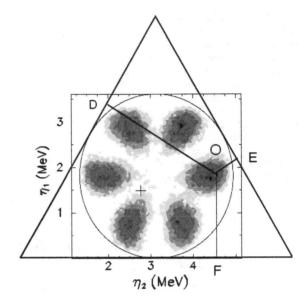

Fig. 6.28 Dalitz projection of the three-α break-up from the 12.71 MeV ($J^{\pi} = 1^+$) state in ^{12}C, with $\eta_1 = E_1$ and $\eta_2 = (E_1 + 2E_2)/\sqrt{3}$. A particular energy correlation is pointed with the vectors OD, OE, OF. The figure is adapted from [134]

$$x = \frac{E_1 + 2E_2}{\sqrt{3}}, \quad y = E_1. \tag{6.20}$$

For broad resonances, the total energy of the particles is normalised in order to fit all the events in a unique triangle. The three-body decay of the 12.71 MeV state in ^{12}C decaying in three α particles is depicted in the Dalitz projection in Fig. 6.28 with a pattern corresponding to the decay of a 1^+ state [134].

Four-particle break-up can be projected in a tetrahedron, with the conversion of the energies in 3D expressed as follows:

$$x = \sqrt{\frac{3}{8}}(E_1 + E_2 + 2E_3), \quad y = \frac{E_1 + 3E_2}{\sqrt{8}}, \quad z = E_1. \tag{6.21}$$

6.6 Computer Codes

A taste of modern experimental cluster physics was given in this chapter. The use of a radioactive beam implies new experimental challenges addressed with new generation detectors, for example γ-tracking arrays or Active Target facilities. Such experimental setups are optimised through detailed simulations in order to make the best use of sophisticated ion beams and detector arrangements. A number of software tools are commonly used in nuclear physics for cross section or energy loss calculations, data analysis and simulation purposes. An overview of some freely available and highly supported codes is given here. This, by all means, is not exhaustive and the reader is invited to explore further as some applications possibly require more specific codes.

LISE: Multipurpose Nuclear Physics and Magnetic Spectrometer Code

LISE is a package combining a number of nuclear physics codes. The initial computer code was primarily designed for Radioactive Ion Beam production and secondary beam transport simulations. Many nuclear physics codes are coupled with the LISE code over a wide range of energies, from 10 keV to 10 GeV using a variety of nuclear physics models. The user can implement some experimental setup and visualise the response of the detectors, for example the expected energy resolution or count rates [135].

SRIM Stopping Power Code

In dealing with cluster decay and charged-particle spectroscopy, calculations of energy loss and multiple scattering are almost every day tasks. Ziegler's [136] tables for stopping power have been shown to be robust over a wide range of energies and ion species, although some discrepancies are reported at very low energy where measurements are not trivial. SRIM [137] can be used with a high level of confidence for energies ranging from 100 keV to GeV per nucleon in multi layered or mixed materials. A large library of useful compound materials met in experimental nuclear physics is also available.

GEANT4 Simulation Code

The GEANT4 code [138] is traditionally used in high-energy physics for modeling complex detectors. Because of its high degree of sophistication, a wide number of applications, for example in low-energy nuclear physics or for radiation detectors currently in orbit around earth, are approached via GEANT simulations. Recent developments of ACTAR and γ-tracking detectors are amongst the best examples of GEANT simulations within the low energy nuclear physics community. The computer code FLUKA [139], also having been developed at CERN over some decades, is oriented towards high-energy particles; however, a wide range of particles are transported with numerous experimental databases for reaction cross sections and nuclear models.

ACTAR Simulation Code

A number of well identified problems have been solved and implemented in the computer code ActaSim [140], for the design of ACtive TARgets. This code allows the assembly of pre-defined geometries, of TPCs under HV and magnetic fields, and simulates the response and read-out of electronic systems for ACTAR. The response of an ACtive TARget can be anticipated and optimised in terms of spatial and energy resolution before any hardware development. Tracking algorithms for active targets are also detailed in Ref. [141].

ROOT Data Analysis Code

A general code for data analysis has been in developement at CERN since the mid 1980s. This code was initially developed under the name PAW, and re-named ROOT in its C++ version [142]. The package is a very versatile software tool that can be used for simple histogram handling to large-scale data analysis of TeraByte datasets in parallel computing architecture. ROOT primarily is a high-energy physics software code, in which nowadays millions of detector channels are involved, but it can be used with ease for smaller scale datasets. Events are stored in optimised ROOT tree database structures for fast analysis. Custom applications are written in C++ and interpreted through the ROOT CINT (C++ interpreter). It comprises a wide range of functions from basic histogram handling to sophisticated fitting applications. ROOT certainly is the most popular software data analysis code within the subatomic physics community. A number of simulation codes (GEANT4) and data analysis software, dedicated to specific experiments, are linked or developed under the ROOT environment by making use of the numerous packages already implemented.

6.7 Concluding Remarks

In these lecture notes, we have presented an overview of a number of well-established methods and techniques, but also of some of the recent developments in the field of experimental nuclear cluster studies. A number of major advances in detection techniques, fine segmentation of semi-conductor detectors, high-density integrated electronics and digital signal processing allows for a high degree of sophistication in the experimental setups. As a result, the increase in efficiency and finer granularity of the silicon detector array enable highly selective and precise measurements. In the field of γ-ray spectroscopy there are very good prospects with the developing $4\pi\gamma$-tracking arrays regarding the investigation of γ-decay events from molecular resonances. In terms of detection techniques, the highly efficient ACtive TARgets were developed mostly in response to the new experimental challenges originating from Rare Ion Beam facilities. A new landscape in the chart is opening towards the neutron-deficient region in the vicinity of and beyond the proton drip line where a number of $2p$ emitters have already been identified. Neutron-rich ion beams are well on their way to more exoticity and a number of interesting cases, such as the ^{11}Li halo nucleus, the neutron-rich Be isotopes where the α–α core persists, or some of the ^{30}Ne and ^{32}Mg nuclei displaying evident cluster structure give a taste of the discoveries ahead of us in the field of nuclear clustering.

References

1. Chatterjee, A., et al.: Phys. Rev. Lett. **101**, 032701 (2008)
2. Ikeda, K., Myo, T., Kato, K., Toki H.: In: Beck, C. (ed.) Clusters in Nuclei. Lecture Notes in Physics vol. 818, p. 165. Springer, Berlin (2010)

3. Horiuchi, H.: In: Beck, C. (ed.) Clusters in Nuclei. Lecture Notes in Physics, vol. 818, p. 57. Springer, Berlin (2010)
4. Hoyle, F.: Astrophys. J. Supp. **1**, 121 (1954)
5. Poenaru, D., Greiner, W.: In: Beck, C. (ed.) Clusters in Nuclei. Lecture Notes in Physics, vol. 818, p. 1 (2010)
6. Price, P.B.: Ann. Rev. Nucl. Part. Sci. **39**, 19 (1989)
7. Price, P.B.: Nucl. Phys. A **502**, 41 (1989)
8. Rose, H.J., Jones, G.A.: Nature **307**, 245 (1984)
9. Gales, S., et al.: Phys. Rev. Lett **53**, 759 (1984)
10. Hourani, E., et al.: Phys. Lett. **60**, 375 (1985)
11. Hourani, E., et al.: Nucl. Instr. Meth. A **264**, 357 (1988)
12. Hourani, E., et al.: Phys. Rev. C **44**, 1424 (1991)
13. Price, P.B., Stevenson, J.D., Barwick, S.W., Ravn, H.L.: Phys. Rev. Lett. **54**, 297 (1985)
14. Shicheng, W., et al.: Phys. Rev. C **36**, 2717 (1987)
15. Bonetti, R., et al.: Eur. Phys. J. A **5**, 235 (1999)
16. Bonetti, R., Guglielmetti, A.: Rom. Rep. Phys. **59**, 301 (2007)
17. Srour, J.R., Lo, D.H.: IEEE Trans. Nucl. Sci. **47**, 1289 (2004)
18. Kutschera, W.: Nucl. Instr. Meth. A **123**, 594 (1997)
19. Libby, W.F.: Phys. Rev. **69**, 671 (1946)
20. Figuera, P., et al.: Z. Phys. A **352**, 315 (1995)
21. Hilscher, D., et al.: Nucl. Instr. Meth. A **414**, 100 (1998)
22. Ortlepp, H.-G., et al.: Nucl. Instr. Meth. A **403**, 65 (1998)
23. Tyukavkin, A.N., et al.: Instr. Exp. Tech. **52**, 508 (2009)
24. Pyatkov Yu., V., et al.: Phys. Atomic Nucl. **73**, 1309 (2010)
25. Rubio, B., Gelletly, W.: Lecture Notes in Physics, vol. 764, p. 99. Springer, Berlin (2009)
26. Severijns, A.: Lecture Notes in Physics, vol. 651, p. 339. Springer, Berlin (2004)
27. Fynbo, H.O.U., et al.: Nature **433**, 136 (2005)
28. Borge, M.J.G., et al.: Physica Scripta T **125**, 103 (2006)
29. Smirnov, D., et al.: Nucl. Instr. Meth. A **547**, 480 (2005)
30. Hyldegaard, S., et al.: Phys. Lett. B **678**, 459 (2009)
31. Riisager, K.: Lecture Notes in Physics, vol. 700, p. 1. Springer, Berlin (2006)
32. Goldansky, V.I.: Nucl. Phys. **19**, 482 (1960)
33. Blank, B., Ploszajczak, M.: Rep. Prog. Phys. **71**, 046301 (2008)
34. Blank, B.: Lect. Notes Phys. **764**, 153 (2009)
35. Giovinazzo, J., et al.: Phys. Rev. Lett. **89**, 102501 (2002)
36. Pfützner, M., et al.: Eur. Phys. J. A **14**, 279 (2002)
37. Mukha, I., et al.: Phys. Rev. C **77**, 061303 (2008)
38. Mukha, I., et al.: Phys. Rev. C **82**, 054315 (2010)
39. Giovinazzo, J., et al.: Phys. Rev. Lett. **99**, 102501 (2007)
40. Huyse, M.: Lect. Notes Phys. **651**, 1 (2004)
41. Alamanos, N., Gillibert, G.: Lect. Notes Phys. **651**, 295 (2004)
42. Morrissey, D.J., Sherril, B.: Lect. Notes Phys. **651**, 113 (2004)
43. Freer, M., et al.: Phys. Rev. C **80**, 041303(R) (2009)
44. Raabe, R., et al.: Phys. Rev. C **80**, 054307 (2009)
45. Raabe, R., et al.: Phys. Rev. Lett. **101**, 212501 (2008)
46. Abgrall, Y., Gabinski, P., Labarsouque, J.: Nucl. Phys. A **232**, 235 (1974)
47. Crannell, H., et al.: Nucl. Phys. A **758**, 399 (2005)
48. Freer, M., et al.: Phys. Rev. C **76**, 034320 (2007)
49. Funaki, Y., et al.: Phys. Rev. Lett. **101**, 082502 (2008)
50. Chernykh, K., et al.: Phys. Rev. Lett. **105**, 022501 (2010)
51. Chernykh, K., et al.: Phys. Rev. Lett. **98**, 032501 (2007)
52. Schatz, H., et al.: Phys. Rev. Lett. **79**, 3845 (1997)
53. Kouzes, R., et al.: Nucl. Phys. A **286**, 253 (1977)

54. Papka, P., et al.: Phys. Rev. C **81**, 054308 (2010)
55. Fortier, S., et al.: Phys. Lett. B **461**, 22 (1999)
56. Cappuzzello, F., et al.: Phys. Lett. B **516**, 21 (2001)
57. Gade, A., Glasmacher, T.: Prog. Part. Nucl. Phys. **60**, 161 (2008)
58. Nociforo, C., et al.: Phys. Lett. B **605**, 79 (2005)
59. Gaelens, M.: Nucl. Instr. Meth. B **204**, 48 (2003)
60. http://www.intds.org
61. Gialanella, L., et al.: Nucl. Instr. Meth. A **522**, 432 (2004)
62. Mohr, P., et al.: Phys. Rev. C **50**, 1543 (1994)
63. Kölle, V., et al.: Nucl. Instr. Meth. A **431**, 160 (1999)
64. Wilmes, S., et al.: Phys. Rev. C **66**, 065802 (2002)
65. Vorobyov, A.A., et al.: Nucl. Instr. Meth. **119**, 509 (1974)
66. Dolégiéviez, P., et al.: Nucl. Instr. Meth. A **564**, 32 (2006)
67. Knowles, P.E., et al.: Nucl. Instr. Meth. A **368**, 604 (1996)
68. Sakurai, H.: Nucl. Phys. A **805**, 526c (2008)
69. Ishimoto, S., et al.: Nucl. Instr. Meth. A **480**, 304 (2002)
70. Dobrovolsky, A.V.: Nucl. Phys. B **214**, 1 (1983)
71. Mizoi, Y.: Nucl. Instr. Meth. A **431**, 112 (1999)
72. Roussel-Chomaz, P.: Eur. Phys. J. A **42**, 447 (2009)
73. Demonchy, C.E., et al.: Nucl. Instr. Meth. A **583**, 341 (2007)
74. Blank, B., et al.: Nucl. Instr. Meth. A **613**, 65 (2010)
75. Laird, A.M.: Nucl. Instr. Meth. A **573**, 306 (2007)
76. Sauli, F.: Nucl. Instr. Meth. A **386**, 531(1997). http://cerncourier.com/cws/article/cern/27921and/29894
77. http://www.nscl.msu.edu/exp/sr/attpc
78. Leo, W.R.: Techniques for Nuclear and Particle Physics Experiments. Springer, Berlin (1987)
79. Knoll, G.F.: Radiation Detection and Measurement. Wiley, New York (1989)
80. Nicolini, R., et al.: Nucl. Instr. Meth. A **582**, 554 (2007)
81. Dorenbos, P., et al.: IEEE Trans. Nucl. Sci **51**(3), 1289–1296 (2004)
82. Paris collaboration, http://paris.ifj.edu.pl
83. Lee, I.-Y: Nucl. Phys. A **834**, 743 (2010)
84. Gelletly, W., Eberth, J.: Lect. Notes Phys. **700**, 79 (2006)
85. Eberth, J., Simpson, J.: Prog. Part. Nucl. Phys. **60**, 283 (2008)
86. Ball, G.C., et al.: Nucl. Phys. A **787**, 118 (2007)
87. Baglin, C.M., et al.: Nucl. Instr. Meth. A **481**, 365 (2002)
88. Mulgin, S.I., et al.: Nucl. Instr. Meth. A **388**, 254 (1997)
89. Yanagimachi, T., et al.: Nucl. Instr. Meth. A **275**, 307 (1989)
90. Ieki, K., et al.: Nucl. Instr. Meth. A **297**, 312 (1990)
91. Bergmann, U.C., Fynbo, H.O.U., Tengblad, O.: Nucl. Instr. Meth. A **515**, 657 (2003)
92. Tengblad, O., et al.: Nucl. Instr. and Meth. A **525**, 458 (2004)
93. Blumenfeld, Y., et al.: Nucl. Instr. Meth. A **421**, 471 (1999)
94. Pollacco, E., et al.: Eur. Phys. J. A **25**, s01, 287 (2005)
95. Mukha, I.: Eur. Phys. J. A **42**, 421 (2009)
96. Stanoiu, M., et al.: Nucl. Instr. Meth. B **266**, 4625 (2008)
97. Montagnoli, G., et al.: Nucl. Instr. Meth. A **547**, 455 (2005)
98. Bachmann, S., et al.: Nucl. Phys. A, 663–664, 1069c (2000)
99. Giomataris, Y., et al.: Nucl. Instr. Meth. A **376**, 29 (1996)
100. Giomataris, Y.: Nucl. Instr. Meth. A **419**, 239 (1998)
101. Drouart, A., et al.: Nucl. Instr. Meth. A **579**, 1090 (2007)
102. Altunbas, C., et al.: Nucl. Instr. Meth. A **490**, 177 (2002)
103. Pagano, A., et al.: Nucl. Phys. A **734**, 504 (2004)
104. Tilquin, I., et al.: Nucl. Instr. Meth. A **365**, 446 (1995)
105. Marqués, F.M., et al.: Phys. Rev. C **65**, 044006 (2002)

106. Skeppstedta, Ö., et al.: Nucl. Instr. Meth. A **421**, 531 (1999)
107. Baumann, T., et al.: Nucl. Instr. Meth. A **543**, 517 (2005)
108. Gerlic, E., et al.: Phys. Rev. C **39**, 2190 (1989)
109. Mihailidis, D.N., et al.: Phys. Rev. C **64**, 054608 (2001)
110. Tamii, A., et al.: Nucl. Instr. Meth. A **605**, 326 (2009)
111. Neveling, R., et al.: Nucl. Instr. Meth. A **654**, 29 (2011)
112. Hutcheon, D.A., et al.: Nucl. Instr. Meth. A **498**, 190 (2003)
113. Gadea, A., et al.: J. Phys. G Nucl. Part. Phys. **31**, S1443 (2005)
114. Salsac, M.-D., et al.: Nucl. Phys. A **801**, 1 (2008)
115. Jenkins, D.G., et al.: Phys. Rev. C **76**, 044310 (2007)
116. Ottini-Hustache, S., et al.: Nucl. Instr. Meth. A **431**, 476 (1999)
117. Pancin, A., et al.: J. Instr. **4**, 12012 (2009)
118. Petri, M., et al.: Nucl. Instr. Meth. A **607**, 412 (2009)
119. Cardella, G., et al.: Nucl. Instr. Meth. A **378**, 262 (1996)
120. Skulski, W., Momayezia, M.: Nucl. Instr. Meth. A **458**, 759 (2001)
121. Litherland, A.E., et al.: Phys. Rev. Lett. **2**, 104 (1959)
122. Alderighi, M., et al.: IEEE Trans. Nucl. Sci. **52**, 1624 (2005)
123. Alderighi, M., et al.: IEEE Trans. Nucl. Sci **53**, 279 (2006)
124. Beck, C., et al.: Phys. Rev. C **80**, 034604 (2009)
125. Ellison, P.A., et al.: Phys. Rev. Lett. **105**, 182701 (2010)
126. Uusitalo, J., et al.: Nucl. Instr. Meth. B **204**, 638 (2003)
127. Julin, R.: Lect. Notes Phys. **651**, 263 (2004)
128. Schmitt, C., et al.: Nucl. Instr. Meth. A **621**, 558 (2010)
129. Ritt, S., Amaudruz, P., Olchanski, K.: http://midas.psi.ch
130. Biedenharn, L.C., Rose, M.E.: Rev. Mod. Phys **25**, 729 (1953)
131. Papka, P., et al.: Phys. Rev. C **81**, 054308 (2010)
132. Muñoz-Britton, T., et al.: J. Phys. G **37**, 105104 (2010)
133. Dalitz, R.H.: Philos. Mag. **44**, 1068 (1953)
134. Fynbo, H.O.U., et al.: Phys. Rev. Lett. **91**, 082502 (2003)
135. Tarasov, O., Bazin, D., Lewitowicz, M., Sorlin, O.: Nucl. Phys. A **701**, 661c (2002). http://www.nscl.msu.edu/lise
136. Ziegler, J.F.: Nucl. Instr. Meth. B **1027**, 219–220 (2004)
137. Ziegler, J.F.: http://www.srim.org/
138. Agostinelli, S., et al.: Nucl. Instr. Meth. A **506**, 250–303 (2003). http://www.geant4.org/
139. Fassò, A., et al.: Proceedings of the Monte Carlo 2000 Conference, p. 955. Springer, Berlin (2001). http://www.fluka.org/fluka.php
140. Alvarez, H.: http://fpsalmon.usc.es/actar/index.shtml
141. Roger, T., et al.: Nucl. Instr. Meth. A **638**, 134 (2011)
142. Brun, R.: Nucl. Instr. Meth. A **389**, 81 (1997). http://root.cern.ch/